Artificial Intelligence (AI)

Artificial Intelligence (AI): Elementary to Advanced Practices

Series Editors: Vijender Kumar Solanki, Zhongyu (Joan) Lu, and Valentina E Balas

In the emerging smart city technology and industries, the role of artificial intelligence is getting more prominent. This AI book series will aim to cover the latest AI work, which will help the naïve user to get support to solve existing problems and for the experienced AI practitioners, it will assist to shedding light for new avenues in the AI domains. The series will cover the recent work carried out in AI and its associated domains, it will cover Logics, Pattern Recognition, NLP, Expert Systems, Machine Learning, Block-Chain, and Big Data. The work domain of AI is quite deep, so it will be covering the latest trends which are evolving with the concepts of AI and it will be helping those new to the fiield, practitioners, students, as well as researchers to gain some new insights.

Cyber Defense Mechanisms
Security, Privacy, and Challenges
Gautam Kumar, Dinesh Kumar Saini, and Nguyen Ha Huy Cuong

Artificial Intelligence Trends for Data Analytics Using Machine Learning and Deep Learning Approaches
K. Gayathri Devi, Mamata Rath and Nguyen Thi Dieu Linh

Transforming Management Using Artificial Intelligence Techniques
Vikas Garg and Rashmi Agrawal

AI and Deep Learning in Biometric Security
Trends, Potential, and Challenges
Gaurav Jaswal, Vivek Kanhangad, and Raghavendra Ramachandra

Enabling Technologies for Next Generation Wireless Communications
Edited by Mohammed Usman, Mohd Wajid, and Mohd Dilshad Ansari

Artificial Intelligence (AI)
Recent Trends and Applications
Edited by S. Kanimozhi Suguna, M. Dhivya, and Sara Paiva

For more information on this series, please visit: https://www.routledge.com/Artificial-Intelligence-AI-Elementary-to-Advanced-Practices/book-series/CRCAIEAP

Artificial Intelligence (AI)

Recent Trends and Applications

Edited by
S. Kanimozhi Suguna, M. Dhivya, and Sara Paiva

CRC Press
Taylor & Francis Group
Boca Raton London New York

CRC Press is an imprint of the
Taylor & Francis Group, an **Informa** business

First edition published 2021
by CRC Press
6000 Broken Sound Parkway NW, Suite 300, Boca Raton, FL 33487-2742

and by CRC Press
2 Park Square, Milton Park, Abingdon, Oxon, OX14 4RN

Library of Congress Cataloging-in-Publication Data

Names: Suguna, S. Kanimozhi, editor.
Title: Artificial intelligence (AI) : recent trends and applications /
edited by S. Kanimozhi Suguna, M. Dhivya, and Sara Paiva.
Description: First edition. | Boca Raton : CRC Press, 2021. | Series:
Artificial intelligence (ai): elementary to advanced practices |
Includes bibliographical references and index.
Identifiers: LCCN 2020048919 (print) | LCCN 2020048920 (ebook) | ISBN
9780367431365 (hardback) | ISBN 9781003005629 (ebook)
Subjects: LCSH: Artificial intelligence.
Classification: LCC Q335 .A8584 2021 (print) | LCC Q335 (ebook) | DDC
006.3--dc23
LC record available at https://lccn.loc.gov/2020048919
LC ebook record available at https://lccn.loc.gov/2020048920

ISBN: 978-0-367-43136-5 (hbk)
ISBN: 978-0-367-75969-8 (pbk)
ISBN: 978-1-003-00562-9 (ebk)

Typeset in Times
by Deanta Global Publishing Services, Chennai, India

Contents

Preface

Artificial Intelligence (AI): Recent Trends and Applications covers AI technologies and highlights the prospects of AI to problems in real-time day-to-day applications. In short, AI is for *everyone, everywhere, and in everything.*

Artificial intelligence has embarked its niche in all fields of engineering and the dissemination is seemingly important in interdisciplinary research. This book serves as a background for covering the basic principles of AI; it elaborates the concepts of artificial intelligence to counterfeit human intelligence, which requires learning and reasoning.

The book encompasses both the theoretical aspects and recent trends of AI incorporated with deep learning strategies towards Smart city, Smart Grid, image processing and computer vision, robotics, instrumentation and automation, etc. The readers will benefit from the exhaustive range of applications in all fields of engineering presented in this book.

Acknowledgements

We would like to recognize CRC Press for providing us with the opportunity and the professionalism shown by teams along the process. Namely, we would like to thank Ms. Erin Harris and Keith Emmanual Arnold for all the support and leadership during the course of publishing this book.

A word of appreciation and recognition to all the authors who have contributed to this book. Also, a special recognition and our gratitude to all the reviewers who made an extraordinary effort to enhance the quality of the final chapters of this book.

Kanimozhi Suguna S has a great privilege to thank the Vice Chancellor, Dean – School of Computing and other fellow colleagues of SASTRA Deemed University, Thanjavur, for their assistance in and encouragement in throughout the progress of the publication. She would like to extend her gratitude to her husband, Shyam Sundar M, son Dharun Krishna S, and parents Gunavathi K and Subramanian G. She would also like to thank her brother Arun Prasad S and all his family members for their continuous support.

Dhivya wishes to thank Dr. Mohan Manghnani, Chairman, New Horizon Educational Institutions and Dr. Manjunatha, Principal, New Horizon College of Engineering, for their whole-hearted cooperation and great encouragement in all endeavours. She also wishes to thank her parents for their continuous support and motivation.

With great respect and honour, Sara Paiva would like to thank the Instituto Politécnico de Viana do Castelo, Portugal and all her colleagues. Special thanks go to her husband Rogério Paiva, children Diana and Leonardo Paiva, and also her parents for all their support.

Obviously, a big word of appreciation goes to our families, our main source of inspiration and strength, namely, in these difficult times we have lived in the last year. Dedicated to The Almighty – The Supreme Power That Drives Us.

Dr. S. Kanimozhi Suguna
Dr. M. Dhivya
Dr. Sara Paiva

Editor biographies

S. Kanimozhi Suguna received her B.Sc., M.C.A., M.Phil., and Ph.D., in 2005, 2008, 2009, and 2015, respectively. She has seven years of teaching experience and is working as Assistant Professor in the Department of CSE, School of Computing, SASTRA University, Thanjavur, India since 2017. Her research interests include image processing, data mining, web mining, and wireless sensor networks. She is an editorial member in IGI: Global and Editorial Advisory Board in various International Conferences. She is a member of IEEE and CSI. She has published 4 book chapters, 12 international journals, and 5 conference presentations.

M. Dhivya received her B.E., M.E., and Ph.D. degree in Electrical Engineering in 2006, 2008, and 2013, respectively. At present, she is working with New Horizon College of Engineering, Bangalore. She has worked as Associate Professor in the Department of Electronics and Communication Engineering at Dr. N.G.P. Institute of Technology. She has received a grant of Rs. 4,00,000 from Texas Instruments for organizing four-day faculty empowerment programme on "Analog, Power, Embedded Systems and Wireless (IoT)." She has guided 30 postgraduate scholars in several disciplines: Applied Electronics, Electric Drives and Embedded Control, Power Electronics and Drives and Control, and Instrumentation. Her research interests are wireless sensor networks, embedded systems, real-time systems, optimization techniques, and machine learning.

Sara Paiva is Associate Professor at the Polytechnic Institute of Viana do Castelo. She received her Ph.D. in Informatics Engineering from University of Vigo. She is also a postdoctoral researcher at the University of Oviedo since January 2018; her field of research is advanced driving assistants and urban mobility. Paiva coordinates ARC4DigiT, the Applied Research Center for Digital Transformation (http://arc4digit.ipvc.pt), created in January 2018. Her main line of research is urban mobility solutions. She has edited five books for Springer and IGI and other six are ongoing. She has contributed more than 40 publications to several international journals, conferences, and books. She is a frequent reviewer of journals and international conferences and supervises several final projects at Bachelor's and Master's levels in her main line of work.

Contributors

Alper Yılmaz
Department of Electrical and Electronics
Bursa Technical University
Turkey

Bettina O'Brien
Department of Computer Science
Pondicherry University
India

Bhuvaneswari T.
Department of CS&A
Queen Mary's College
Chennai, India

Dimitrios Alexios Karras
General Department
National and Kapodistrian University of
  Athens
Evia, Greece
and
Computer Engineering Department
Epoka University
Tirana, Albania

Fernando Olivera-Domingo
UnidadProfesionalInterdisciplinaria de
  Ingeniería Campus Zacatecas
InstitutoPolitécnicoNacional
Zacatecas, México

Girija Narasimhan
Department of Information Technology
Higher College of Technology
Muscat, Oman

Gökay Bayrak
Department of Electrical and
  Electronics
Bursa Technical University
Turkey

Indu Manimaran
Deaprtment of Computer Science
California State University
Long Beach, California

Jari Porras
Department of Software Engineering
LUT University
Lappeenranta, Finland

Jayden Khakurel
Department of Child Psychiatry
University of Turku
Turku, Finland

José Manuel Ortiz-Rodríguez
Laboratorio de Innovación y
  DesarrolloTecnológicoenInteligencia
  Artificial
Universidad Autónoma de Zacatecas
Zacatecas, México

Lalitha, R.
Department of Computer Science and
  Engineering
Rajalakshmi Institute of Technology
Chennai, India

Ma.del Rosario Martínez-Blanco
Laboratorio de Innovación y
  DesarrolloTecnológicoenInteligencia
  Artificial
Universidad Autónoma de Zacatecas
Zacatecas, México

Nabanita Dutta
Department of Energy and Power
  Electronics
VIT University
Vellore, India

Papazoglou, P. M.
General Department
National and Kapodistrian University of
    Athens
Evia, Greece

Rakesh K.
Department of Computer Science
Pondicherry University
India

Sai Charan Bharadwaj
Department of Energy and Power
    Electronics
VIT University
Vellore, India

Sathiya S.
Department of Instrumentation and
    Control Engineering
Dr. B R Ambedkar National Institute of
    Technology
Jalandhar, India

Sindhu Rajendran
Department of Electronics and
    Communication
R. V. College of Engineering
Bangalore, India

Srividya P.
Department of Electronics and
    Communication
R. V. College of Engineering
Bangalore, India

Sweta Saraff
Department of Behavioral Sciences
Amity Institute of Psychology and
    Allied Sciences
Amity University Kolkata
India

Teodoro Ibarra-Pérez
Laboratorio de Innovación y
    DesarrolloTecnológicoenInteligencia
    Artificial
Universidad Autónoma de Zacatecas
Zacatecas, México

Thulasi M. Santhi
Deaparment of Instrumentation and
    Control Engineering
Dr. B R Ambedkar National Institute of
    Technology
Jalandhar, India

Udendhran Mudalyar
Department of Computer Science and
    Engineering
Bharathidasan University
Tiruchirappalli, India

Uma V.
Department of Computer Science
Pondicherry University
India

Uma Shankar Subramaniam
Renewable Energy Lab
Prince Sultan University Riyadh
Saudi Arabia

Venkateshkumar M.
Department of EEE
Amrita Vishwa Vidyapeetham
    University
Chennai, India

Vijayalakshmi A.
Department of Computer Science
Christ(Deemed to be University)
Bangalore, India

1 Advances in Large-Scale Systems Simulation Modelling Using Multi-Agent Architectures Optimized with Artificial Intelligence Techniques for Improved Concurrency-Supported Scheduling Mechanisms with Application to Wireless Systems Simulation

P.M. Papazoglou and Dimitrios Alexios Karras

CONTENTS

1.1   LITERATURE REVIEW

1.1.1   Simulation Methodologies Applied in Wireless Communication Systems (WCS)

1.1.1.1   Simulation of WCS

A real wireless network's efficiency and behaviour can be tested using simulation systems without the need for field experiments and prototype creation. The simulation solutions give the opportunity to grow to a desired wireless network channel allocation schemes, network architectures, etc. The simulation software development approach becomes a very critical issue influencing the resulting network model and efficiency, due to the complexity of real wireless networks. A big challenge for wireless network simulation is the discovery of a way to tackle the actual actions of the network and not just speed up execution time using parallel machines. The simulation model and environment structure affect the performance of simulated wireless networks, and for this reason the design and development of such systems is studied thoroughly. Modern simulation tools provide network engineers with the opportunity to develop and test wireless communication systems at low cost very quickly. There are three major simulation techniques (Chaturvedi, A., et al. 2001): discrete event simulation (DES), system dynamics, and multi-agents. The most widely known simulation tools are based on the DES concept and use various model architectures to implement. A more accurate and reliable simulation environment can be developed with the help of efficient model architectures (Chaturvedi, A., et al. 2001; Liu, W., et al. 1996; Zeng, X., et al. 1998; Bajaj, L., et al. 1999; Kelly, O.E., et al. 2000; Liu, et al. 2001; Boukerche, A., et al. 2001; Bononi, L. and D'Angelo, G., 2003) to speed up execution times. Rapid development of parallel systems with multiple processors has contributed to more effective simulation execution. On the other hand, the programming technology, and especially the multithreading systems, provides an alternative approach to simulation model implementation. Inside the simulation model, the actual network activities must be realistically modelled. In the case of cellular networks, such operations are the essential network functions such as new call entry, reallocation (handoff) (Krishna, S., et al. 2016), device movement, and call termination. The simulation model will include sub-models for three types of basic components:

• Facilities by network
• Network operating parameters (e.g. number of cells, base station locations, allocation channel schemes, etc.)

• Mathematical network models (e.g. propagation models, statistical distributions, signal computations, etc.)

The sequential simulation is a very time-consuming method due to the high complexity of the models used (Liu, W., et al. 1996) and thus the parallelization is a critical issue particularly for wireless networks of large scale (Liu, W., et al. 1996; Zeng, X., et al. 1998; Bajaj, L., et al. 1999; Kelly, O.E., et al. 1999; Liu, et al. 2001; Boukerche, A., et al. 2001; Bononi, L. and D'Angelo, G. 2003). The key goal of parallel simulation is the minimization of the execution time (Zeng, X., et al. 1998; Bajaj, L., et al. 1999). In the process of parallelization, sets of network entities have to be mapped to multiple processors to achieve load balance (Zeng, X., et al. 1998). Multiple-processor load distribution is a key issue (Liu, W., et al. 1996; Zeng, X., et al. 1998; Bajaj, L., et al. 1999). Several features, procedures, and entities, such as geographical area (Kelly, O.E., et al. 1999; Boukerche, A., et al. 2001), radio channels (Kelly, O.E., et al. 1999; Boukerche, A., et al. 2001), interference calculations (Kelly, O.E., et al. 1999), mobile hosts (MH) (Boukerche, A., et al. 2001), and network cells (Boukerche, A., et al. 2001) can be parallelized in parallel implementations. There are important problems that need to be tackled effectively in order to achieve optimum parallelization, such as synchronization of processors (Zeng, X., et al. 1998; Bajaj, L., et al. 1999; Boukerche, A., et al. 2001), load balancing control (Liu, W., et al. 1996; Zeng, X., et al. 1998), and cycle mapping to processors (Liu, W., et al. 1996; Zeng, X., et al. 1998). The above-mentioned studies have resulted in two basic points for further investigation:

• How to achieve full parallelization for sophisticated computer systems
• How to model the current network environment and behaviour more effectively using advanced programming principles

Advanced computer and programming technologies must be used to build an effective simulation model, tailored to the actual network (Papazoglou et al. 2008–2016).

1.1.1.2 Discrete Event Simulation

1.1.1.2.1   *The Main Concepts of Discrete Event Simulation*

DES (Subramania, S., 2001; Tolk, A. 2012; Barr, R., 2004; Goh, R.S.M. and Thng, I., L-J, 2003; Reddy, D., et al. 2006; Schriber, T.J., and Brunner, D.T., 1997; Misra, J., 1986; Tropper, C., 2002; Balakrishnan, V., et al 1997; Jefferson, D.R., and Barnes, P.D., 2017) represents the most well-known methodology of simulation, in particular for communication systems. According to the principle of DES, events occur within the simulation time at discrete points in time. Simulation time progresses based on sequence of events. These events express the basic actions within the physical network, such as new call arrival. Each event is created with an exclusive time stamp which is later involved when running the event. A scheduler defines the processing of events over simulation time (Mehta et al. 2010; Siow, R., et al. 2004; Chung, K., et al. 1993; Whitaker, P., 2001; Naoumov, V., and Gross, T., 2003) that picks out events with a minimum time stamp associated with maximum priority. Therefore, a priority queue is the foundation of the entire scheduling process. Characterized DES systems can be sequential or parallel.

Sequential DES systems are the most common in science community. In such systems, in a three-step cycle, the scheduling mechanism can be analyzed:

- *Dequeue:* an incident of minimum time stamp is deleted from the queue
- *Run:* diagnosis of dequeued case
- *Enqueue:* entrance into the queue of a newly created event

ns-2, ns-3 (Fall, K., and Varadhan, K., 2015), widely accepted by the scientific community (Kurkowski, S., et al 2005), use this scheduling mechanism for event execution. In ns-2, ns-3 (Fall, K., and Varadhan, K., 2015), the scheduler selects the next earliest incident, executes it until ending it up, and returns to run the next event of highest priority or with least time stamp among the outstanding events. In ns-2, ns-3 (Fall, K., and Varadhan, K., 2015), only one operation can be executed at any given time, and so it is a single-threaded simulator. If two or more events are scheduled to take place (to execute) at the "same time," their execution is performed in a first come first scheduled scheme – first dispatched, based on the time stamp of each incidence and thus, the simulation model's adaptability to the actual network behaviour, including all relevant network physical activities, is strongly based on the scheduling algorithm. On the other hand, parallel implementation of DES systems was introduced in the literature primarily to achieve speed up of execution. Due to the fact that the main goal is load distribution and speed-up execution, the parallel DES systems do not modify the principle of the event scheduling found in sequential DES. The system performance is thus a critical point in terms of execution time (Liu, W., et al 1996; Zeng, X., et al 1998; Bajaj, L., et al 1999; Kelly, O.E, et al 1999; Liu, et al. 2001; Boukerche, A., et al 2001; Bononi, L. and D'Angelo, G., 2003). Parallel efficiency plays a major role in the performance of the entire system (Liu, W., et al. 1996; Zeng, X., et al. 1998; Kelly, O.E, et al. 1999; Boukerche, A., et al. 2001) and therefore synchronization of multiple processors is important (Liu, W., et al. 1996; Zeng, X., et al. 1998).

1.1.1.2.2 Simulation Time Definition

The time can be defined in different ways with respect to the real time and simulation time. There are three known definitions of time (Fujimoto, R.M. 2000):

- Physical time length
- Simulator period
- Official clock time

Physical time is the time in the physical (real) system; simulation time is the modelled physical time in the simulation system; and wall-clock time is the simulation program's runtime.

The definition of simulation time is given in Fujimoto (2000) as follows:

«Simulation time is defined as a totally ordered set of values in which each value represents an instant of time in the modelled physical system. Furthermore, for any two simulation time values T1 representing physical time P1, and

T2 representing P2, if T1 < T2, then P1 occurs before P2, and (T2 − T1) is equal to (P2 −P1) K for some constant K. If T1 < T2 then T1 is said to occur before T2 and if T1>T2, then, obviously T1 happened after T2.»

1.1.1.2.3   Efficient Usage of Computer Technology for Supporting Parallel DES Systems

The availability of computer technology and tools among academic and research institutes led to the development of advanced simulation systems that effectively confront the complexity and time of execution. Scientific community and industry have a great interest in that field. In the case of speeding up the simulation execution time, INTEL (Liu, W., et al. 1996) and DARPA (Zeng, X., et al. 1998; Bajaj, L., et al 1999) supported several projects in educational institutions such as UCLA, but according to Fujimoto et al. (2003), the effective simulation of large-scale networks is still an unresolved problem. In general, parallel computing can be categorized according to system features and capabilities (e.g. architecture). Multiprocessing, computer cluster, parallel supercomputer, and distributed computing are the parallel computing-based categories. In parallel implementation of DES systems, two different strong technologies were used: clusters and multiple processor machines. A cluster usually consists of two or more workstations connected via a local network (Baker M., et al 2000). It is possible to view this group of loosely coupled computers as one single computer. The most common cluster types are high-availability clusters and high-performance clusters. High-availability clusters are used to improve a cluster's availability of the services it offers. High-performance clusters are used mainly by splitting a computational function into more than one processing unit to provide improved performance. One of the most common cluster projects is the Beowulf project used in over 50 educational institutes and research centres. The parallel computing system forms a computer of multiple processors. The process to solve a problem in this scheme can be divided into smaller tasks performed by different processors. There are several critical issues in parallel computing such as synchronization and parallel overhead which affect the performance of the whole system. Many strategies for execution time minimization were applied based on the model's thorough decrease (Liu, W., et al. 1996). If the parallelization is sufficiently effective, the reduction of details of the model is not necessary. Also, in a parallel system with $N=P$ ($N=$partitions of the layout, $P=$number of processors), there must be efficient synchronization between processors. Effective parallelization can be accomplished by eliminating global variables and unnecessary model synchronizations, effective partitioning (locality of exploitation and balance of loads), etc. A range of workstation processors with a total of 16 processors is used in simulation experiments of (Liu, W., et al. 1996) and (Zeng, X., et al. 1998). The use of 16 processors for over 200 network nodes provided major acceleration. The partitioning scheme plays a major role in the resulting speedup according to Zeng et al. (1998). In Kelly et al. (1999), interference calculations are parallelized and sub-geographic points and sub-matrix zones for path losses are mapped as entities in one processor (Kelly, O.E., et al. 1999). Multiple channel zones are mapped to multiple processors according to Kelly et al. (1999), and significant speedup is achieved using up to eight processors. For the effective simulation of wireless and mobile networks, a cluster of workstations is used (Boukerche, A., et al. 2001). Quite impressive speedup is accomplished using up to 16 processors based on Mobile Host partitioning. For parallel DES, a networked cluster of PCs use a set of physical execution units (Bononi, L. and D'Angelo, G., 2003).

1.1.1.3 Event Scheduling

1.1.1.3.1 The Importance of Event Scheduling

Events, being the main entities of a DES system, express the physical activities of a real-world wireless network. With respect to the aforementioned network facilities, each service could be considered an incidence for a particular mobile user (MU). During simulation time, an incidence generator creates events as, for instance, new call arrivals in the system. Such an event is marked with a time stamp (within simulation time) that specifies the starting execution point in time. A critical entity within the simulation system, namely, the scheduler, is executing by selecting the next earliest event, based on its associated time stamp, running it until ending it up, and returning to execute the next event expressed by the next least time stamp. The scheduling system is a standard tool within the actual wireless network for the event service incidence sequence control. When an operation is not executed, the pending operation set (PES) (Goh, R.S.M. and Thng, I., L-J, 2003; Tan, K.L., and Thng, L.-J., 2000; Siangsukone, T., et al. 2003) is the set of all non-simulated and processed tasks generated during simulation time. Consequently, PES corresponds to a priority queue controlling the stream of events simulation based on the contemporary minimum time stamp, that is the task of highest priority (Goh, R.S.M. and Thng, I., L-J, 2003). The selected scheduling method imposes how the actual network activities that occurred can be reflected in the simulation model in a realistic way. Therefore, scheduling could be viewed as a generalized mapping method within the simulation time of the DES algorithm for the real network events-activities execution.

1.1.1.3.2 The Current Standard Event Scheduling Mechanism: Calendar Queue (CQ)

The major application of priority queuing is the realization of the PES sets within the DES simulation systems (Blackstone, J.H., et al 1981; Henriksen, J.O., 1977; Kingston, J. H., 1984; McCormack, W.M. and Sargent, R.G., 1981). The CQ concept was first introduced by Brown (1988), has been analyzed in Siangsukone et al. (2003), and even nowadays is established as the most popular scheduling scheme among the implemented DES systems such as ns-2, ns-3 (Berkeley, USA) (Fall, K., and Varadhan, K., 2015), Ptolemy II (Berkeley, USA) (Whitaker P. 2001), Jist (Cornell University, USA) (http://jist.ece.cornell.edu/javadoc/index.html?jist/runtime/Scheduler.Calendar.html). Numerous other variations of the CQ aiming at increasing performance of the queuing system, as for instance by optimally resizing the queue, have been proposed in the literature such as DSplay (Siow, R. et al. 2004), MList (Goh, R.S.M. and Thng, I., L-J, 2003), Markov hold model (Chung, K., et al. 1993), and SNOOPY CQ (Tan, K.L., and Thng, L.-J., 2000). The CQ concept is derived from the common desktop calendar of one-page entry every day. Each incident is scheduled on the relevant calendar page. Each event's schedule timing defines its priority. If an event on the calendar is enqueued, then this event is scheduled for future running. The earliest event within the calendar is dequeued by searching it within the relevant page of the current date and removing it from that page (Brown, R. 1988).

A CQ scheme is considered to consist of an array of lists. Each such list queue contains future incidences. According to the CQ principle, a large queue of N events is partitioned to M shorter queues called Buckets. Each bucket is associated with a specific time range corresponding to future events executions. Any incidence bound to an occurrence time $t(e)$ is associated with the mth bucket in year y ($y=0, 1, 2, \ldots$) if and only if

$$t(e) \in \left[\left(\left(yM + m \right) \delta \right), \left(yM + m + 1 \right) \delta \right] \tag{1.1}$$

In order to find the associated relevant bucket number $m(e)$ where an incidence e will occur at time $t(e)$, the following type is involved:

$$m(e) = \left\lfloor \frac{t(e)}{\delta} \right\rfloor \bmod M \tag{1.2}$$

Consider that $N = 10$, $\delta = 1$, $M = 8$, and $t(e) = 3.52$ (Figure 1.1) for a new incidence e.
Using Equation 1.2, the relevant bucket number for the incidence e is $m(e) = 3$.
An analytical study of the CQ scheme can be found in Erickson et al. (1994).

1.1.1.3.3   Application of CQ in Sequential DES

Figure 1.2 demonstrates a complete CQ operation in a sequential system of DES. The CQ initially includes six events (step 0) that are to be executed at a later time unit. The event e1 (NC – New Call), with time stamp 3.62, will be dequeued first as it has the highest priority among the queue buckets, having acquired the minimum time stamp. A novel incidence with time stamp 9.98 is enqueued in step (3), and in step (4) the event e2 (RC – reallocation of a call) is dequeued. The entire CQ function follows the Dequeue-Run-Enqueue protocol. In any step of the entire operation, this rule can be evaluated (e.g. Dequeue{step(9)}-Run{step(10)}- Enqueue{step(11)}).

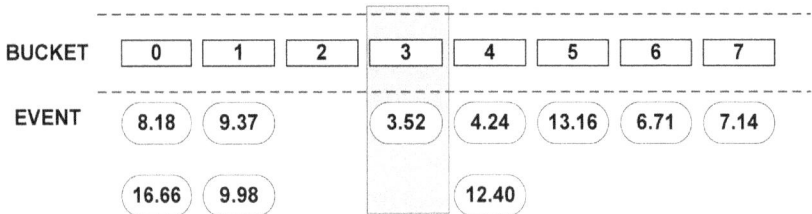

BUCKET	0	1	2	3	4	5	6	7
EVENT	8.18	9.37		3.52	4.24	13.16	6.71	7.14
	16.66	9.98			12.40			

FIGURE 1.1   Illustration of a CQ implementation with eight buckets.

Buckets

| 0 | 1 | 2 | 3 | 4 | 5 | 6 | 7 |

Event Generator — (0) (1) — Dequeued Events

9.98 | 8.18 9.37 | 3.62 4.24 | 6.71 7.14 | 3.62 | e₁ NC

12.40 | 8.18 9.37 | (4) 4.24 | 6.71 (7) | 4.24 | e₂ RC
9.98 (3)

8.18 9.37 (6) | 12.40 | 6.71 7.14 | 6.71 | e₃ MC

8.18 9.37 | 12.40 | 7.14 (9) | 7.14 | e₄ NC
9.98

13.16 | 8.18 9.37 | 12.40 13.16 (11) (12) | 8.18 | e₅ RC

16.66 | 9.98
16.66 9.37 (14) | 12.40 13.16 (15) | 9.37 | e₆ RC
9.98

16.66 9.98 | 12.40 13.16 (17) | 9.98 | e₇ NC

16.66 | 12.40 13.16 (19) | 12.40 | e₈ FC

Enqueue | 16.66 | (21) 13.16 | 13.16 | e₉ MC

16.66 | 16.66 | e₁₀ FC
(23)

Execution Sequence

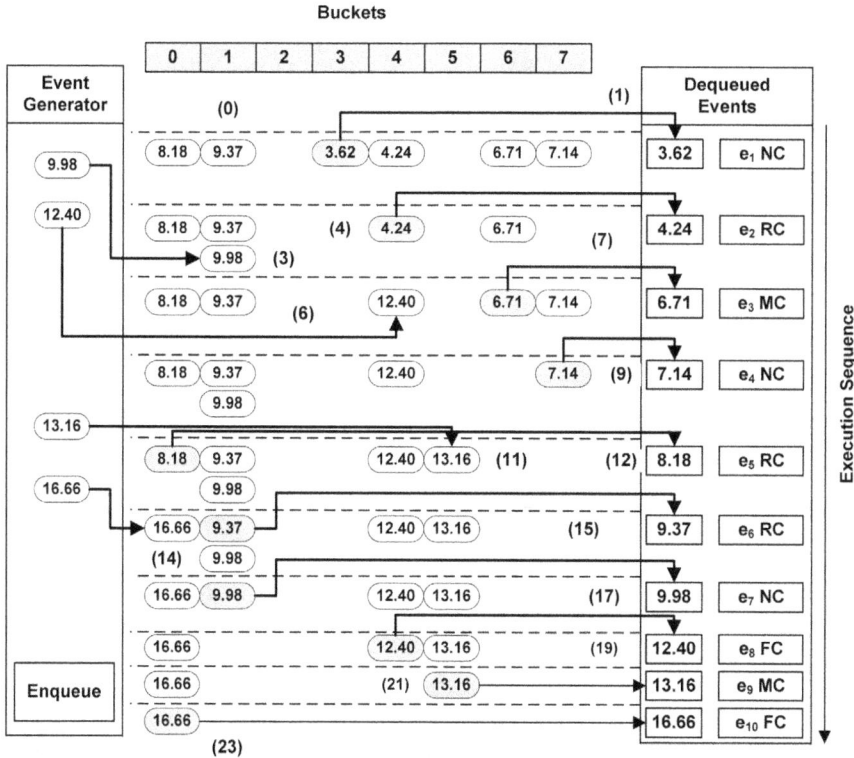

FIGURE 1.2 Sample of a Sequential DES operation (events/facilities: NC – new call arrival, RC – call reallocation/handoff, MC – movement of user running a call, FC – terminated call).

The generator of the incidences creates events with corresponding time signs. The dequeued incidences are running either in rising priority order or in increasing time stamp order (Figure 1.3). The final processes running is sequential oversimulation time except for events of nearly identical time stamps.

1.1.1.3.4 Application of CQ in Parallel DES

Where a key goal is to speed simulation execution, parallel models and implementations are used.

Speeding up can be performed in a parallel DES system if the work load of the total processing is spread over individual processors. Figure 1.4 shows a scenario with a cellular network that is divided into four clusters. Each cluster contains seven uniquely frequently used cells. A separate processor is assigned to the processing load for simulation of all possible MU resources in each cluster.

In parallel DES, the implementation of just one central CQ is an incorrect selection since, given the multiple processor availability, the time stamps of the enqueued events compel the entire system to be sequential. This is because all activities in any cell must be carried out in a particular order based on their associative time stamps.

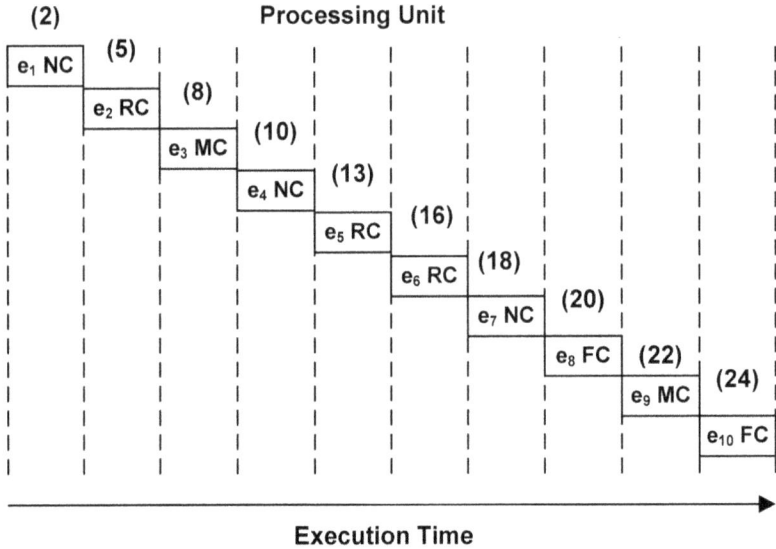

FIGURE 1.3 Event running following the dequeuing order of the processing unit (events/facilities: NC – new call arrival, RC – call reallocation/handoff, MC – movement of user running a call, FC – terminated call).

The events belonging to various clusters can be executed in parallel according to Figure 1.5. All the events are running serially within each processor and its associated cluster. Many more processors are required to obtain true competition. Parallel sets of sequential events can run in parallel within this scheme. Therefore, only selected rivalry can be accommodated within this framework, and only for services that belong to different network partitions.

1.1.2 CHANNEL ASSIGNMENT IN WCS

A cellular system's capability may be defined in terms of the number of channels available, or the number of users that the network can support at the same time. The total number of channels made available to a system depends on each channel's allotted spectrum and bandwidth. The frequency spectrum available is limited and the number of MUs is increasing day by day, therefore the channels have to be reused as much as possible to increase the system capacity. One of the basic resource management issues in a mobile communication system is allocating channels to cells or mobile phones. A channel allocation scheme has the function of allocating channels to cells or MUs in a way that minimizes:

(A) the probability of discontinuing incoming calls;
(B) the probability of dropped calls, and
(C) the probability of any call carrier-to-interference ratio falling below the prespecified value (Papazoglou et al. 2008–2016).

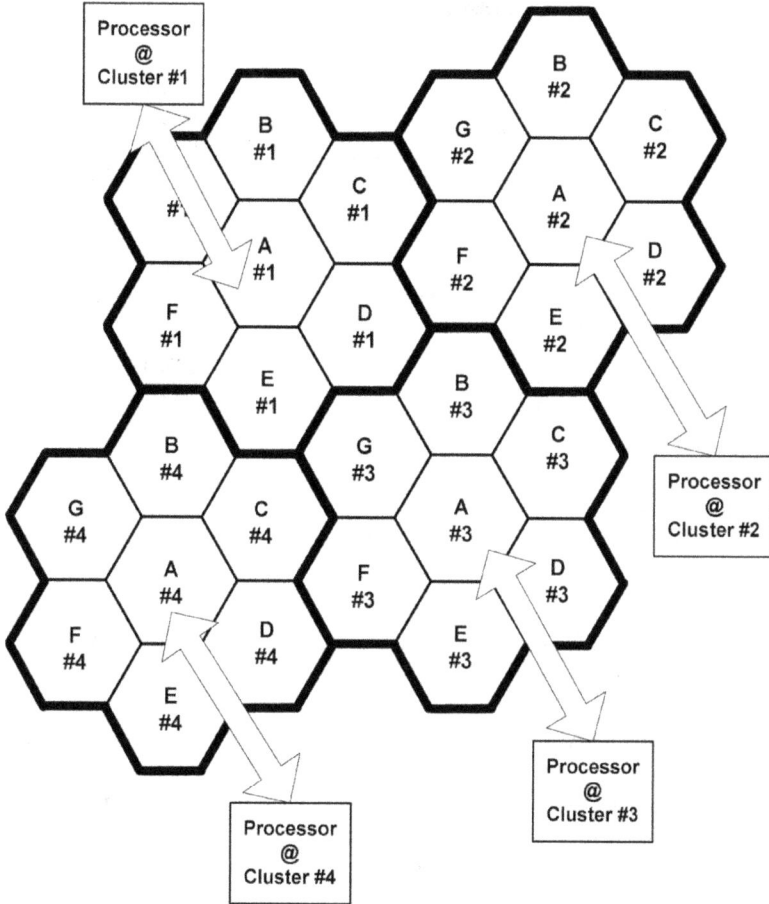

FIGURE 1.4 Excerpt of spatial partitioning/parallelization.

Many channel allocation schemes were widely invested in literature with a goal of maximizing reuse of frequency. Channel allocation schemes are classified into three strategies: fixed channel allocation (FCA) (Zhang, M. and Yum, T.S., 1989; Lai, W.K. and Coghill, G.C., 1996; MacDonald, V.H., 1979; Elnoubi, S.M., et al 1982; Xu, Z. and Mirchandani, P.B., 1982); dynamic channel allocation (DCA) (Zhang, M. and Yum, T.S., 1989; Cimini L.J. and Foschini, G.J., 1992; Cox, D.C. and Reudink, D.O., 1973; Del Re, E, Fantacci, R, and Giambene, G. 1996; Kahwa, TJ, and Geor-gans, ND. 1978; Papazoglou, PM, and Karras, DA. 2016; Papazoglou, PM, Karras, DA, and Papademetriou, RC. 2008a; Papazoglou, PM, Kar-ras, DA, and Papademetriou, RC. 2008b; Papazoglou, PM, Karras, DA, and Papademetriou, RC. 2009; Papazoglou, PM, Karras, DA, and Papademetriou, RC. 2011; Papazoglou, PM, Karras, DA, and Papademetriou, RC. 2016; Sivarajan, KN, McEliece, RJ, and Ketchum, JW. 1990; Subramania Sharma, T, and Thazhuthaveetil, MJ. 2001) and HCA (Hybrid Channel Allocation).

Buckets

| 0 | 1 | 2 | 3 | 4 | 5 | 6 | 7 |

Cluster #1 Scheduler: 8.18   19.38   4.24   15.00

4.24	e_1 NC	Processor @ #1
8.18	e_2 RC	
15.00	e_3 MC	
19.38	e_4 NC	

Cluster #2 Scheduler: 9.32   5.23

| 5.23 | e_1 MC | Processor @ #2 |
| 9.32 | e_2 FC | |

Cluster #3 Scheduler: 8.55   1.26   4.11

1.26	e_1 FC	Processor @ #3
4.11	e_2 RC	
8.55	e_3 RC	

Cluster #4 Scheduler: 9.26   2.69   6.04   15.66

2.69	e_1 MC	Processor @ #4
6.04	e_2 RC	
9.26	e_3 NC	
15.66	e_4 NC	

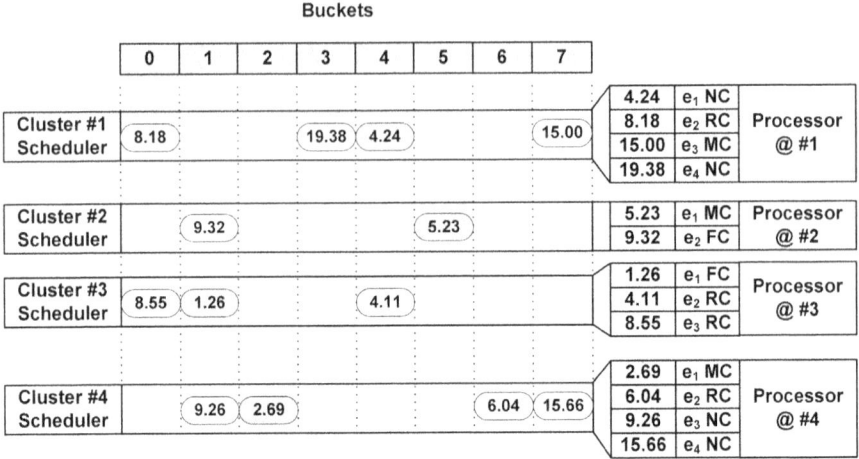

FIGURE 1.5 Excerpt of different CQs based on network spatial partitioning (events/facilities: NC – new call arrival, RC – call reallocation/handoff, MC – movement of user running a call, FC – terminated call).

In FCA, every cell is permanently assigned a set of channels based on a pre-estimated traffic intensity. No permanent allocation of channels to cells occurs in DCA. Rather, all the channels available are open to all the cells, and the channels are allocated in a dynamic manner on a call-by-call basis. The FCA scheme is simple, but does not adapt to changing traffic and user delivery conditions. In addition, frequency planning in a microcellular environment becomes more difficult because it is based on accurate knowledge of traffic and interference conditions. DCA overcomes these shortcomings however; under heavy load conditions, FCA outperforms most known DCA schemes. HCA combines the features of both the FCA and DCA techniques to overcome the drawbacks of FCA and DCA.

1.1.3 MULTI-AGENT SYSTEMS IN WCS

1.1.3.1 Agent and Multi-Agent Systems

An agent is considered to be a computational machine that autonomously interacts with its environment and works for the reason for which it was programmed (Maes, P., 1995), (Papazoglou et al. 2008–2016). Agents may be approached as organizations committed to a specific task and smaller than a standard application (Smith, D.C., et al. 1994). An agent also perceives the conditions of the world, behaves according to these conditions, interprets perceptions, and addresses issues (Hayes-Roth, B., 1995). Mobile agents are used in a distributed environment such as the internet and offer flexible networks customized to MU needs, and use the available bandwidth effectively. There are certain basic attributes which distinguish agents from other programs. Such qualities of greatest significance are as follows:

- Adaptability that reflects a learning and development potential centred on growing experience. This attribute is, in other words, about the shift of agent according to external or internal events (Splunter, S., et al. 2003; Russell, S., and Norvig, P., 2002).
- Independence in the acts needed to achieve predefined goals. Growing agent has the power over its own behaviour according to this characteristic (Huhns, M., and Singh, M., Eds. 1998; Norman, T., and Long, D., 1995; Ekdahl, B., 2001).
- Partnering with other stakeholders to reach mutual goals.
- Interactivity with surroundings.

According to Jennings (2000), a multi-agent system (MAS) has a number of agents that interact with each other through communication. These agents function in an environment and have different areas of environmental control. Many areas of influence can be coincided within the environment. In modern problems, the modelling of all system aspects requires a number of agents, and so the MASs emerged. MASs can be described as a loosely connected network of problem-solver entities (Sycara, K., 1998) working together with the shared aim of solving the whole problem beyond each individual entity's solving capabilities. Agent technology has been used in the management of telecommunications networks in many instances (Hayzelden, A., and Bigham, J., 1999; Bodanese, E.L, 2000; Iraqi, Y. and Boutaba, R., 2000; Jiang, Y., and Jiang, J., 2008). In these cases, the architecture is straightforward. Modelling of cellular systems and particularly the methodology of simulation have not been presented in terms of MAS so far. Their simulation is based primarily on sequential or parallel models (for faster execution) but not on models of the agents. According to Wooldridge (2004), negotiation approaches are based on other components, such as negotiation sets (possible agents' proposals), protocols (definition of legal proposals), tactics selection (usual private), etc.

Growing agent has its own view of the environment and also personal objective and resource claim in a competitive or collaborative environment. Negotiation between the agents is a required re-requirement for achieving the goal and satisfying the specified con-stresses. Negotiations may be challenged as competitive or cooperative (Zhang, X., et al. 2001). The negotiators are self-interested in successful deals to achieve their own objectives. The negotiators are working together in joint discussions to find a compromise for a shared target.

1.1.3.2 Multi-Agent Systems in WCS

For the issue of resource distribution, the MAS technology has been used in many studies (Papazoglou et al. 2008–2016). Different network elements, such as Base Stations (BSs), cells, etc. were modelled as agents within the established models. An overview of communication systems agent technology is presented in Hayzelden and Bigham (1999). This overview is focused on software agents used for the management of communications. More specifically, agents can be used to deal with some important issues, such as complexity of the network, mobility of the MU, and network management. A MAS for the control of resources in multimedia wireless mobile networks is provided in Iraqi and Boutaba (2000). The call dropping likelihood is small based on the proposed MAS (Iraqi, Y. and Boutaba, R., 2000), although the wireless network makes a high average usage of the bandwidth. According to Iraqi and Boutaba (2000), the final decision to accept a call is based on neighbouring cells participating. Therefore, every cell or BS has an agent working. A cooperative agreement in a MAS is defined in Jiang and Jiang (2008) to help real-time load balancing of a mobile cellular network. Within the current MAS, agents are used (a) for market representation of different service providers and (b) for network operators handling the radio capital of different network regions. According to Jiang and Jiang (2008), the coordination of agents is done by messages, and negotiations are focused on the final agreements. Negotiation is an intelligent control strategy in the above study to dynamically change the shape and size of the cells and to manage the traffic load over the network.

To suggest a distributed channel allocation scheme using intelligent software agents, a detailed simulation model for wireless cellular networks was developed by Bodanese (2000). In the above analysis, smart collaborative software agents offer BSs autonomy, enhance network robustness, enable network resource negotiation, and improve resource allocation. To this end, several aspects of infrastructure and operation of the cellular network have been modelled. According to Bodanese (2000), the basic features implemented in the model are call set-up, handoffs based on signal strength and defined threshold, forced call termination emulation (due to insufficient signal), Mobile Station self-location based on signal strength from nearest BSs, Mobile Stations with arbitrary trajectories, cell boundaries, number of Mobile Stations per unit. Four specific aspects of the configuration and function of the entire network were modelled in Bodanese (2000). Those aspects include (a) the network model defining the network layout (cell structure, BSs, MSs), (b) the mobile station (signal calculations, call generation inter-arrival time, call length), (b1) call server (SNR measurements), (b2) handoff, (b3) call dropping, (b4) call termination, (b5) filter (sending packets from the radio link to the server), (b6) trajectory tracking, which is responsible for Mobile Station movement (random cyclical, driving on a highway, etc.), (c) BS which includes SNR measurements, handoff (monitoring the received voice packets), etc., (d) MTSO (Mobile Telephone Switching Office) which offers services such as HLR (Home Location Register), VLR (Visitor Location Register), etc., (d1) keeps track of each active MS. All the implemented network models in Bodanese (2000) do not support multimedia real-time services. A novel modelling methodology for supporting wireless network services is also described in the current report, based on MAS technology.

1.1.4 THE CONCEPT OF CELLULAR NETWORK

The cellular principle divides the geographic area covered into a series of smaller service areas called cells. That cell is fitted with a BS and a number of mobile terminals (e.g. mobile phones, palms, laptops, etc.). The BS is fitted with reception and radio transmission equipment. The mobile terminal inside a cell communicates with the cell-related BS via wireless links. A number of BSs are linked via microwave links or dedicated leased lines to the Base Station Controller (BSC). The BSC provides logic for the management of the base stations' radio resources under its control. As a MU moves from cell to cell, it is also responsible for transferring an ongoing call from one base station to another. A mobile terminal must first obtain a channel from the BS in order to establish a communication with a BS. A channel consists of a pair of signal transmission frequencies between the BS and MU.

1.1.5 SIMULATION LANGUAGES (SLs)

The great interest in simulation applications led to the development of specialized Simulation Languages (SLs) that offered the possibility of a developed model being simulated quickly. The biggest advantage of the SLs is the significant reduction in time for programming the model. This advantage, contrasting with the programming time, helps the developer to give more time for model analysis and development. Modern SLs provide design guidance over production and study. The SLs most known offer several useful features:

- Are process or event oriented
- Mechanism simulating time flow
- Generation of random numbers and various distributions
- Fit for debugging devices

On the other hand, SLs have several drawbacks:

- High procurement and maintenance costs
- Time cost for the initial language learning
- High resource requirements in contrast with conventional programming languages (computer power, memory)

In order to pick every SL correctly, multiple variables need to be taken into account. Those characteristics include:

- SL-accessibility as a (free) commercial or academic resource
- SL technical features such as (a) how the SL encounters incidents, activities, and processes, (b) problem-solving usability, (c) data collection and analysis capabilities, (d) model extension support, (e) random number and sample generation, (f) debugging procedures
- The programming effort required
- Computer time
- Specifications for software resources
- The mobile model

Most approaches to simulation models and applications have allowed us using general-purpose languages, general SLs, and packages for special-purpose simulations. Languages of general use have the following principal characteristics:

- Bottom cost
- Are available
- No supplementary training is required
- Every model starts from the ground up
- Hard to check
- Reusability of code is minimal
- The business process period is a long one

General SLs deliver uniform modelling features, shorter development cycle, and simple verification. These languages have higher costs, restricted portability, and require more instruction.

Languages such as Simula (Nierstrasz, O., 1989), GPSS (Karian, Z.A., and Dudewicz, E.J., 1998), SimScript (Russell, E.C., 1999), Parsec (Subramania, S., 2001; Zeng, X., et al. 1998; Takai, M., 1999; Reddy, D., et al. 2006; Abar, S., et al. 2017), modsim (West, J. and Mullarney, A., 1998), slam (Miller, J.O., et al. 1988), GlomoSim (Zeng, X., et al. 1998), and other special purposes of high-flexibility simulation devices. Other software tools such as MATLAB have plenty of scientific libraries and capabilities for the creation of portable script code but suffer from low-performance execution. A large-scale simulation model includes a scalable network environment with advanced features such as portability, internetworking, and high network activity adaptability.

The invention and development of new languages for simulation introduces important drawbacks:

- The new languages are specific to the domain and rarely adapted by the scientific community.
- The respective libraries require designers to adapt their applications to specific needs.
- Designers can't achieve high simulation kernel adaptability in their applications.

Java language is the most appropriate to develop scalable, portable, and high-performance network applications, and is adopted by the majority of the scientific community. On the other hand, Java supports multiple threads that mean manipulation of the concurrent tasks. This functionality is very critical for supporting network events at the same time.

1.2 THE PROPOSED SIMULATION MODEL

1.2.1 NETWORK STRUCTURE

The cellular network consists of N cells and each of the cells includes a BS in the centre position (Figure 1.6).

Each cell is divided into a mesh of spots where users may exist from the start of the call or because of handoff situations.

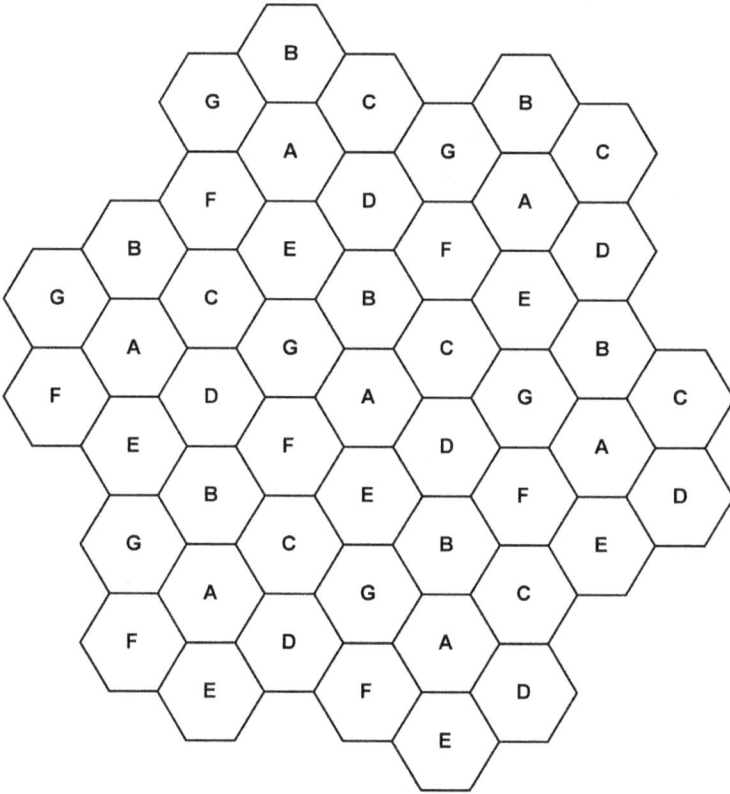

FIGURE 1.6   Structure of cellular network model.

   The basic parameter for mesh building is the fineness, which means the distance between the spots.

1.2.1.1   Operational Parameters

A set of parameters determine the conditions of network operation, such as signal propagation, signal rates, and MU behaviour. The parameter basics are: CNR (carrier-to-noise ratio) on cell edge (dB), CNIR (carrier-to-noise plus interference ratio) threshold (dB), average call arrival rate (calls/hour, lambda), average call hold time (seconds), simulation time, cell channels, path loss factor (alpha), standard shadowing deviation (sigma), cell mesh fineness, etc.

1.2.2   MODELLED NETWORK SERVICES AND CHANNEL ALLOCATION

1.2.2.1   Network Services

1.2.2.1.1   New Call Arrival (NC)

The number or MUs is increasing, the calls by each MU are restricted, and so the call arrivals can be considered as random and independent. In the simulation system, the new calls result from a random or a Poisson distribution with regard to a pre-defined regular pattern.

1.2.2.1.2 Call Termination (FC)

For each new MU, the call length is based on an exponential function. The call keep-ing time for later review is applied to current simulation time. The related protocol checks the phased call time for linked MUs.

1.2.2.1.3 Reallocation Management-Handoff (RC)

The computations are based on the intensity of the signal and how it affects adjacent cells from other related MUs. If the CNIR threshold is not reached by a MU signal, the procedure may attempt to find another suitable source. First, the algorithm calcu-lates the strength of the signal between MU and BS and then calculates any interfer-ence from connected MUs in later time. If an approved channel is identified, then the new MU is allocated, otherwise the call will be dropped.

1.2.2.1.4 MU Movement (MC)

The algorithm locates the MUs which are related and adjusts their current positions. A movement based on the Gaussian distribution is generated by the MU. This distri-bution is also used in simulation models similar to that found in literature (Dixit et al. 2001). The formula of Gaussian distribution probability density function is

$$f(x) = \frac{1}{\sigma\sqrt{2\pi}} e^{-(x-\mu)^2/2\sigma^2} \tag{1.3}$$

where σ is the standard deviation and μ is the baseline expected value. The Gaussian distribution applied has mean value 0 and standard deviation 1. Consequently, the probability density function is

$$f(x) = \frac{1}{\sqrt{2\pi}} e^{-x^2/2} \tag{1.4}$$

In the MU movement procedure, firstly a Gaussian number is generated, e.g. x_1 and the corresponding $f(x_1)$ when $x_1 \in [-0.5, 0.5]$. If another Gaussian number, e.g. x_2, where $|x_2| \geq f(x_1)$, a MU move is generated. The new position of any MU is maxi-mum two cells distance from the initiated cell.

1.2.2.2 Channel Allocation

1.2.2.2.1 Channel Allocation Criteria

Any channel allocation procedure is successful only if the following criteria are met:

- Channel availability
- Carrier strength (between MU and BS)
- CNR (signal purity)
- Signal to noise plus interference ratio CNIR (interference from other con-nected MUs)

1.2.2.2.2   Channel Allocation Procedure and Calculation

After a new call arrival, several actions take place in turn:

(a) Check if the maximum MU capacity in the cell neighbour has been reached.
(b) Calculate a random MU position in the mesh.
(c) Place the new MU according to the cell's BS position and mesh spot.
(d) Calculate the signal strength between BS and new MU in the call-initiated cell.

Firstly, the shadow attenuation (Molisch, A., 2012; Lee, W.C.Y., 1995) is derived. Next

$$sh = 10^{\frac{\sigma \cdot n}{10}} \tag{1.5}$$

where σ is the shadowing standard deviation and n is the associated number from the normal distribution. The distance attenuation dw can be derived from the shadow attenuation and distance between MU and BS. Through MU and BS, the CNR could be computed (Molisch, A., 2012; Lee, W.C.Y., 1995):

$$cn = 10^{\frac{cnedge}{10}} \cdot dw \tag{1.6}$$

where *cnedge* is the CNR on cell edge (dB).

(e) Calculate interference among the new MU and other MUs that use the same channel.
(f) Check if the CNIR ratio is acceptable according to a predefined threshold.
(g) If CNIR is acceptable, establish the new call and update UR, otherwise use any alternative selected DCA variation.

1.2.2.3   Traffic Generation

When a new call arrival occurs with regard to Poisson distribution, a random number x is generated from Poisson distribution and the corresponding $P(x)$ is calculated as follows:

$$P(x) = \frac{e^{-\lambda} \cdot \lambda^x}{x!} \tag{1.7}$$

where λ represents the new call arrival rate. This rate is analogous to the time of the day. A fixed set of simulation time steps represents a whole day. Each day is divided into five zones according to traffic conditions. Table 1.1 illustrates the five zones with corresponding λ parameter. Zone 1 represents the hours 12–9 a.m. ($\lambda = 1$) and zone 3 the hours 1–4 p.m. ($\lambda = 5$).

TABLE 1.1

Daily Zones

Zone	Zone Percentage (Approximately)	λ	Description
1	34%	1	Least busy zone
2	21%	4	
3	12%	5	Most busy zone
4	17%	3	
5	17%	2	

1.2.3 THE MULTI-AGENT/MULTILAYERED MODEL

The basic events like new call delivery, user movement, call status control etc. will arise concurrently in a real-world cellular network. As soon as a new user attempts to be connected to the network through a call, another call could be terminated at the matching time, another user is moved or reassigned to a new channel, etc. Hence, these things need to be simultaneously modelled. The main idea is that each basic event is independent of each other, acting independently, interacting with the environment, and being able to be implemented as an agent. The entire structure is divided into three layers in the proposed architecture (Figure 1.7).

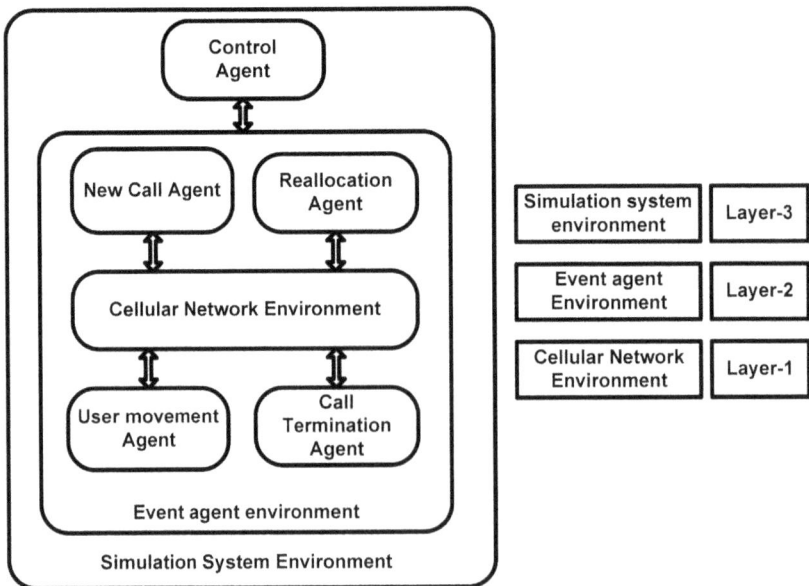

FIGURE 1.7   The proposed layered agent architecture.

NC Agent #3	RC Agent #3
MC Agent #3	FC Agent #3

NC Agent #1	RC Agent #1
MC Agent #1	FC Agent #1

NC Agent #2	RC Agent #2
MC Agent #2	FC Agent #2

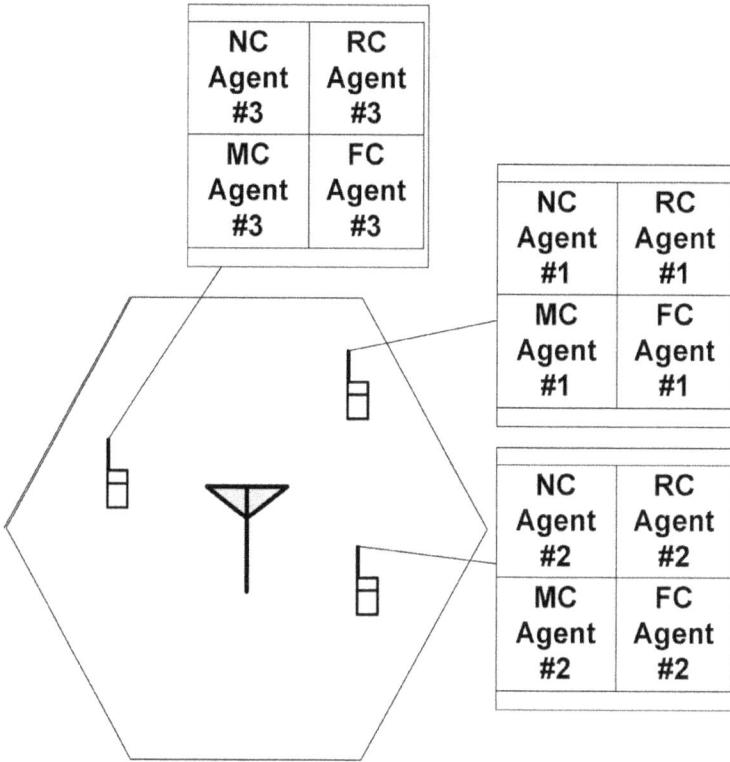

FIGURE 1.8 Three agent sets for three individual MUs.

The fundamental layer (layer 1) embodies the structure of the cellular network, where the basic events (new call, call rearrangement, user movement, call termination) take place. Layer 2 is made up of four agents that implement the synchronized events and describe the behaviour of the network. The control agent (main agent) that exists in the third layer activates the system with corresponding event agents. The set of four agents is recurring for every individual MU (Figure 1.8).

The main agent includes a clock logic which is responsible for the layer 2 agents' activation and ensures that the additional procedures are triggered in the correct order. The agents communicate for the activation status through layers 2 and 3 and exchange information (messages) etc. Event agents are independent entities that react with the appropriate environment and are coordinated by the control agent. The cellular network environment includes parameters such as information about the signal power, user characteristics, network status, etc. Every agent is implemented as a Java thread and is defined in relation to layered architecture by state, actions, and position. Agents in layer 2 are concurrently executed according to signals from the control node. The simulation time changes only when the layer 2 agents have completed the concurrent acts. Layer 1 uses shared data objects that are protected from the event agents in order to simultaneously access (read or write). Each event agent keeps the control agent informed about the status of its execution. The control agent is active, while the time for simulation is not over and so the rest of the agents. This agent has a sequential step-value clock (e.g. 1,2,3, etc.). A corresponding action is turned on in each value. All the acts (agents and supplementary procedures) are carried out as threads. The requisite supplementary activities (preparation tasks) are triggered in the first phase of the clock, while the other agents and procedures are disabled. In the second step of the clock, event agents are turned on while other procedures are disabled. Control agent prevents simultaneous activation within simulation time of the event agents, initial and final simulation procedures.

1.2.4 THEORETICAL ANALYSIS OF AGENTS ADAPTED TO MODELLED NETWORK SERVICES

1.2.4.1 Network Agent Definition

According to Wooldridge and Jennings (1995), "An agent is a computer device located in any setting that is capable of autonomous action in this environment to meet its design goals."

An agent communicates with its environment, obtains input information from it, and carries out a reasonable appropriate behaviour that affects this environment (Figure 1.9).

NC and RC agents (NCA, RCA) interconnect with the cellular network (environment), obtain input information (blocking the probability-network performance, decreasing RCA probability), and perform certain actions (giving priority to new calls or handovers for different MUs) affecting network efficiency (Figure 1.10).

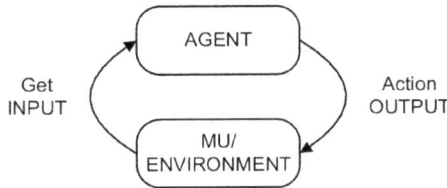

FIGURE 1.9   Agent definition.

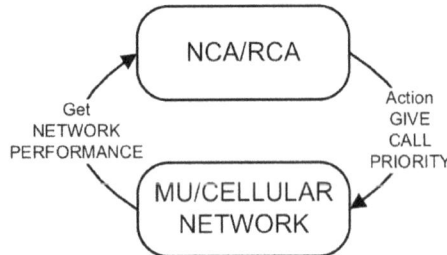

FIGURE 1.10   Network agent definition.

The agents' main features can be concisely outlined as follows:

- **Responsiveness**. An agent perceives its environment and reacts to fulfil the design goals. NCA perceives network efficiency and (a) prioritizes new calls and (b) negotiates the best deal with RCA to achieve its design goal of reducing blocking risk (network performance optimization). RCA works similarly but with probability dropping.
- **Pro-activeness**. Takes the initiative to demonstrate the goal-oriented behaviour to achieve the design goals. To get performance benefits for the network, network agents send messages to other MU agents (NCA to RCA and vice versa) in order.
- **Socio-power**. Interaction with other MU agents to attain the concept goals. NCA works with RCA to reach design goals.

1.2.4.2   Architecture of the Intelligent Network Agents

Let's assume that the conceivable discrete states of the network setting can be designated by the set E as

$$E = \{e, e', \ldots\} \tag{1.8}$$

Moreover, let's assume that the probable discrete states of the wireless network setting up can be defined by the set E as

$$E = \{LL, LH, HL, HH\} \tag{1.9}$$

where the memberships of E signify the network performance, L signifies low level, and H the high level. The pairs are linked to blocking and dropping probability, respectively.

On the other hand, it is adopted that each agent has a set of possible interactions with this set-up. These interactions change the environmental status and are determined by the set

$$Ac = \{\alpha_{k0}, \alpha_{k1}, \ldots\} \tag{1.10}$$

In the case of network set-up, the matching actions are determined by the set

$$Ac = \{INP_{NC}, INP_{RC}, DeP_{NC}, DeP_{RC}, DN_{NC}, DN_{RC}\} \tag{1.11}$$

where INP is the action "Increase Priority," DeP represents the action "Decrease Priority," and, finally, DN is the action "Do Nothing."

The external network set-up changes its state with regard to the above-mentioned activities. A sequence of actions results in a sequence of environmental state changes. Thus, a run r of an agent within the environment can be defined as

$$r : e_0 \xrightarrow{\alpha_0} e_1 \xrightarrow{\alpha_1} e_2 \xrightarrow{\alpha_2} e_3 \xrightarrow{\alpha_3} \cdots \xrightarrow{\alpha_{u-1}} e_u \tag{1.12}$$

A run r for the cellular network might develop

$$r : HL \xrightarrow{INP_{NC}, DN_{RC}} LL \xrightarrow{DN_{NC}, DN_{RC}} LH \xrightarrow{DN_{NC}, INP_{RC}} LL \xrightarrow{DN_{NC}, DN_{RC}} \cdots \xrightarrow{\alpha_{u-1}} e_u \qquad (1.13)$$

Let, also, consider the following sets:

   R: Set of possible finite sequences (over E and Ac)
   R^{AC}: Subset of R that ends with an action
   R^{E}: Subset of R that ends with an environment state

A state transformer function has been made known by Fagin et al. (1995) in order to depict the effect of an agent activity on the network set-up:

$$\tau : R^{Ac} \to \gamma\left(E\right) \qquad (1.14)$$

The previously mentioned function associates a run to a set of probable environmental states. Assuming that no successor state exists to r, $\tau(r)$ is developed as

$$\tau\left(r\right) = \varnothing \qquad (1.15)$$

The entire environment is defined (states, transformer function) as

$$Env = \left\langle E, e_0, \tau \right\rangle \qquad (1.16)$$

where E is the state set, e_o is the initial state, and τ is the transformer function. For modelling agents, it is adopted that an agent characterizes a function for mapping runs to activities and thus:

$$Ag : R^{E} \to Ac \qquad (1.17)$$

In other words, an agent decides about actions (to select an appropriate action to perform) based on the past events of the system. For expressing the entire system (agents, environment), a suitable set is developed:

$$R\left(Ag, Env\right) \qquad (1.18)$$

Finally, the sequence $(e_0, \alpha_0, e_1, \alpha_1, e_2, \ldots)$ expresses a run of an agent Ag (in the network set-up) $Env = \left\langle E, e_0, \tau \right\rangle$, if:

$$\alpha_0 = Ag\left(e_0\right) \qquad (1.19)$$

A run of agent NCA or RCA is expressed for instance by the sequence $(HL < INP_{NC} >, LL < DN >, LH <, \ldots)$, if $IP_{NC} = Ag\left(HL\right)$

and for $u > 0$,

$$e_u \in \tau\left(e_0, \alpha_0, ..., \alpha_{u-1}\right) \tag{1.20}$$

$$\alpha_u = Ag\left(e_0, \alpha_0, ..., e_u\right) \tag{1.21}$$

In the case of network set-up and for $u > 0$

$$e_u \in \tau\left(HL < INP_{NC}, DN_{RC} >, ..., \alpha_{u-1}\right) \tag{1.22}$$

$$\alpha_u = Ag\left(HL < INP_{NC}, DN_{RC} >, ..., e_u\right) \tag{1.23}$$

where Ag corresponds to NCA or RCA.

1.2.4.3 Network Agent Interface

As mentioned above, an agent perceives environment and acts on it. These two distinct actions are expressed by two functions separately (Figure 1.11).

Correspondingly, NCA and RCA, perceive network set-up and react on it. These two distinct actions for NCA and RCA are expressed by two functions, respectively (Figure 1.12).

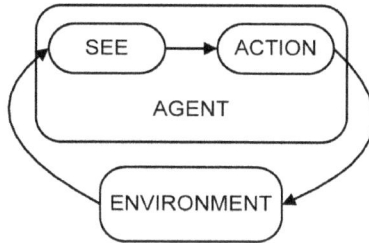

FIGURE 1.11   Agents interface.

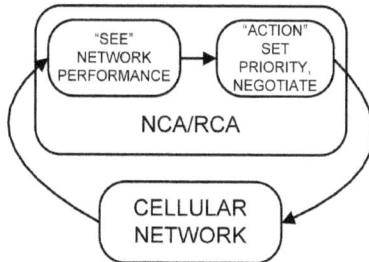

FIGURE 1.12   Network agents interface.

The "*perceive*" function maps environment states to observations and "*activity*" maps sequences of perceptions to actions.

As a concrete example of the above framework, let x express the statement "*metric M_1 is adequate*" and let y represent the statement "*metric M_2 is adequate.*" Thus, the set E contains four combinations of x and y. This set can be expressed as follows:

$$E = \left\{ \{\bar{x}, \bar{y}\}, \{\bar{x}, y\}, \{x, \bar{y}\}, \{x, y\} \right\} \tag{1.24}$$

with

$$e_1 = \{\bar{x}, \bar{y}\}, \quad e_2 = \{\bar{x}, y\}, \quad e_3 = \{x, \bar{y}\}, \quad e_4 = \{x, y\} \tag{1.25}$$

Network behaviour is evaluated through two basic statistical metrics which are the blocking and dropping likelihoods. Thus, (1.24) and (1.25) will be expressed in terms of the above metrics as follows:

$$E = \left\{ \{\bar{B}, \bar{D}\}, \{\bar{B}, D\}, \{B, \bar{D}\}, \{B, D\} \right\} \tag{1.26}$$

with

$$e_1 = \{\bar{B}, \bar{D}\}, \quad e_2 = \{\bar{B}, D\}, \quad e_3 = \{B, \bar{D}\}, \quad e_4 = \{B, D\} \tag{1.27}$$

where B represents the statement "*Blocking possibility is adequate*" and D represents the statement "*Dropping possibility is adequate.*" Now, the set E contains four mixtures of B and D.

With regard to (1.24) and (1.25), the "perceive" function of the agent, will have two observations in its range, P1 and P2 that indicate if the metric M1 is acceptable or not. The behaviour of the "perceive" function can be described as follows:

$$see(e) = \left\{ \begin{array}{llll} p_1 & \text{if} & e = e_1 & \text{or} \quad e = e_2 \\ p_2 & \text{if} & e = e_3 & \text{or} \quad e = e_4 \end{array} \right\} \tag{1.28}$$

According to (1.26) and (1.27), the "perceive function" of the NCA will have two observations in its range P1 and P2 that indicate if the blocking possibility is acceptable or not. The behaviour of the "perceive" function can be outlined as follows:

$$see(e) = \left\{ \begin{array}{llll} P_1 & \text{if} & e = e_1 & \text{or} \quad e = e_2 \quad \text{bad} \\ P_2 & \text{if} & e = e_3 & \text{or} \quad e = e_4 \quad \text{good} \end{array} \right\} \tag{1.29}$$

Identically for RCA, the perceive function $see(e)$ is formulated as follows:

$$see(e) = \left\{ \begin{array}{llll} P_1 & \text{if} & e = e_1 & \text{or} \quad e = e_3 \quad \text{bad} \\ P_2 & \text{if} & e = e_2 & \text{or} \quad e = e_4 \quad \text{good} \end{array} \right\} \tag{1.30}$$

With two given environmental states $e \in E$ and $e' \in E$, then $e \sim e'$ can be written only if $see(e) = see(e')$. An agent has perfect perception if the different environment states are equal to distinct observations. In such case

$$|\sim| = |E| \tag{1.31}$$

In contrast with (1.15), when the perception of an agent does not exist,

$$|\sim| = 1 \tag{1.32}$$

1.2.4.4 Network Agents Which Maintain State

In many cases, an agent maintains state. In such a case, there are two additional objects within the internal structure (Figure 1.13). In the proposed MA model, each of the two basic agents (NCA, RCA) can be in one of the following states: (a) active (in cooperation), (b) active (in competition), (c) inactive (Figure 1.14).

The agent (internal) state is updated by the "next" function. This function can be expressed as $next(i_0, see(e))$ and the corresponding action is $action(next(i_0, see(e)))$, where i_0 is the internal initial state.

1.2.4.5 Network Agent Utility Functions

A different utility function can be assigned to each agent (NCA, RCA) in order to "measure" how good the corresponding outcome is (blocking/dropping). This

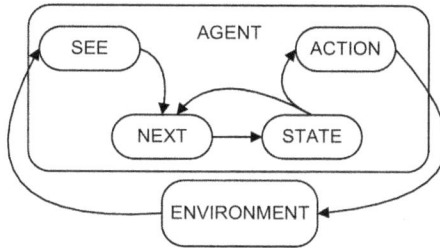

FIGURE 1.13 Agent with state.

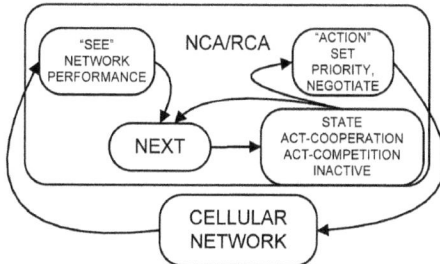

FIGURE 1.14 Network agent with state.

function assigns a real number to each outcome, indicating how good the outcome is for the selected agent. In other words, this utility function defines the preference ordering over all the outcomes.

Let ω and ω' be two possible outcomes in Ω set (where $\Omega = \{\omega 1, \omega 2, \ldots\}$), with utility functions that give

$$u_i(\omega) \geq u_i(\omega') \tag{1.33}$$

Based on (1.16), agent i prefers better the ω outcome.

1.2.4.6 Multi-Agent Encounters

In a large-scale environment with more than one agent, except the environment itself, the agent encounters also have to be modelled. Assume that there are two possible actions by two agents such as "C", which means "Cooperate", and "D", which means "Defect" and let the action set $Ac = \{C,D\}$. Based on the above, the environment behaviour can be modelled with the function

$$\tau : Ac_i \times Ac_j \rightarrow \Omega \tag{1.34}$$

where Ac_i and Ac_j represent the two agents i and j, respectively.

As an example, let the environment function

$$\tau(D,D) = \omega_1, \quad \tau(D,C) = \omega_2, \quad \tau(C,D) = \omega_3, \quad \tau(C,C) = \omega_4 \tag{1.35}$$

The two possible actions for NCA and RCA are the Cooperation (CR) and Competition (CT) and so the action set becomes $Ac = \{CP, CT\}$. Now the expression (1.35) is as follows:

$$\tau(CT,CT) = \omega_1, \tau(CT,CR) = \omega_2, \tau(CR,CT) = \omega_3, \tau(CR,CR) = \omega_4 \tag{1.36}$$

According to (1.35), each action is mapped to a different outcome.

Using utility functions for the ω in (1.35), let

$$
\begin{array}{llll}
u_i(\omega_1) = 1 & u_i(\omega_2) = 1 & u_i(\omega_3) = 4 & u_i(\omega_4) = 4 \\
u_j(\omega_1) = 1 & u_j(\omega_2) = 4 & u_j(\omega_3) = 1 & u_j(\omega_4) = 4
\end{array}
\tag{1.37}
$$

The corresponding utility functions (NCA, RCA) in (1.18a) are

$$
\begin{array}{llll}
u_{NC}(\omega_1) = 1 & u_{NC}(\omega_2) = 1 & u_{NC}(\omega_3) = 4 & u_{NC}(\omega_4) = 4 \\
u_{RC}(\omega_1) = 1 & u_{RC}(\omega_2) = 4 & u_{RC}(\omega_3) = 1 & u_{RC}(\omega_4) = 4
\end{array}
\tag{1.38}
$$

From (1.35) and (1.37), the outcomes for every action combination of the two agents i,j can be expressed as follows:

$$
\begin{array}{llll}
u_i(D,D) = 1 & u_i(D,C) = 1 & u_i(C,D) = 4 & u_i(C,C) = 4 \\
u_j(D,D) = 1 & u_j(D,C) = 4 & u_j(C,D) = 1 & u_j(C,C) = 4
\end{array}
\tag{1.39}
$$

Combining now (1.36) and (1.38) for NCA and RCA, the outcomes can be expressed as follows:

$$u_{NC}(CT,CT) = 1 \quad u_{NC}(CT,CR) = 1 \quad u_{NC}(CR,CT) = 4 \quad u_{NC}(CR,CR) = 4$$
$$u_{RC}(CT,CT) = 1 \quad u_{RC}(CT,CR) = 4 \quad u_{RC}(CR,CT) = 1 \quad u_{RC}(CR,CR) = 4 \quad (1.40)$$

The only possible agent actions for the given outcomes can be rewritten (see (1.39)):

$$C,C \geq_i C,D >_i D,C \geq_i D,D \qquad (1.41)$$

In the same way, the possible NCA, RCA actions based on (1.20a) are

$$CR,CR \geq_{NC} CR,CT >_{NC} CT,CR \geq_{NC} CT,CT \qquad (1.42)$$

It is obvious from (1.39), (1.40), and (1.41) which action will be selected from each agent, and thus each agent knows exactly what to do.

1.2.5 EVENT INTERLEAVING AS SCHEDULING TECHNIQUE BASED ON REAL-TIME SCHEDULING THEORY

1.2.5.1 Real-Time Scheduling Algorithms for Implementing Synchronized Processes or Events

Real-time scheduling is the most commonly recognized method for handling and running simultaneous tasks where response time is a critical problem. The principle of real time scheduling can be used to establish an alternative scheduling of events compared to the state-of-the-art scheduling mechanism (CQ) used in DES systems to accommodate concurrent events.

Real-time (RTS) systems are commonly used in a range of vital applications, including robotics, avionics, telecommunications, process control, etc. In several embedded systems, RTSs are also used. The common aspect of the above programs is the tight deadlines for production tests. In other words, the timing of the results is critical and in some cases even harmful. According to Burns and Wellings (2001), "Real-time systems are those whose correctness depends not only on the rationality of the computational results, but also on their precise timing." According to Nahas (2014), the interaction of the real-time control system with the environment requires different components, such as sensors that read the external environment state. The software architectures are common RTS implementation. Based on Nahas (2014) and Locke (1992), these architectures can be classified into (a) cyclic executives, (b) event-activated simultaneous task systems, (c) message-passing systems, and (d) client–server systems. A specialized scheduler is required in concurrent task systems (Nahas, M., 2014) in order to take scheduling decisions as the notion of a task is retained in real time. The real-time scheduling theory is one way of predicting the timing behaviour of complex multitasking software. Several developments have been made since the initial formulation of the scheduling theory, such as rate monotonic analysis in Liu and Layland (1973) and deadline monotonic analysis in Audsley et al. (1992). These two improvements were incorporated into the scheduling theory of fixed priorities (Audsley, et al. 1995). Some very important characteristics of the RTSs found in Buttazzo (2011) are (a) repetitive contact with the environment, (b) multiple actions concurrently transitioning rapidly to events and requiring a high degree of competition, (c) competition for shared resources, (d) actions caused externally by events (within the system) or after time development, (e) must be stable in overload, and (f) must be maintainable and of scalable open architecture.

1.2.5.2   Process Life Span in a Real-Time Scheduling Set-Up

The process life span in an RTS can be investigated by utilizing the single-queue methodology combined with the pre-emption scheme.

An RTS consists of three basic components (Fidge, C.J., 2002): (a) a collection containing the computational processes to be performed (typically subroutines with a private control thread), (b) a run-time scheduler (e.g. dispatcher) specifying the process execution sequence, and (c) a collection of common resources to be accessed. Figure 1.15 demonstrates the standard life span of a job execution through a queue from event to termination.

The processes that are facilitated by the system can be characterized by the time features of arrival as (a) periodic (non-changing static arrival intervals) (Buttazzo, G.C, 2011), (b) aperiodic (random arrivals) (Sprunt, B. 1990), and (c) infrequent (they have only a non-changing minimum inter-arrival time) (Sprunt, B. 1990).

1.2.5.3   Scheduling Concurrent Events in WCS

In a wireless set-up, where events occur synchronously, the performance of the network depends not only on the rationality of the computational results, but, also on the correctness of the time sequence in which the results are encountered, implying that the logical sequence of the various network procedures is also related to efficient bandwidth management. Network operations such as new call entry, reallocation (handoff), call termination, and so on can be regarded as tasks the network needs to perform. A major role for the success of the resulting network is the scheduling mechanism that manages the process under service. That is, this framework determines how, in terms of network performance metrics, the network can precisely express the synchronized processes (incidents) in the most efficient way. The progress and actions of the event service may be seen by applying the response time analysis as a first step. The sequential execution of an event is due to the existence of one processor in conventional computers. The most common simulation tools, such

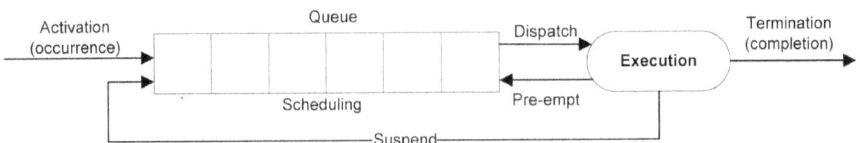

FIGURE 1.15   Task lifecycle.

as ns-2/ns-3, are focused on the scheduling of the CQ form, which reflects the serial logic. Within this conceptual framework the execution sequence of each generated event is defined from the associated time stamps. CQ retains information about activities for future executions. Such time stamps can also be used as goals. The computational multi-processing approach can be implemented when the network events are addressed as synchronized. In a simultaneous model, another MU is moving or attempting for reallocation (handoff) when a MU is under network processing. Thus, the execution of the mission (event) has to be incomplete to accommodate the waiting synchronized MUs. This reasoning leads to the inter-leaving scheme that can normally be enforced using a multi-threading framework. Figure 1.16 demonstrates the queuing models for the two approaches (consecutive and concurrent).

In a concurrent set-up, events can be viewed as multiple processes that must be facilitated by the available computing unit (Figure 1.17).

1.2.5.4 Response Time Analysis

Owing to the frequency of tasks, the whole system can be interpreted as a multi-processing device in real time when other tasks are being performed. That approach leads to sophisticated control and interleaving of events. Multi-processing concepts

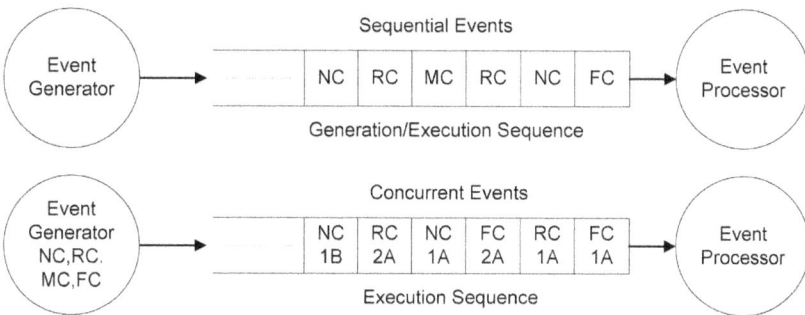

FIGURE 1.16 Queuing models for serial and synchronized processes.

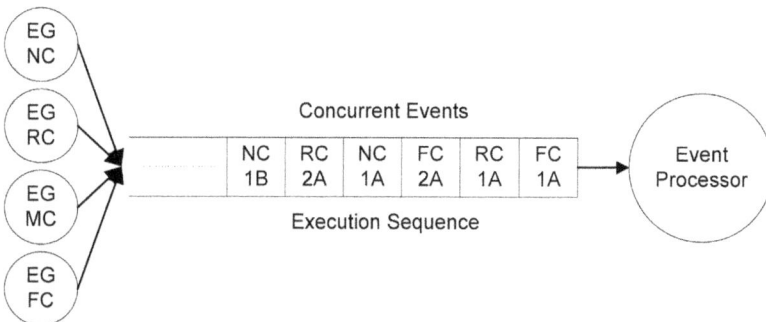

FIGURE 1.17 Multiple processes (events) generators related to one server.

and an analysis of schedulability (Fidge, C.J. 2002), reflect events behaviour that could be analyzed as follows:

Any event e, with $e \in \{NC, RC, MC, FC\}$ has a release time r:

$$r_e = (n \cdot T_e) - J_e \tag{1.43}$$

where n is the nth event, T_e is the event period, and J_e is the associated Jitter. J_e is computed as

$$J_e = re(\max) - re(\min) \tag{1.44}$$

In the computations, $J_e = 0$.

Computational time of an event is

$$C = (n+1) \cdot C_e \tag{1.45}$$

Let R_i be the response time for process i, C_i be the worst-case calculation time, and I_i be the interference, due to higher priority processes. The response time for process i is

$$R_i = C_i + I_i \tag{1.46}$$

Assuming that all priorities are unique, the interference for process i, is

$$I_i = \sum_{j \in hp(i)} \left\lceil \frac{R_i}{T_j} \right\rceil C_j \tag{1.47}$$

where $hp(i)$ is the set of all processes that have greater priority than i.

From Equations 1.46) and 1.47, the worst-case response time of process i can be computed through the following formula:

$$R_i = C_i + \sum_{j \in hp(i)} \left\lceil \frac{R_i}{T_j} \right\rceil C_j \tag{1.48}$$

Equation 1.48, can be solved iteratively (Audsley et al. 1993), commencing with the initial estimate for R_i equal to 0. Knowing the xth value, the $(x+1)$ value can be approached as follows:

$$R_i^{x+1} = C_i + \sum_{j \in hp(i)} \left\lceil \frac{R_i^x}{T_j} \right\rceil C_j \tag{1.49}$$

The main and useful conclusion derived from Equation 1.49 is process i schedulability, which means that the deadlines required can be met. For safe conclusions, the above equation must be re-evaluated to see if it converges to an R_i value such that $R_i < D_i$ (D_i = the given process i time limit). Instead of the existence of specialized

tools to evaluate the above equations, such tools do not necessarily give the programmer any insight into the dynamic conduct of each process (Fidge, C.J., 2002).
Simulation is the most effective method for demonstrating the timing actions of the
processes in consideration (Fidge, C.J., 2002).

The event goals must be set for conducting full calculations. Let PFC, PRC, PMC,
and PNC be the priorities for the event types FC, RC, MC, and NC with PFC > PRC,
PRC > PMC, PMC > PNC, respectively. The corresponding $hp(j)$ sets involving the
definitions above can be defined as follows:

$$hp(FC) = \{\varnothing\} \tag{1.50}$$

$$hp(RC) = \{FC\} \tag{1.51}$$

$$hp(MC) = \{RC, FC\} \tag{1.52}$$

$$hp(NC) = \{MC, RC, FC\} \tag{1.53}$$

Equation 1.49 can be solved iteratively (Audsley et al. 1993). Thus, the response time
for next event $n+1$ can be calculated as follows:

$$R_e^{n+1} = C + \sum_{j \in hp(e)} \left\lceil \frac{R_e^n}{T_j} \right\rceil \cdot C_j \tag{1.54}$$

Formulating the above equations to network procedures, the resulted response times
are as follows:

Assuming that $R_e^0 = 0$,

$$R_{FC}^{n+1} = C_{FC} + \sum_{j \in hp(FC)} \left\lceil \frac{R_{FC}^n}{T_j} \right\rceil \cdot C_j \Rightarrow R_{FC} = C_{FC} \tag{1.55}$$

Since $hp(FC) = \{\varnothing\}$

$$R_{RC}^{n+1} = C_{RC} + \left\lceil \frac{R_{RC}^n}{T_{FC}} \right\rceil \cdot C_{FC} \tag{1.56}$$

$$R_{MC}^{n+1} = C_{MC} + \left\lceil \frac{R_{MC}^n}{T_{RC}} \right\rceil \cdot C_{RC} + \left\lceil \frac{R_{MC}^n}{T_{FC}} \right\rceil \cdot C_{FC} \tag{1.57}$$

$$R_{NC}^{n+1} = C_{NC} + \left\lceil \frac{R_{NC}^n}{T_{MC}} \right\rceil \cdot C_{MC} + \left\lceil \frac{R_{NC}^n}{T_{RC}} \right\rceil \cdot C_{RC} + \left\lceil \frac{R_{NC}^n}{T_{FC}} \right\rceil \cdot C_{FC} \tag{1.58}$$

Based on the above rules, a pre-emptive event scheduling can be applied.

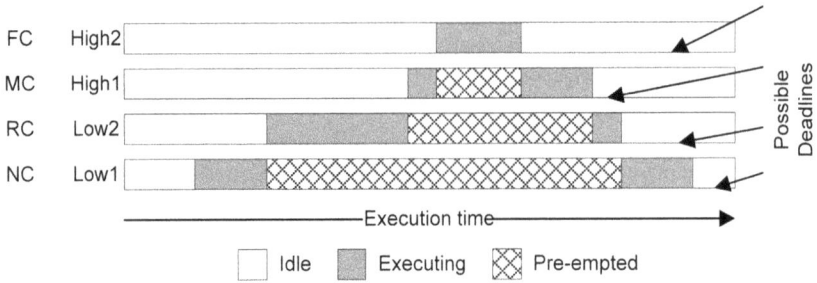

FIGURE 1.18 Pre-emptive stationary priority scheduling (tasks/events: NC – new call, RC – call reallocation/handoff, MC – movement call, FC – finished call).

1.2.5.5   Pre-emptive Stationary Priority Scheduling (PSPS)

In PSPS scheduling, each task has a non-changing priority that is constant during execution time. The main rewards of this strategy are (a) faster facilitations of processes with higher priority (Fidge, C.J., 2002; Pont et al. 2007), (b) simpler realization, (c) stability in overloads and spiking behaviour. In a real network set-up, while an MU is under processing, another MU might be arrived.

Figure 1.18 illustrates four processes with different priorities (Low1 < Low2 < High1 < High2) during a time window of execution.

For example, while an NC type of event is under processing, a new event (RC) is arrived (Figure 1.18).

1.2.6   Supported DCA Variations

For the assessment procedure, we used various DCA variations:

* Unbalanced version reflecting traditional DCA
* A conventional balanced version based on a less congested neighbourhood algorithm)
* A conventional Best CNR channel assignment algorithm that is based on the best measured CNR between current MU and local or neighbouring BS
* A traditional round blocking algorithm where the adjacent channels are in turn checked

All the above algorithms assume sequential search only within the neighbouring cells between the current user and available channels.

The following proposed algorithm assigns channels inside neighbouring cells simultaneously, provided that N agents/users emerge at a given time.

* Artificial intelligence-based variation controlled and good CNR DCA for concurrent network assignment

1.2.6.1   The Conventional Unbalanced Variation (Classical DCA)

Unbalanced variation represents the classical DCA scheme where only one attempt (in call initiating cell) for connection is done based on the CNIR criterion (Figure 1.19).

Compute CNR from Current BS

Find available channel

Compute Interference from other users

CNIR > CNIR Threshold ?

YES : Success =1

NO : Success = 0

FIGURE 1.19 Unbalanced algorithm.

1.2.6.2 The Conventional Balanced Variation (Min Cell Congestion)

The unbalanced DCA algorithm assigns the first channel which meets the conditions and the balanced DCA uses the least congestion algorithm to complete the channel assignment.

The balanced approach uses the least congested algorithm, as we have already stated. Initially we presume that the cell where the new call occurs is the least congested. The algorithm checks all the neighbouring region's cells sequentially and identifies the least congested cell (Figure 1.20).

1.2.6.3 The Conventional Best CNR Variation

The best CNR method goes around the initiated cell's six neighbour cells and calculates the CNR between the base stations for the user and the neighbour cells. The software tries to create a new channel from the corresponding cell until the strongest CNR is identified. Initially, the algorithm assumes that the strongest CNR occurs

Current Congestion = Congestion of Current cell (%)
Least Congested = Current cell (number of cell)
Min Congestion = Current Congestion (%)
START_LOOP
   Get first/next Neighbour cell
   Current Congestion = Congestion of Neighbour
   Current Congestion < Min Congestion ?
YES :
      Least Congested = Current Neighbour cell
      Min Congestion = Current congestion
NO : -
Have all Neighbour cells checked ?
NO :GOTO START_LOOP
YES :Break
END LOOP

FIGURE 1.20 Min cell congestion algorithm.

Max CNR cell = First Neighbour cell
Current cell = First Neighbour cell
Max CNR = CNR of Current cell
START_LOOP
Reach first/next Neighbour cell
   Current cell = Neighbour cell
   Calculate CNR for Current cell
   Current CNR greater than Max CNR ?
YES :
          Max CNR cell = Current Neighbour cell
          Max CNR = CNR of Current Neighbour cell
NO :
Have all Neighbour cells reached and checked ?
NO :GOTOSTART_LOOP
YES :Break
END LOOP

FIGURE 1.21   Best CNR algorithm.

from the first neighbour cell and proceeds to measure CNR sequentially from the rest of the neighbour cells until the strongest CNR is identified. Figure 1.21 illustrates the structure and logic of the algorithm, respectively.

1.2.6.4   The Conventional Round Blocking Variation

The round blocking algorithm checks in neighbouring cells, and the algorithm stops when a successful channel assignment is made. If the call establishment in the last neighbouring cell is not successful, then the call is blocked. The round blocking algorithm and its rationale are shown in Figure 1.22, respectively.

START_LOOP
   Get first/next Neighbour cell
   Compute CNR with respect to the Current BS
Check channels availability
   Calculate Interference from other users
   CNIR > CNIR Threshold ?
YES : Success =1
NO : Success = 0
   Last Neighbour reached OR Successful linkestablishment?
YES :Break
NO :GOTOSTART_LOOP
END LOOP

FIGURE 1.22   Round blocking algorithm.

1.2.6.5   The Proposed Novel Artificial Intelligence Based Balanced and Best CNR DCA Variation for Concurrent Channel Assignment

Suppose at any given moment t, N agents/users emerge in a given neighbour cell under consideration. Also suppose that $C(j)$ is the number of available cells per cell j in this neighbour. The problem is how each agent/user i out of the total N agents/users should be assigned to a channel in an optimal way according to the least congestion criterion and best CNR criterion. If this is done sequentially, then it is not obviously optimal. In order that this assignment problem could be solved in an optimal way, we should involve either dynamic programming or a global optimization methodology. In this research, we have employed genetic algorithms which can lead to an acceptable solution, especially when the problem is of fair dimensionality.

The chromosome in the proposed algorithm is as follows

agent_1	agent_2	agent_3	...	.....	....	.....	....	agent_N-1	agent_N
Ch(1)	Ch(2)	Ch(3)						Ch(N-1)	Ch(N)

Therefore, the chromosome comprises the possible channels (out of the available ones) to be assigned to every agent/user. If there are no available channels for all N agents/users we put zeros in the corresponding positions in the chromosome. An exhaustive search of all possible combinations of C in total available channels in groups of N agents/users, $\left(\dfrac{C}{N} \right)$, requires factorial time increase as it is known and makes impossible the task in real time. Genetic algorithms in this case can provide an affordable near optimal solution.

The fitness function used is comprised of the following parts:

1. $\displaystyle\sum_{j=1}^{all\ cells} \sum_{i=1}^{C(j)} CNR(i,j)$ should be maximized. This means that for all channels

   $C(j)$ of cell j assigned, let's say, to $C(j)$ agents/users out of N and for all cells the total CNR should be maximized.

2. $\displaystyle\sum_{i} \sum_{j} \left(congestion(i) - congestion(j) \right)^2$ should be minimized for all cell

   pairs (i,j).

3. Each available channel $ch(i)$ assigned to agent i should be used only once.

4. $\displaystyle\sum_{i} ch(i,j) \le C(j)$, that is, all channels of cell j assigned to agents/users i

   should be less than or equal to the available channels $C(j)$ of channels j.

The above fitness function and the chromosome architecture define completely the genetic algorithm (GA) solution

1.2.7 IMPLEMENTATION ARCHITECTURES

We have implemented and tested three different approaches of agent-oriented architectures using Java threads. Each approach is characterized by the parallelization grade that influences the simulation efficiency and the reliability related to network behaviour.

1.2.7.1 Conventional Model

The conventional model consists of one thread and the event processing is based on the CQ mechanism (Figure 1.23). According to this model, any generated event is served through a three-step procedure which includes enqueue, dequeue, and execution. Based on the type of the highest priority event, the corresponding network service is activated.

1.2.7.2 Concurrent Models

As we mentioned before, in a real network, processes take place simultaneously. A model that supports concurrent architectures approaches in a more reliable way the real network. We have developed and tested two architectural models with different

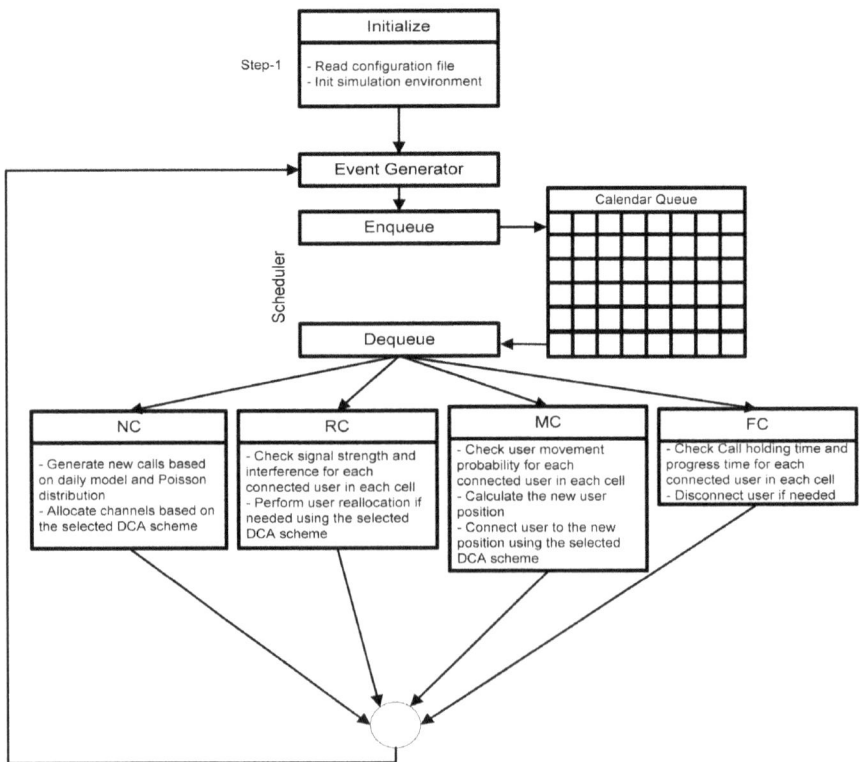

FIGURE 1.23   Conventional model.

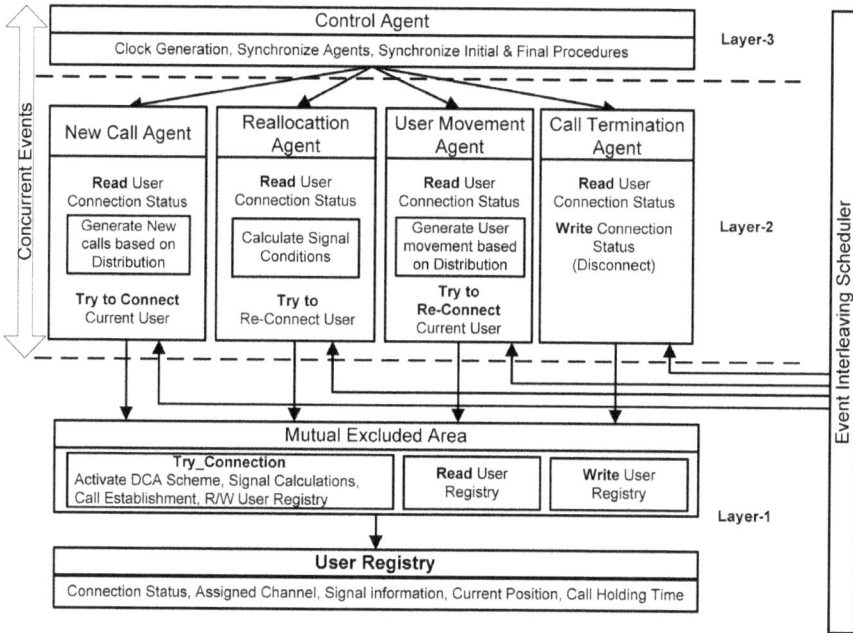

FIGURE 1.24 Basic concurrency.

levels of concurrency. These models rely on the mentioned layered agent architecture using Java multi-threading techniques for the implementation.

1.2.7.2.1 Basic Concurrency

The concurrent version of the model must prevent simultaneous access to the user registry where critical data for the MUs are stored. Figure 1.24 illustrates the structure of the implemented architecture. Level 0 represents the control agent that synchronizes the execution of the concurrent tasks (New call, reallocation, etc.) in combination with the thread scheduler and administers the simulation time. Concurrent tasks are activated within the level 1 and share the connection establishment methods. The main network processes like new call arrival, reallocation check, user movement, etc. execute common tasks such as access to user registry and connection establishment. Connection establishment and user registry access (read, write) are implemented with three mutual excluded methods (level 2).

These methods belong to the same thread. The main agents act simultaneously and try to finish for each MU in each cell. A new connection based on reallocation, user movement, or new call arrival requires signal calculations and other DCA operations. These actions are implemented using the try_connection method. All the threads work simultaneously except the above method and access to user registry that is used by one thread at a time preventing the data corruption of user registry.

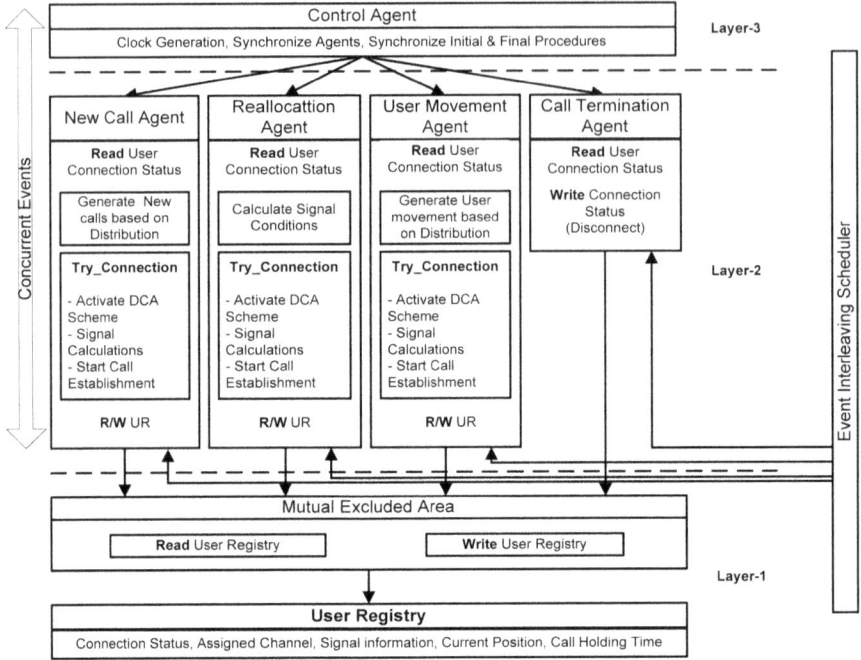

FIGURE 1.25   Concurrency with multiple synchronized blocks.

1.2.7.2.2   Concurrency with Multiple Synchronized Blocks

In this model, every main agent (new call, reallocation, etc.) has a separate instance of the try_connection method. Mutual exclusion is now moved to user registry access (Figure 1.25). Each agent can complete its connection establishment independently from the others. The resource competition between concurrent threads is restricted to user registry access and not in try_connection method like the previous mentioned model. Thus, we achieve a higher level of concurrency.

1.2.7.2.3   Thread Usage

The main implementation concept is based on thread logic. Each user/network operation is implemented as a thread. The Java programs with regard to the above-mentioned concurrent architectures consist of four basic threads (call termination, reallocation, new call arrival, user move) and three additional (clock synchronization, initial/final procedures). As the concurrency grade increases, the Java threads become more competitive for accessing the user table (read/write tasks). Thus, the possibility for deadlocks is rising. In order to face effectively the deadlock conditions, synchronized blocks instead of synchronized methods can be used. Synchronized blocks set locks for shorter periods than synchronized methods and so the deadlock possibilities decrease. Table 1.2 shows the threads and synchronized methods/blocks that are used in the proposed architectures.

TABLE 1.2
Thread Usage by Architectural Model

Concurrency Architecture	Basic Concurrency	Concurrency with Multiple Synchronized Blocks
No. of threads	7	7
Thread tasks	Clock, initial loop procedures, final loop procedures, call termination, reallocation check, new call arrival, user movement	
No. of synchronized methods	3	0
Synchronized methods	Read User Registry Write User Registry Try_connection	–
No. of synchronized blocks	0	18 (Read/Write User registry)

In the first approach, the most of the access points for user registry exist within the try_connection method that is synchronized. Due to that synchronization, the competition for user registry access is low. Increasing concurrency, the try_connection is concurrent among the four main threads and so only the access methods (read/write to user registry) are synchronized. With this approach, we achieve high concurrency but the deadlocks seem to be a difficult situation. Using synchronized blocks in contrast with synchronized methods, we faced successfully the deadlock conditions. Extending the above models for avoiding deadlocks, conditional synchronization can be applied. With conditional synchronization, the user registry access is synchronized only when two or more threads try to access data for the same MU.

1.3 SIMULATION MODEL EVALUATION

1.3.1 NETWORK BEHAVIOUR

One of the most important features for characterizing the performance of a cellular network is the probability of blocking. When a new call arrival happens and a channel cannot be allocated by the network, then we say this call is blocked. The probability of blocking $P_{blocking}$ is determined from the proportion.

$$P_{blocking} = \frac{number\ of\ blocked\ calls}{number\ of\ calls} \qquad (1.59)$$

The risk of a call dropping is also an additional and a very critical cellular network feature. If a call is ongoing and the required quality conditions are not met, then this call is mandatorily led to termination. The probability of dropping P_{fc} is computed from the following proportion:

$$P_{fc} = \frac{number\ of\ forced\ calls}{number\ of\ calls - number\ of\ blocked\ calls} \qquad (1.60)$$

1.3.2 Monte Carlo Simulation Method

Monte Carlo (MC) methods are stochastic techniques, meaning they are based on the use of random numbers and probability statistics to investigate problems. For MC simulations, the processes are random; so each time it is run, it will come up with slightly different results. This method gives us a way to model complex systems that are often extremely hard to investigate with other types of techniques. Moreover, MC method is used in order to estimate parameters of an unknown distribution (e.g. network behaviour) by statistical simulation (Fishman, G.S., 1995).

1.3.3 Simulation Model Behaviour

The stability of the simulation model is derived by sequential Monte Carlo simulations from the progressive standard deviation of selected network properties and metrics.

A further metric for evaluating the performance of the simulation model is periodicity, a metric dependent on outliers. As we have already mentioned, simulation time is based on a regular model. The periodicity is determined within the simulation time using all available pairs of days. Assume that the likelihood of blockage within a two-day set of 48 hours is expressed by the set a and set b matrices, respectively. First, we calculate the $BPDk$ blocking likelihood differences within a pair of days for every hour:

$$\vdots$$

$$BPD_k[i] = \left| seta\left[i \right] - setb\left[i \right] \right| \qquad (1.61)$$

where k is the selected pair and i is the i-element of each set. The difference $seta[i]$ – $setb[i]$ is calculated only if the $seta[i]$ and $setb[i]$ values are not outliers; otherwise, the data pairs are rejected. We then compute the expected value M_k for all possible such successive pairs:

$$M_k[] = \frac{\sum\limits_{i=1}^{n} BPD_k\left[i \right]}{n} \qquad (1.62)$$

Third, we calculate periodicity by measuring the mean value between the mean M_k. When the difference between elements tends to be zero between pairs of different days, the pairs tend to be equal; so the factor of periodicity tends to be zero for the given pair. When more pairs have a low periodicity factor, then the mean factor value of more pairs is also low. Smaller periodicity factor means higher periodicity and simulation model behaviour is more stable in simulation time as opposed to the regular model.

1.3.4 Results Accuracy

Outliers and other common statistical metrics are used to measure the accuracy of results. Outliers are unusual data values which may distort the results obtained

(Barnett, V. and Lewis, T., 1994). When out of an agreed area, we describe a value as outlier. This region's boundaries are defined as follows:

$$L = m \pm 0.95 stdX \tag{1.63}$$

where $stdX$ is the standard deviation of the likelihood of blocking or link dropping, and m is the mean simulation time value.

To investigate how the system design and implementation affects the accuracy and reliability of the test, we employed several overall statistical metrics with regard to blocking and dropping probabilities throughout all simulation runs such as expected value of standard deviation, mean value, variance, etc.

1.3.5 REFERENCE ANALYSIS MODEL EMPLOYING ONE CELL ONLY

Initially, a conceptual level reference model was developed to check the accuracy of the event interleaving approach compared to established CQ approaches. Due to the conceptual model's simplicity, only one cell is presumed to support MUs within the coverage area. Figure 1.26 shows the cell offering n channels to support new call entries, reallocation, and movement of MUs. This model includes no advanced

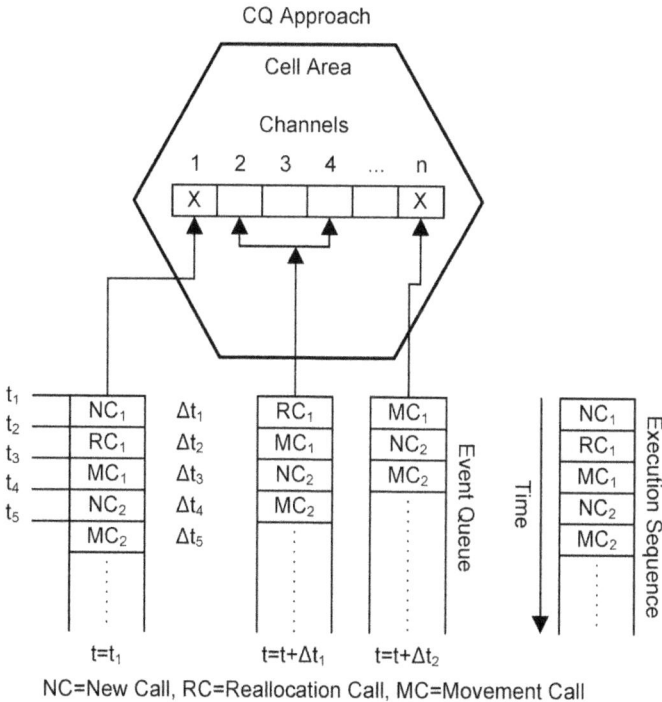

FIGURE 1.26 MU service operation based on CQ approach.

mathematical model (e.g. for signal propagation) since it is based on how the channels express the clients (MUs).

Both models were initially tested for event-generation with deterministic data. Figure 1.26 shows how some of the five events (NC1, NC2, RC1, MC1, RC2) perform. The scheduler restores the event with the least time stamp (highest priority) for subsequent execution from the queue. When the event interleaving system manages the same events queue (Figure 1.27), the execution sequence remains identical, while the scheduler time slice (time slice width – TSW) is greater than or equal to the event computational time (ECT). In other words, if $TSW > ECT$, then one slice of time in the predetermined sequence is enough to complete execution of each generated event. And the two conceptual models yield similar results. When the TSW is less than the ECT for finishing execution of the event, or the TSWs are assigned asymmetrically to active threads, the results are totally different from the traditional CQ approach. This is due to the competitiveness of running threads, associated with individual MUs, for specific resources management and the relevant scheme of channel allocation at different time units.

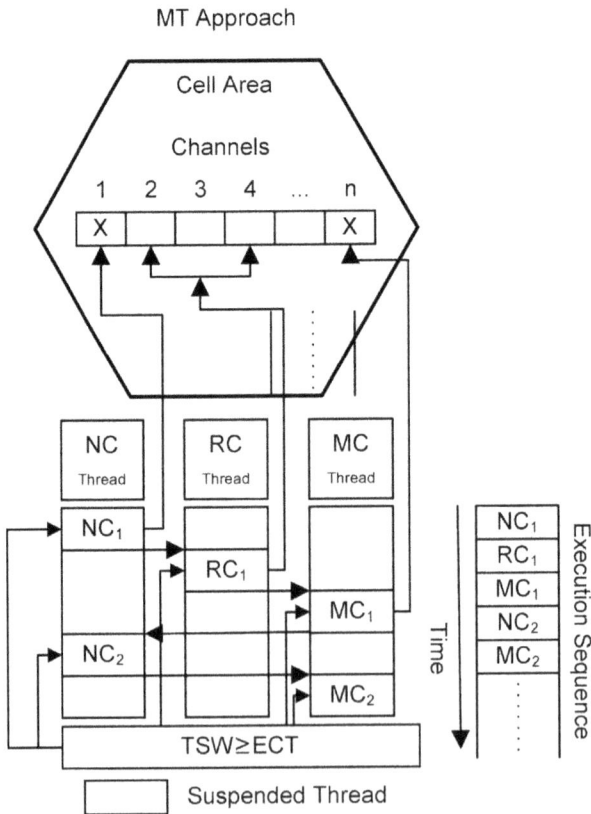

FIGURE 1.27   MU service operation based on the Event Interleaving scheme.

1.4 EXPERIMENTAL RESULTS

1.4.1 INDICATIVE RESULTS BASED ON FIVE DAYS OF NETWORK OPERATION

The simulation results are based on unbalanced and balanced DCA schemes where the wireless network consists of 19 cells with capacity of 50 users per cell and 32 channels per cell. Figures 1.28–1.30 show a typical blocking probability graph for the given network using the daily traffic model (five days of network operation) and based on the proposed architectures.

1.4.2 MODEL BEHAVIOUR BASED ON ARCHITECTURAL VARIATIONS

It is obvious from Figures 1.31–1.36 that the selection of the concurrency affects positively the simulation model behaviour.

1.4.3 SCHEDULING MECHANISM COMPARISON

Figures 1.37 and 1.38

1.4.4 RESPONSE TIME ANALYSIS RESULTS

Figures 1.39–1.42 show the time behaviour of two tested event interleaving schemes. In Pre-emptive Event Interleaving (PEI) scheme, when a new event arrives with higher priority, the current under processing event is pre-empted and so the PEI is

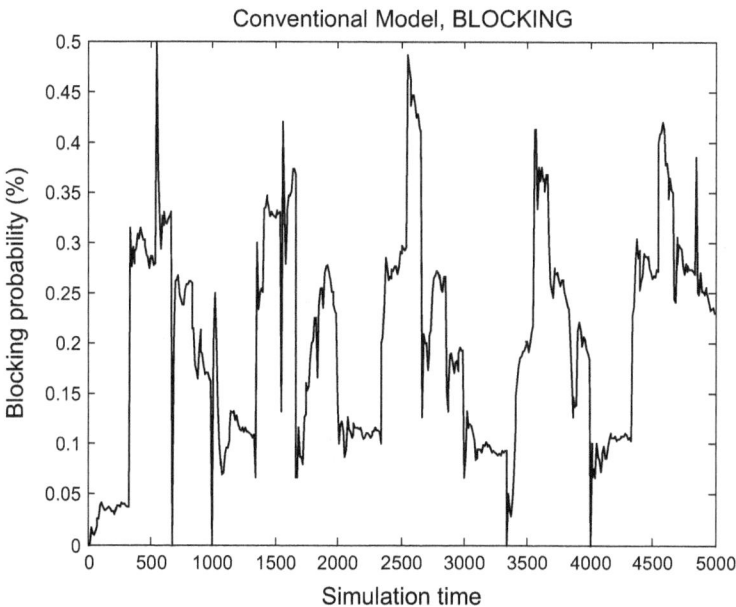

FIGURE 1.28 Conventional model, blocking probability.

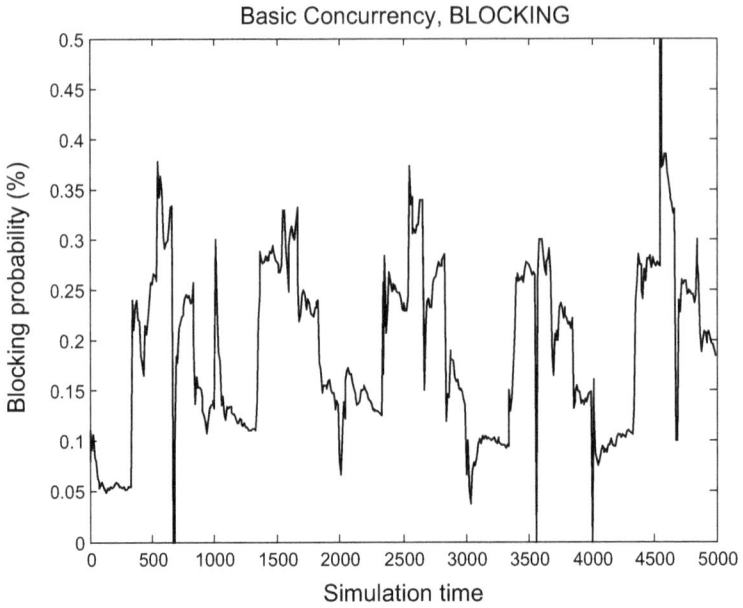

FIGURE 1.29   Basic concurrency, blocking probability.

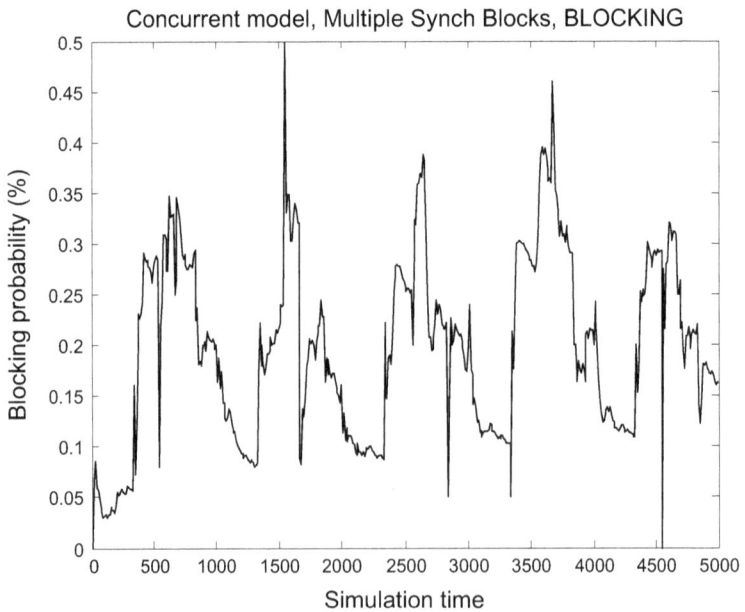

FIGURE 1.30   Concurrency with multiple synchronized blocks, blocking probability.

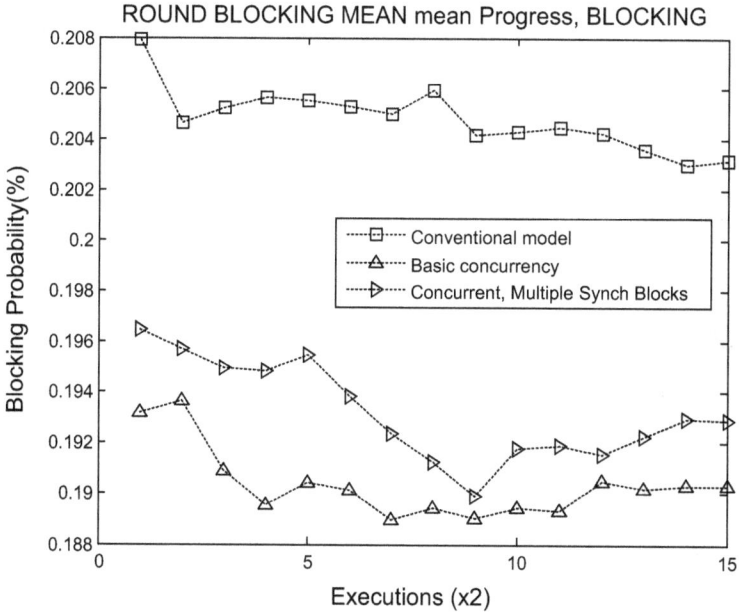

FIGURE 1.31 Proposed balanced DCA and best CNR based on GAs, blocking probability.

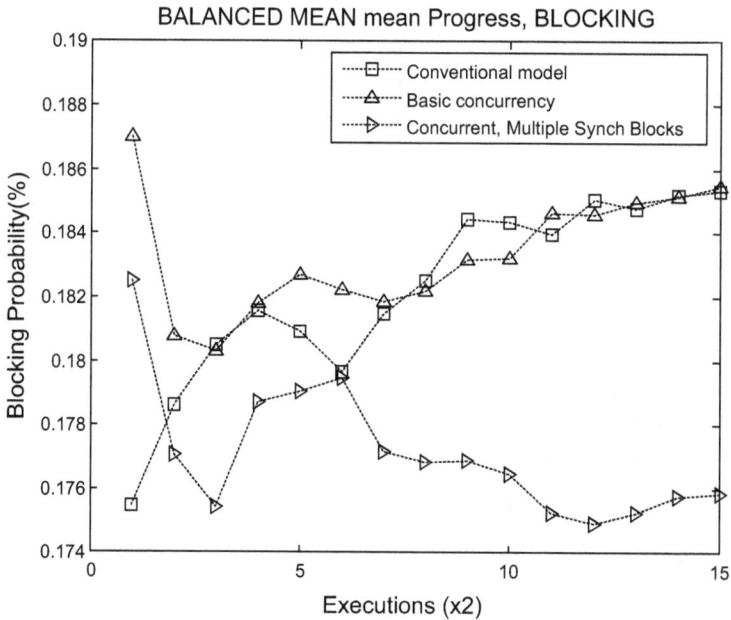

FIGURE 1.32 Conventional balanced DCA, blocking probability.

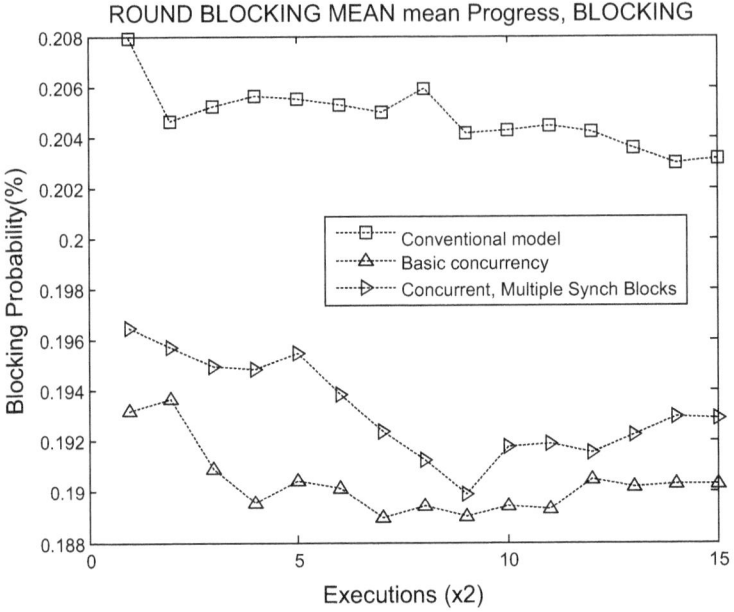

FIGURE 1.33   Best CNR DCA, blocking probability.

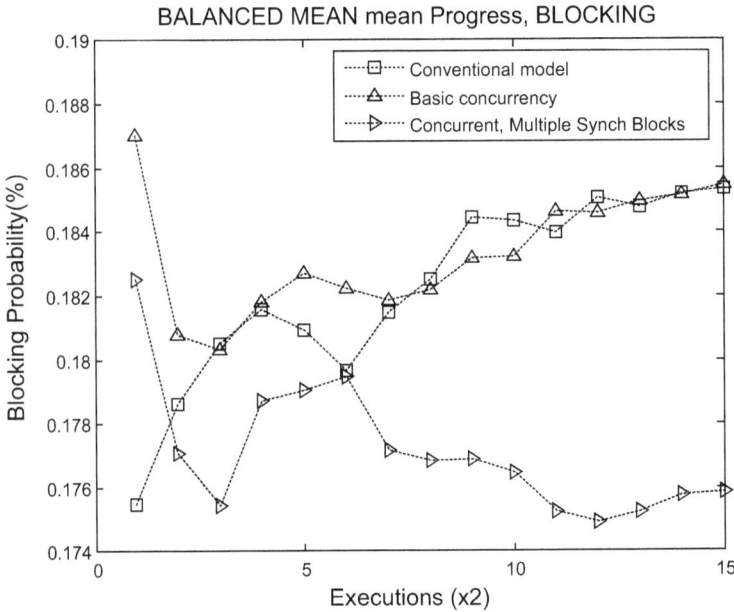

FIGURE 1.34   Proposed balanced DCA and best CNR based on GAs, dropping probability variance.

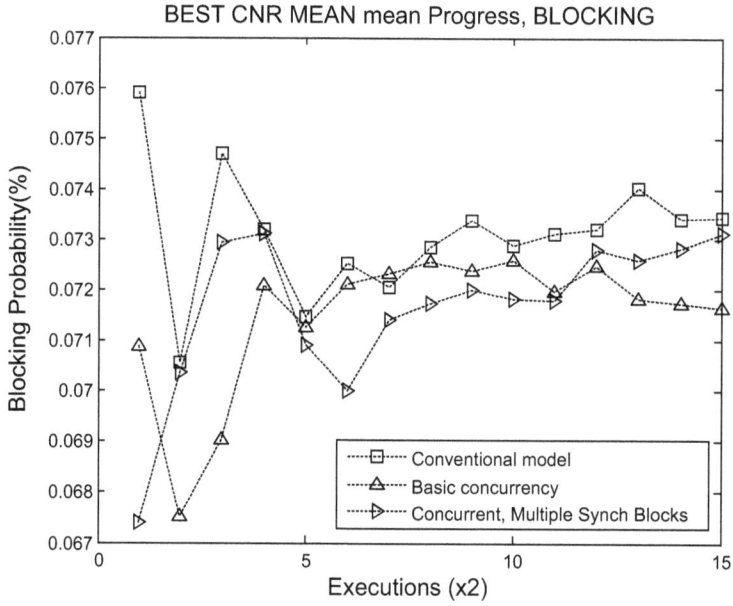

FIGURE 1.35 Unbalanced DCA, dropping probability periodicity.

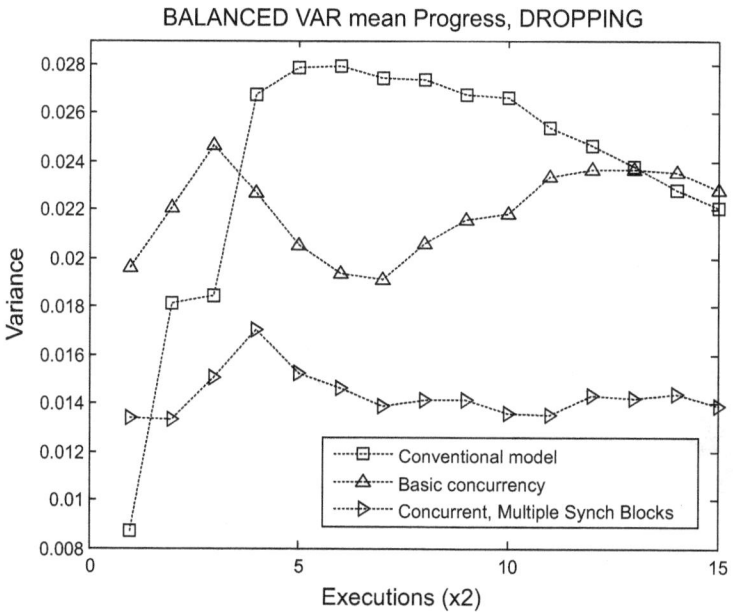

FIGURE 1.36 Proposed balanced DCA and best CNR based on GAs, dropping probability periodicity.

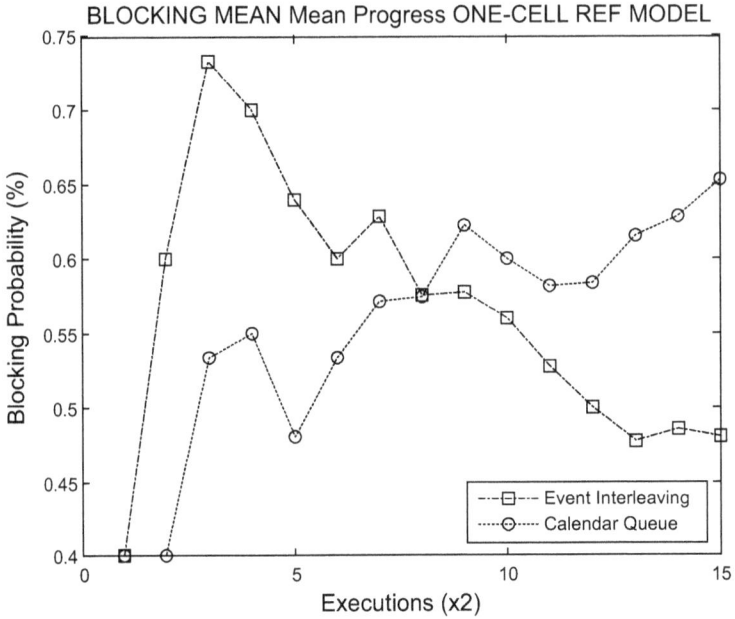

FIGURE 1.37   CQ blocking probability and Event Interleaving techniques (MC runs, reference model of one cell).

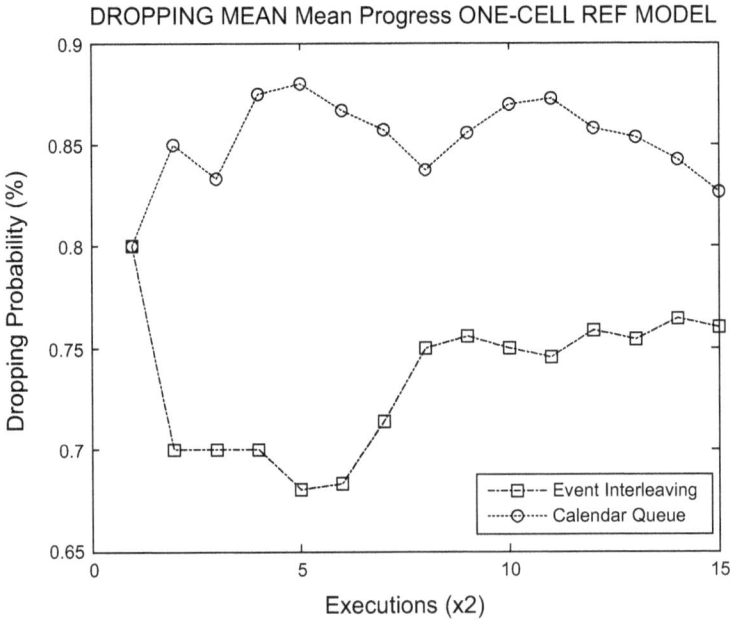

FIGURE 1.38   CQ dropping probability and Event Interleaving techniques (MC runs, reference model of one cell).

ARRIVAL-DEPARTURE MEAN Mean Progress
ONE-CELL REF model (Tnc=Trc=Tmc=Tfc)

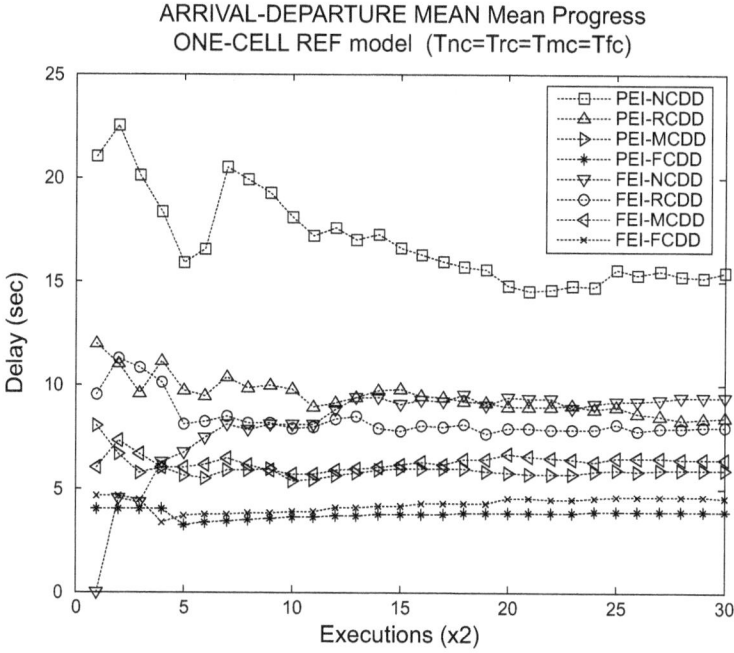

FIGURE 1.39 Delay calculations between events arrival and departure time units (equal arrival time units).

ARRIAVAL-DEPARTURE MEAN Mean Progress
ONE-CELL REF model (Tnc#Trc#Tmc#Tfc)

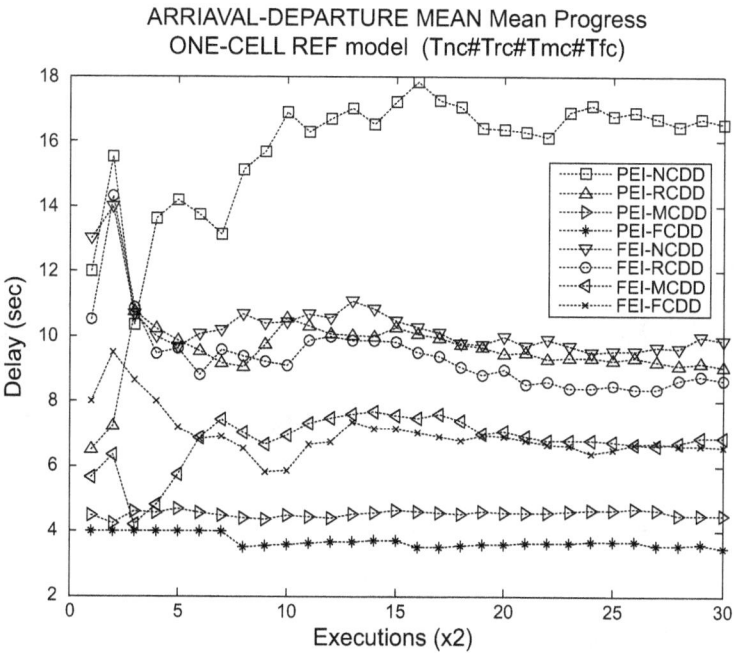

FIGURE 1.40 Delay calculations between events arrival and departure time units (aperiodic and random arrival times).

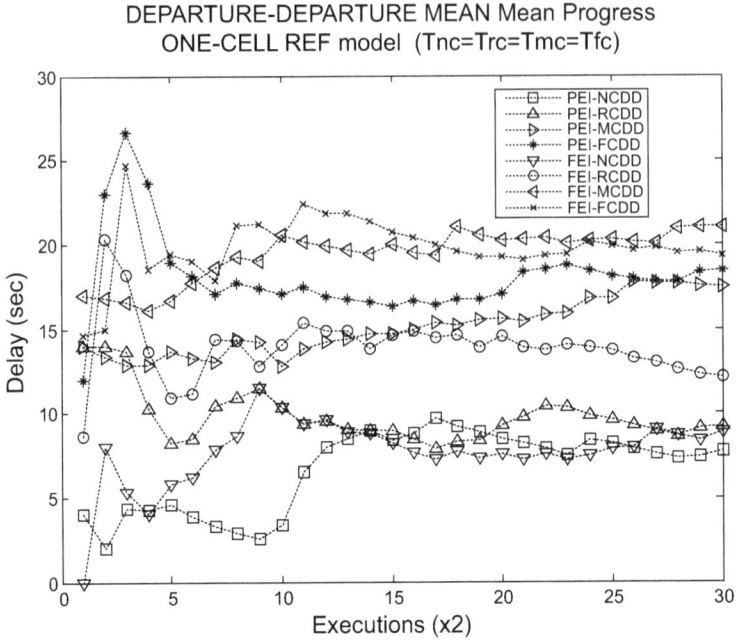

FIGURE 1.41   Delay measurements between events departure times (equal arrival times).

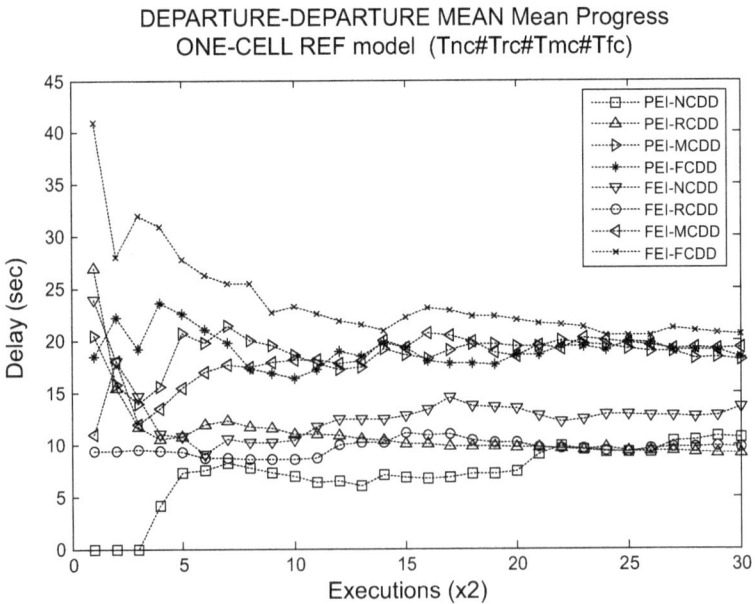

FIGURE 1.42   Delay measurements between events departure times (aperiodic arrival times).

based clearly on the real-time scheduling concept. On the other hand, a variation of PEI called Fair Event Interleaving (FEI) is also tested. In FEI, each event is executed in turn based on priority settings, and the system gives equal computational periods for each event under processing.

1.5 CONCLUSIONS AND FUTURE WORK

This research work investigates scheduling issues in multi-agent systems modelling wireless communications systems. Several contributions have been herein introduced by analyzing conventional scheduling methods and the more suitable concurrent scheduling methodologies in case of synchronized agent models. Moreover, a comprehensive simulation model for simulating large-scale wireless networks is proposed based on multi-agent analysis. A novel modelling methodology of large-scale network services and the corresponding theoretical analysis and implementation based on the multi-agent concept is initially presented. Based on this modelling, an optimized channel allocation algorithm through artificial intelligence techniques, namely, a genetic algorithm model, has been presented with success when concurrent events emerge in the system, which is the actual real-world case. Involving this modelling approach, a novel and alternative to the state-of-the-art event scheduling mechanism using real-time scheduling theory for supporting improved concurrent network events modelling is also presented. The simulation results show the effectiveness of the concurrent multi-agent-based models using artificial intelligence optimization techniques as well as the efficiency of the event interleaving for concurrent network events. However, such promising results should be verified in larger scale networks, especially in wireless sensor networks, where the parameter of power consumption plays a critical role together with congestion and best CNR control.

REFERENCES

Abar, S, Theodoropoulos, GK, and Lemarinier, P. 2017. Agent based modelling and simulation tools: A review of the state-of-art software. *Computer Science Review*, 24, 13–33.

Audsley, NC, Burns, A, Richardson, MF, and Wellings, AJ. 1992. Deadline monotonic scheduling theory. In *Proceedings 18th IFAC/IFIP Workshop on Real-Time Programming (WRTP'92)*, June.

Audsley, N, Burns, A, Richardson, M, Tindell, K, and Wellings, A. 1993. Applying new scheduling theory to static priority pre-emptive scheduling. *Software Engineering Journal*, 8(5), 284–292.

Audsley, NC, Burns, A, Davis, RI, Tindell, KW, and Wellings, AJ. 1995. Fixed priority pre-emptive scheduling: An historical perspective. *Real-Time Systems*, 8, 173–198.

Bajaj, L, Takai, M, Ahuja, R, and Bagrodia, R. 1999. Simulation of large-scale heterogeneous communication systems. In *Proceedings of IEEE Military Communications Conference (MILCOM'99)*.

Baker, Mark, Rajkumar Buyya, and Dan Hyde. "Cluster computing: A high-performance contender." arXiv preprint cs/0009020 (2000), pp. 79–83, https://arxiv.org/ftp/cs/paper s/0009/0009020.pdf

Balakrishnan, V, et al. 1997. A framework for performance analysis of parallel discrete event simulators. In *Proceedings of the 29th Conference on Winter Simulation*.

Barr, R. 2004. *An Efficient, Unifying Approach To Simulation Using Virtual Machines.* Cornell University, PhD Dissertation

Beowulf Project. http://www.beowulf.org.

Blackstone, JH, Hogg, CL, and Phillips, DT. 1981. A two-list synchronization procedure for discrete event simulation. *Communications of the ACM*, 24(12), 825–629.

Bodanese, EL. 2000. *A Distributed Channel Allocation Scheme for Cellular Networks Using Intelligent Software Agents.* PhD thesis, University of London.

Bononi, L, and D'Angelo, G. 2003. A novel approach for distributed simulation of wireless mobile systems. *PWC 2003*, LNCS 2775, 829–834.

Boukerche, A, Das, SK, and Fabbri, A. 2001. SWiMNet: A scalable parallel simulation testbed for wireless and mobile networks. *Wireless Networks*, 7, 467–486.

Brown, R. 1988. Calendar queues: A fast O(1) priority queue implementation for the simulation event set problem. *Communications of the ACM*, 31(10), 1220–1227.

Burns, A, and Wellings, AJ. 2001. *Real-Time Systems and Programming Languages: Ada 95, Real-Time Java, and Real-Time POSIX.* Pearson Education, 3rd edition, ISBN-13: 978-0201729887, ISBN-10: 0201729881, https://www.amazon.com/Real-Time-Systems-Programming-Languages/dp/0201729881/ref=sr_1_1?dchild=1&keywords=9780201729887&linkCode=qs&qid=1608488674&s=books&sr=1-1

Buttazzo, GC. 2011. *Hard Real-Time Computing Systems: Predictable Scheduling Algorithms and Applications*, Vol. 24. Springer Science & Business Media, LLC 2011, New York, doi: 10.1007/978-1-4614-0676-1.

Chaturvedi, A, Dickieson, J, Dolk, DR, and Scholl, J. 2001. Introduction to agent-based simulation and system dynamics minitrack. In *Proceedings of the 34th Hawaii International Conference on System Sciences.*

Chung, K, Sang, J, and Rego, V. 1993. A performance comparison of event calendar algorithms: An empirical approach. *Software—Practice and Experience*, 23(10), 1107–1138.

Cimini, LJ, and Foschini, GJ. 1992. Distributed algorithms for dynamic channel allocation in microcellular systems. In *IEEE Vehicular Technology Conference*, 641–644.

Cox, DC, and Reudink, DO. 1973. Increasing channel occupancy in large scale mobile radio systems: Dynamic channel reassignment. *IEEE Transanctions on Vehicular Technology*, VT-22, 218–222.

Del Re, E, Fantacci, R, and Giambene, G. 1996. A dynamic channel alloca-tion technique based on hopfield neural networks. *IEEE Transanctions on Vehicular Technology*, VT-45(1), 26–32.

Dixit, S, Yile Guo, Antoniou Z, Resource management and quality of service in third generation wireless networks, IEEE Communications Magazine (Volume: 39, Issue: 2, Feb. 2001)

Ekdahl, B. 2001. How autonomous is an autonomous agent? In *Proceedings of the 5th Conference on Systemic, Cybernetics and Informatics (SCI 2001)*, July 22–25, Orlando, USA.

Elnoubi, SM, Singh, R, and Gupta, SC. 1982. A new frequency channel assignment algorithm in high capacity mobile communication systems. *IEEE Transactions on Vehicular Technology*, VT-21(3), 125–131.

Erickson, KB, Ladner, RE, and LaMarca, A. 1994. Optimizing static calendar queue. *Annual IEEE Symposium on Foundations of Computer Science*, 35, 732–743.

Fagin, R. et al. 1995. *Reasoning About Knowledge.* MIT Press, Cambridge, MA.

Fall, K, and Varadhan, K. 2015. The ns manual (2011). http://www.isi.edu/nsnam/ns/doc/ns doc. pdf.

Fidge, CJ. 2002. *Real-Time Scheduling Theory, Technical Report.* University of Queensland, No. 02-19, https://espace.library.uq.edu.au/data/UQ_10543/svrc_02_19.pdf?dsi_vers ion=53e60a44db45e9802571ffd4a16a4c33&Expires=1608490365&Key-Pair-Id=AP

KAJKNBJ4MJBJNC6NLQ&Signature=RQTLaQ~jI7v5b6glxW3O3ctcm0PGv~h
78x5yuyTLLD8bhKV8Imuzsy7OP1Usyh8-yU9EwkQb0-hnzDkzYvpZEJ1gzJbNRlJU
YwFmOR9fgOqIPjr8UJnLRjgWV6n2fbxMwwrPXi0IodLe~vQsrsMZXRFgAAe2
Dq3vwFIb-2bS7fd1v3Zl~tsujYUk0JZXCqe9eT9fUz6XH0q-dmnyORhO0WgELFnj6
ett-IHXq7KbeTXAFnCMu0-Ls6IXKAqFs0pRtXTcbqDpj5X1oa3jHypGDBvPQ
2TCvZ0jlRuJPRmh2d0OiEBJVw-JJsPkhCM6AKALnNmZA8Dr2ZeeBf072~UZb
Q__

Fishman, GS. 1996. *Monte Carlo: Concepts, Algorithms, and Applications.* New York: Springer-Verlag, Springer Science + Business Media.

Fujimoto, RM. 2000. *Parallel and Distributed Simulation Systems.* (Vol. 300). New York: Wiley-Interscience.

Fujimoto, RM, et al. 2003. Large-scale network simulation: How big? How fast? In *IEEE, International Symposium on Modeling, Analysis, and Simulation of Computer and Telecommunications Systems MASCOTS.*

Goh, RSM and Thng, IL-J. 2003. MLIST: An efficient pending event set structure for discrete event simulation. *International Journal of Simulation*, 4(5–6), 66–77.

Hayes-Roth, B. 1995. An architecture for adaptive intelligent systems. *Artificial Intelligence: Special Issue on Agents and Interactivity*, 72, 329–365.

Hayzelden, A. and Bigham, J. 1999. *Software Agents for Future Communications Systems.* Springer-Verlag, Berlin.

Henriksen, JO. 1977. An improved events list algorithm. In *Proceedings of the 1977 Winter Simulation Conference (Gaithersburg. Md., Dec. 5–7).* IEEE, Piscataway, NJ, 547–557. http://jist.ece.cornell.edu/javadoc/jist/runtime/Scheduler.Calendar.html.

Huhns, M, and Singh, M. Eds. 1998. *Agents and Multi Agent Systems: Themes, Approaches, and Challenges.* Readings in Agents, Chapter 1, Morgan Kaufmann Publishers, USA, 1–23.

Iraqi, Y, and Boutaba, R. 2000. A multi-agent system for resource management. In *Wireless Mobile Multimedia Networks, LNCS 1960.* Springer-Verlag, Berlin Heidelberg, 218–229.

Jefferson, DR, and Barnes, PD. 2017. Virtual time III: Unification of conservative and optimistic synchronization in parallel discrete event simulation. In *2017 Winter Simulation Conference (WSC).* IEEE.

Jennings, NR. 2000. On agent-base software engineering. *Artificial Intelligence*, 117, 277–296.

Jiang, Y, and Jiang, J. 2008. Contextual resource negotiation-based task allocation and load balancing in complex software systems. *IEEE Transactions on Parallel and Distributed Systems*, 20(5), 641–653.

Kahwa, TJ, and Georgans, ND. 1978. A hybrid channel assignment schemes in large-scale, cellular structured mobile communication systems. *IEEE Transactions on Communications*, 26, 432–438.

Karian, ZA, and Dudewicz, EJ (1998). *Modern Statistical, Systems, and GPSS Simulation*, 2nd edition, CRC Press.

Kelly, OE, Lai, J, Mandayam, NB, Ogielski, AT, Panchal, J, and Yates, RD. 2000. Scalable parallel simulations of wireless networks with WiPPET: Modeling of radio propagation, mobility and protocols. *Mobile Networks and Applications*, 5(3), 199–208, Springer.

Kingston, JH. 1984. *Analysis of Algorithms for the Simulation Event List.* Ph.D. thesis, Basser Dept. of Computer Science, Univ. of Sydney, Australia. July.

Krishna, S, Goldsmith, A, and Carlton, M. Seamless handoff, offload, and load balancing in integrated Wi-Fi/small cell systems. *US Patent 9,510,256, 2016 - Google Patents.*

Kurkowski, S, Camp, T and Colagrosso, M. MANET Simulation Studies: The Current State and New Simulation Tools, 2005.

Lai, WK, and Coghill, GC. 1996. Channel assignment through evolutionary optimization. *IEEE Transactions on Vehicular Technology*, 45(1), 91–96.

Lee, WCY. 1995. *Mobile Cellular Telecommunications*. McGraw-Hill Professional. ISBN:978-0-07-038089-9, https://dl.acm.org/doi/book/10.5555/555761

Liu, CL, and Layland, JW. 1973. Scheduling algorithms for multiprogramming in a hard real-time environment. *Journal of the ACM*, 20(1), 46–61.

Liu, W, Chiang, CC, Wu, HK, Jha, V, Gerla, M, and Bagrodia, R. 1996. Parallel simulation environment for mobile wireless networks. In *Proceedings of the 1996 Winter Simulation Conference, WSC'96*, 605–612.

Liu, J, Perrone, LF, Nicol, DM, Liljenstam, M, Elliott, C, and Pearson, D, 2001, June. *Simulation Modeling of Large-Scale ad-hoc Sensor Networks*. In European Simulation Interoperability Workshop (Vol. 200, No. 1).

Locke, CD. 1992. Software architecture for hard real-time applications: Cyclic executives vs. fixed priority executives. *The Journal of Real-Time Systems*, 4, 37–53.

MacDonald, VH. 1979. The cellular concepts. *The Bell System Technical, Journal*, 58, 15–42.

Maes, P. 1995. Artificial life meets entertainment: Life like autonomous agents. *Communications of the ACM*, 38(11), 108–114.

McCormack, WM, and Sargent, RG. 1981. Analysis of future event-set algorithms for discrete event simulation. *Communications of the ACM*, 24(12), 801–812.

Mehta, S, Kwak, KS, and Najnin, S. 2010. Network and System Simulation Tools for Next Generation Networks: A Case Study. *Modelling, Simulation and Identification*, Azah Mohamed (Ed.), ISBN: 978-953-307-136-7, InTech, Available from: http://www.inte chopen.com/books/modelling--simulation-andidentification/network-and-system-simu lation-tools-for-wireless-networks-a-case-study

Miller, J, Weyrich, O, and Suen, D. 1988. A software engineering oriented compari-son of simulation languages. In *Proceedings of the 1988 Eastern Simulation Conferences: Tools for the Simulationists*. Orlando, FL, April.

Misra, J. 1986. Distributed discrete-event simulation, *ACM. Computing Surveys, (CSUR)*, 18(1), 39-65, March.1

Molisch, AF. 2012. *Wireless Communications*, Vol. 34. John Wiley & Sons, UK, 2nd edition, ISBN: 1118355687, 9781118355688

Nahas, M. 2014. Studying the impact of scheduler implementation on task jitter in real-time resource-constrained embedded systems. *Journal of Embedded Systems*, 2(3), 39–52.

Naoumov, V, and Gross, T. 2003. Simulation of Large Ad Hoc Networks, *MSWiM'03*. In Proceedings of the 6th ACM international workshop on Modeling analysis and simulation of wireless and mobile systems (pp. 50–57), September, San Diego, CA.

Nierstrasz, O. 1989. A survey of object-oriented concepts. In W Kim and F Lochovsk, editors, *Object- Oriented Concepts, Databases and Applications*. ACM Press, Addison Wesley, 3–21.

Norman, T, and Long, D. 1995. Goal creation in motivated agents. In *Proceedings of the Workshop on Agent Theories, Architectures, and Languages on Intelligent Agents*.

Papazoglou, PM, and Karras, DA. 2016. A conceptual multi-agent modeling of dynamic scheduling in wireless sensor networks. *Lecture Notes in Electrical Engineering*, 348, 385–398. Publisher Springer-Verlag, ISSN:1876-1100.

Papazoglou, PM, Karras, DA, and Papademetriou, RC. 2008a. An improved multi-agent simulation methodology for modelling and evaluating wireless communication systems resource allocation algorithms. *Journal of Universal Computer Science*, 14(7), 1061–1079.

Papazoglou, PM, Karras, DA, and Papademetriou, RC. 2008b. On a new generation of event scheduling algorithms and evaluation techniques for efficient simulation modelling of large scale cellular networks bandwidth management based on multitasking theory. *WSEAS Transactions on Communications*, 7(10), 1024–1034, Publisher: World Scientific and Engineering Academy and Society (WSEAS) Press, ISSN: 1109-2742.

Papazoglou, PM, Karras, DA, and Papademetriou, RC. 2009. On the efficient implementation of a high performance multi-agent simulation system for modeling cellular communications involving a novel event scheduling. *International Journal of Simulation Systems, Science & Technology*, Special Issue on: Performance Engineering, 10(1).

Papazoglou, PM, Karras, DA, and Papademetriou, RC. 2011. Evaluating novel DCA variations for efficient channel assignment in cellular communications through a generic Java simulation system. In *Proceedings of 2011 Intern. Conference on Computer Science and Network Technology*. Harbin, December 24–26.

Papazoglou, PM, Karras, DA, and Papademetriou, RC. 2016. On integrating natural computing based optimization with channel assignment mining and decision making towards efficient spectrum reuse in cellular networks modelled through multi-agent system schemes. *Lecture Notes in Electrical Engineering*, 348, 783–798. Publisher Springer-Verlag, ISSN:1876-1100.

Pincus, R. (1995). Barnett, V., and Lewis T.: Outliers in Statistical Data. J. Wiley & Sons 1994, XVII. 582 pp.,£ 49.95. *Biometrical Journal*, 37(2), 256–256.

Pont, MJ, Kurian, S, Wang, H, and Phatrapornnant, T. 2007. Selecting an appropriate scheduler for use with time-triggered embedded systems. In *EuroPLoP*, 595–618.

Reddy, D, et al. 2006. Measuring and explaining differences in wireless simulation models. In *14th IEEE International Symposium on Modeling, Analysis, and Simulation*. IEEE.

Russell, EC. 1999. *Building Simulation Models with Simscript II.5*. CACI Products Co., CACI Products Company, 3333 North Torrey Pines Court La Jolla, California 92037, https://www.cosc.brocku.ca/Offerings/4P94/sii5/zbuildin.pdf

Russell, S, and Norvig, P. 2003. *Artificial Intelligence: A Modern Approach*, 2nd ed. Upper Saddle River, NJ: Pearson Education.

Schriber, TJ, and Brunner, DT. 1997. Inside discrete-event simulation software: How it works and why it matters. In *Proceedings of the 1997 Winter Simulation Conference*.

Siangsukone, T, Aswakul, C, and Wuttisittikulkij, L. 2003. Study of optimised bucket widths in calendar queue for discrete event simulator. In *Thailand's Electrical Engineering Conference (EECON-26)*.

Siow, R, Goh, M, and Thng, L-J. 2004. DSplay: An efficient dynamic priority queue structure for discrete event simulation. In *Simtect Simulation Conference*.

Sivarajan, KN, McEliece, RJ, and Ketchum, JW. 1990. Dynamic channel assignment in cellular radio. In *IEEE 40th Vehicular Technology Conference*, 631–637.

Smith, DC, Cypher, A, and Spohrer, J. 1994. KidSim: Programming agents without a programming language. *Communications of the ACM*, 37(7), 55–67.

Splunter, S, Wijngaards, N, and Brazier, F. 2003. Structuring agents for adaptation. In E Alonso et al. editors. *Adaptive Agents and Multi-Agent Systems, LNAI*, Vol. 2636, 174–186, Heidelberg: Springer, Berlin.

Sprunt, B. 1990. *Aperiodic Task Scheduling for Real-Time Systems*. PhD dissertation. Carnegie Mellon University.

Subramania Sharma, T, and Thazhuthaveetil, MJ. 2001. TWLinuX: Operating system support for optimistic parallel discrete event simulation. *High Performance Computing–HiPC 2001*, 262.

Sycara, K. 1998. Multi-agent systems. *Artificial Intelligence Magazine*, 19(2), American Association for Artificial Intelligence, 0738-4602-1998.

Takai, M, Bagrodia, R, Lee, A, and Gerla, M (1999, August). Impact of channel models on simulation of large scale wireless networks. In Proceedings of the 2nd ACM international workshop on Modeling, analysis and simulation of wireless and mobile systems (pp. 7–14).

Tan, KL, and Thng, L-J. 2000. Snoopy calendar queue. In *Proceedings of the 2000 Winter Simulation Conference.*

Tolk, A. 2012. Verification and validation. *Engineering Principles of Combat Modeling and Distributed Simulation*, pp. 263–294, published by John Wiley and Sons, NJ, USA, ISBN: 978-0-470-87429-5

Tropper, C. 2002. Parallel discrete-event simulation applications. *Journal of Parallel and Distributed Computing*, Volume 62, Issue 3, March 2002, 327–335, Elsevier

West, J, and Mullarney, A. 1998. ModSim: A language for distributed simulation. In *Proceedings SCS Multi Conference on Distributed Simulation.* San Diego, CA, February, 155–159.

Whitaker, P. 2001. *The Simulation of Synchronous Reactive Systems in Ptolemy II.* Electronics Research Laboratory, College of Engineering, University of California, https://ptolemy.berkeley.edu/publications/papers/01/sr/sr.pdf

Wooldridge, M. 2009. *An Introduction to Multi-Agent Systems*, 2nd edition. John Wiley & Sons, UK, ISBN: 978-0-470-51946-2

Wooldridge, M, and Jennings, NR. 1995. Intelligent agents: theory and practice. *The Knowledge Engineering Review*, 10(2), 115–152.

Xu, Z, and Mirchandani, PB. 1982. Virtually fixed channel assignment for cellular radio-telephone systems: A model and evaluation. In *IEEE International Conference on Communications, ICC'92*, Chicago, Vol. 2, 1037–1041.

Zeng, X, Bagrodia, R, and Gerla, M. 1998. GloMoSim: A library for parallel simulation of large-scale wireless networks. In *Proceedings of the 12th Workshop on Parallel and Distributed Simulations.*

Zhang, M, and Yum, TS. 1989. Comparisons of channel assignment strategies in cellular mobile telephone systems. *IEEE Transactions on Vehicular Technology*, 38(4), 211–215.

Zhang, X, et al. 2001. A proposed approach to sophisticated negotiation. In *AAAI Fall Symposium on Negotiation Methods for Autonomous Cooperative Systems.* November.

2 Let's Find Out
Why Do Users React Differently to Applications Infused with AI Algorithms?

Jayden Khakurel, Indu Manimaran, and Jari Porras

CONTENTS

2.1 INTRODUCTION

In recent years, researchers and application developers have increasingly begun to infuse applications with machine learning or natural language processing algorithms that offer robust empirical performance to tackle real human problems and enhance the quality of life in various domains (Khakurel et al. 2018; Inkpen et al. 2019). Russell, Moskowitz, and Raglin (2017) point out, "Humans' interaction with information will only increase in the future, and this interaction will likely be facilitated by artificial intelligent proxies" (p. 33). This supports the idea that applications infused with artificial intelligence (AI) algorithms will become ubiquitous in our lifetimes, and humans will both interact and integrate with these programs.

Eventually, these programs will replace human–human with human–machine communication (endowed with more or less artificial intelligence) (Hilbert and Aravindakshan 2018). Evidence has emerged that there are similarities (e.g., user's alignment) and differences (e.g., less engagement) between human and artificial intelligence interactions; however, what causes users to behave differently when interacting with AI is still unknown (Rzepka and Berger 2018). Further, while some studies (Yannakakis and Togelius 2018; Amershi et al. 2019) show that apps with more or less AI are used to facilitate entire or partial automated communication, for example, via Chatbot, personal assistant, or, in some scenarios, video games controlled by non-player characters, the AI may demonstrate unpredictable behaviours that can be disruptive, confusing, offensive, and even dangerous. These unpredictable behaviours during human–AI interactions may backfire and induce anger, anxiety, and/or feelings of discomfort among individuals, resulting in negative service evaluations, customer dissatisfaction, and increased switch intentions (Feng et al. 2019). To examine this phenomenon, we explored the following research questions: (i) What causes users to behave differently toward an app using AI? (ii) How can we minimize unpredictable behaviours and improve user engagement with these apps and technologies?

To look at AI–human interactions, it is crucial to look at emotions in general. Here, emerging insights into discrete state emotions (DSEs) indicate that each discrete state emotion, such as excitement, happiness, anger, relaxation, and desire, has a unique mechanism that causes a unique mental state with measurable outcomes and has an influence on human decisions; indeed, DSEs are important in understanding and predicting a user's decisions and behaviours (Angie et al. 2011; Harris and Isaacowitz 2015; Harmon-Jones et al. 2016). Therefore, in this study, we formulate the main hypothesis that either the lack or presence of DSEs may cause differences in interaction behaviours when users work with AI-infused apps. To test the main hypothesis, five sub-hypotheses were formulated, which are explained in Section 2.2. The testing of these sub-hypotheses was carried out in two sessions using the complete counterbalancing technique to control the order effect which is generally recommended for conditions ($K!$) less than 4, $K \leq 4$ (Allen Mike 2017). For instance, in this study, there are two conditions ($k!\ 2$): condition A: apps game experience; and condition B: dyadic game experience. Using all possible orders, two different combinations – AB, BA – are generated. Two groups were formulated in which group 1 (14 participants) interacted with conditions A and then B. During the app game experience, participants held their phone with one hand and interacted with an android game app based on the traditional game "rock–paper–scissors," also known as "roshambo," but infused with machine learning algorithms to execute roshambo moves. During dyadic game interactions, players who participated in the first session played the roshambo game against other participants.

After each session, DSEs were collected using the Discrete Emotions Questionnaire (DEQ) (Harmon-Jones et al. 2016). Having data from separate sessions enabled us to identify the variation in DSEs between the two sessions. Further, we analyzed the collected data using the statistical data analysis language R and the descriptive statistical analysis functions available in R core (R Core Team 2017) and the psych library (Revelle 2017). The results expand the existing research on the influence of DSEs on users' interaction behaviours, providing informal guidelines that can be used to minimize unpredictable behaviours and improve user engagement with apps or technologies.

The remainder of this chapter is structured as follows. Section 2.2 provides related work and hypothesis formulation. Section 2.3 then provides the methodology applied in the study. We discuss the findings in Section 2.4, and finally, we present the discussion in Section 2.5 and promising avenues for further research in Section 2.6.

2.2 RELATED WORK AND HYPOTHESIS FORMULATION

We focus in particular on the interactions between users and AI, looking at users' experiences, needs, and challenges. Rzepka and Berger (2018) conducted a literature review and aggregated the knowledge regarding human–AI interactions. Their review revealed that (i) users' interactions with AI systems trigger contradictory behavioural responses; (ii) AI systems trigger perceptions of threat among users; and (iii) users assign humanness and social characteristics to AI systems.

Väänänen et al. (2019) conducted a pilot survey and adopted a Geneva emotion wheel to understand users' positive and negative experiences and descriptions of expected or potential experiences of AI in their everyday lives; they found that the participants experienced some sort of negative experiences, such as a sense of anger, disappointment, and irritation. Further, the participants expressed that they did not want AI to feel too humanized. On the other hand, in terms of positive experiences, a sense of control and trust/reliability were the most reported wanted experiences in addition to relief, feeling of safety, efficiency, satisfaction, contentment, and pleasure. Subsequently, the authors suggested that understanding, collaborating, and sensitive interactions could strengthen the human–AI relationship.

Emotion, the key component that drives human behaviour, plays an essential role in human cognitive processes, problem-solving competence, and decision-making (de Freitas et al. 2005; Nass and Brave 2010; Khakurel 2018). Psychological studies of emotions have been structured into two levels: the *dimensional level* that posits emotions come in the form of dimensions such as arousal (high/low intensity), valence (positive/negative), and motivation direction (approach/avoid) (Corener Paul 2013; Harmon-Jones et al. 2016, 2017), and the *discrete level*, which posits emotions come in the form of discrete states such as anger, fear, joy, and so on. According to Harmon-Jones et al. (2017), "each primary process or 'basic' discrete emotion is posited to evoke a specific response tendency that will address a specific evolutionarily important need (e.g., protection from harm by fear, rejection of harmful substances by disgust)" (p. 2).

Given the influence of a user's discrete emotions, we derive the main hypothesis that either the lack or presence of DSEs may cause differences in interaction behaviours when users work with AI-infused apps. To establish the main hypothesis, a set of sub-hypotheses were formulated – H1, H2, H3, H4, and H5 – that focus on game-based apps and that are based on the DSE states.

2.2.1 EXCITEMENT

Excitement refers to an individual's response to a situation which increases the adequacy; it supplements the routine modes of responses which appear inadequate (Stratton 1928). We propose that when a player interacts with opponents, there appear to be surprises (i.e., don't know what to expect in the next move), which usually excite the other player more than when interacting with precise and pragmatic games, which are more dry in nature unless the players are playing for the sake of playing them, addicted to, or wanting to win the games. Such excitement may cause other players to cry if the excitation is strong or can result in wild laughter. Therefore, we hypothesize that playing with a human as opponent will lead to excitement, whereas playing against AI-infused apps will lead to less excitement, affecting a user's interaction behaviour. H1: Playing with AI-infused apps will lead to less excitement and thus generate different behavioural responses.

2.2.2 ANGER

Williams (2017) and Mill et al. (2018) refer to the term "anger" as a multifaceted construct that entails a negative activation and can be perceived as a threat to an individual's emotional well-being. In line with Charlton (2009) and Hadlington and Scase (2018), we propose that a poor game app experience with an AI app, including the app's self-efficacy and related anxiety, can be a detrimental factor that may frustrate users, triggering an anger experience. We hypothesize that playing against AI gaming apps triggers a stronger anger experience than when playing against a human opponent, which reflects what Mill et al. (2018) refer to as the "appraisal of an anger-provoking or frustrating situation triggers the anger experience which, in turn, generates different behavioral responses" (p. 739). In contrast, we propose that while playing with a human, individuals will use more coping strategies, and anger will hence be less intense because both players could solve the matter with a wide range of discussion. H2: Playing with AI-infused apps will lead to frustration among users, triggering anger experiences and thus generating different behavioural responses.

2.2.3 DESIRE

Zalta (Schroeder 2017) refers to desire as a state of mind that is commonly associated with a number of different effects. Furthermore, the author also points out that a person with a desire tends to act in certain ways, feel in certain ways, and think in certain ways. In the context of the current study, during a game experience, players may have the desire to win the game, regardless of who they are playing against. Previous studies have shown that when it comes to desire in game experiences, individuals take relative gains more vigorously than personal gains because they either have the desire to defeat, or create a flow of experiences during the game, or do not like receiving less than others (Messick and Thorngate 1967; Meyer-Parlapanis et al. 2017). However, we propose that when playing against AI-infused apps, the desire to win is heightened more because of the collision between two perceptual factors, i.e., time pressure and achieving the highest score. Therefore, we hypothesize that playing with an AI game app increases the desire to win compared with playing against a human opponent, leading the user to behave differently. H3: Increased feelings of the desire "to win" while playing against AI-infused apps will evoke different behavioural responses.

2.2.4 HAPPINESS

Previous studies have conceptualized happiness as the positive emotions formed due to general interaction between internal (endogenic) and external (exogenic) factors, which are associated with and precede numerous successful outcomes (Lyubomirsky et al. 2005; Dfarhud et al. 2014; Harmon-Jones et al. 2016). In the context of the current study, we propose individuals are less happy during and after the game experience with the AI because of (i) social disengagement (Baym et al. 2004), (ii) presence of parameters related to game addiction (i.e., salience, mood modification, tolerance, withdrawal, conflict, and relapse [Griffiths 2008; Hull et al. 2013]), and (iii) frustrating experiences (i.e., error messages, dropped network connections, long download times, and hard-to-find features [Ceaparu et al. 2004]). In line with Mehl et al. (2010), in face-to-face interactions (i.e., conversations while having the game experience), excitement may actually make individuals happier than the actual game experience with the AI app. Therefore, we hypothesize that interpersonal communication and other DSE entities, namely excitement, may cause individuals to remain happier while playing against a human opponent. H4: The interpersonal communication and other DSE entities resulting from the dyadic game experience will influence happiness, generating more positive behavioural responses than the game experience with an AI app.

2.2.5 RELAX

Previous studies (Smith 2007) have organized relaxation states into four groups: basic relaxation, core mindfulness, positive energy, and transcendence. These relaxation states are essential in creating a relaxation response (Benson et al. 1974). In the context of the current study, we propose that a face-to-face game experience will let players disclose personal items, build impressions, and compare values (Baym et al. 2004), which will eventually lead to basic relaxation. Therefore, we hypothesize that playing a game against a human opponent makes a player more relaxed than playing against an AI app, leading the user to behave differently. H5: Having a human opponent during the game experience makes players more relaxed than having an AI opponent.

2.3 METHODOLOGY

The following sections elaborate on the methodological approach used in the current study.

2.3.1 PARTICIPANTS

A total of 26 participants (19 males, 7 females; mean age: 25.6 years; standard deviation: 2.5; with adequate knowledge about technology; 12 participants were outgoing and enjoyed being with a lot of people; 15 participants enjoyed more solitude) had mobile devices using an Android operating system (OS) were voluntarily recruited at the university by word-of-mouth. The sample size is sufficient based on the recommendation from Macefield (2009), where the author states that a group size of 3–20 participants is typically valid, with 5–10 participants demonstrating a sensible baseline range in experimental studies. All of the participants were presented with an informed consent form detailing participants' right to confidentiality, risks, data storage, the use of anonymized data, the voluntary nature of participation, and that no health-related data would be collected. No incentives were provided in exchange for participation.

2.3.2 PROCEDURE

We adopted the complete counterbalancing technique to control the order effect, which is generally recommended for conditions ($K!$) less than 4, $K \leq 4$ (Allen Mike 2017). For instance, in this study, there are two conditions ($k!$ 2), condition A: apps game experience, and condition B: dyadic game experience. Using all possible orders, two different combinations – AB, BA – were generated. Two groups (a total of 26 participants) were formulated in which group 1 (14 participants) interacted with conditions A and then B. Figure 2.1 shows the complete counterbalancing for an experiment with two conditions adopted in this study.

During the first session, group 1 (14 participants) interacted with condition A, where participants were asked to download an android-based game app based on the traditional game "rock–paper–scissors"; this app used machine learning algorithms. The participants were instructed to play exactly 30 rounds. While playing, players made a hand gesture for rock, paper, or scissors before the timer ran out. The winner of the round was decided based on the standard rules. The players then reported the DSEs that they experienced while interacting with the game through the DEQ adopted from Harmon-Jones et al. (2016). In a survey, the participants were asked to rate on a 7-point Likert scale (1 = not at all to 7 = an extreme amount): "While playing with a mobile game with integrated AI algorithms, to what extent did you experience

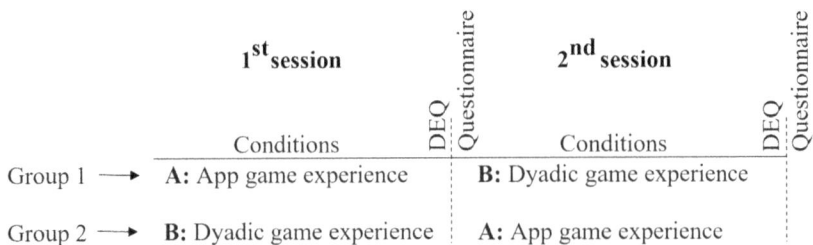

	1st session	DEQ Questionnaire	2nd session	DEQ Questionnaire
	Conditions		Conditions	
Group 1 →	A: App game experience		B: Dyadic game experience	
Group 2 →	B: Dyadic game experience		A: App game experience	

FIGURE 2.1   Complete counterbalancing for an experiment with two conditions.

these emotions?" Furthermore, the participants were asked to provide qualitative feedback of their experience.

Similarly, group 2 (12 participants) interacted with condition B (see Figure 2.1). During the dyadic game experience, each pair was asked to play "rock–paper–scissors" with another participant for a total number of 30 rounds. Hence, a player must guess what their opponent will choose and pick the appropriate object to beat them (University of Stirling 2017). Each player then reported the DSEs separately that they experienced while interacting with the game through the DEQ adopted from Harmon-Jones et al. (2016). In a survey, the participants were asked to rate on a 7-point Likert scale (1 = not at all to 7 = an extreme amount): "While with your partners, to what extent did you experience these emotions?" Furthermore, the participants were asked to provide qualitative feedback on their experience.

During the second session, group 1 (14 participants) interacted with condition B, whereas group 2 (12 participants) interacted with condition A to eliminate order effects. The participants performed a similar task, filled the survey, and were provided feedback after the first session.

Quantitative data collected from the two sessions were analyzed using the statistical data analysis language R and the descriptive statistical analysis functions available in R core (R Core Team 2017) and the psych library (Revelle 2017). We first used the Mann–Whitney U test (Wohlin et al. 2012) to analyze the difference in distributions between the data sets. A continuity correction was enabled to compensate for non-continuous variables (Bergmann and Ludbrook 2000). The Bonferroni correction was used to adjust the p-value to compensate for the family-wise error rate in multiple comparisons (Abdi 2007). We calculated the effect size r using the guidelines by Tofan et al. (2016) for the Mann–Whitney U test. We evaluated the effect size as proposed by Cohen (1994): in r, a large effect is 0.5, a medium effect is 0.3, and a small effect is 0.1.

2.4 FINDINGS

2.4.1 Descriptive Statistics and Hypothesis Testing Outcomes

The results indicate that participants remained excited ($M = 6.54$, SD = 0.859), had desire ($M = 6.62$, SD = 0.571), were happy ($M = 5.77$, SD = 0.704), and were relaxed ($M = 6.50$, SD = 1.273) while playing against a human opponent compared with the AI app. The results also indicate that individuals were angrier ($M = 2.58$, SD = 1.79) while playing against the AI compared with a human opponent. To establish the main hypotheses, the sub-hypotheses formulated in Section 2.3 were tested, they are summarized in Appendix A. The results indicate that all the sub-hypotheses are correct. The results show DSEs, namely excitement ($p < 0.001$; $r = 0.74$), happiness ($p < 0.001$; $r = 0.59$), and relaxation ($p < 0.001$; $r = 0.71$), were felt during the dyadic experience, whereas there were increased feelings of the desire "to win" when playing against the AI-infused app. However, anger caused because of frustration ($p < 0.05$; $r = 0.38$) was found to have medium effect on the app's game experience. The research findings support that either a lack or presence of DSEs may cause differences in interaction behaviours when users play with AI-infused apps.

2.4.2 QUALITATIVE FEEDBACK

We further analyzed our qualitative data that measured the overall experience after playing the games to see what essential challenges affected the emotions of users while interacting with the AI app. Many participants commented that the interaction between humans was more fun, more competitive, and quicker than with the AI app. For example, P6 commented, "Human is more fun definitely, but AI could be an alternative when you want to play and no one is around." The comment shows that users may use the mobile app as an alternative method when lonely. Similarly, P7 stated, "Human was more fun, AI takes more time to recognize the gesture, so I feel that communication is easier and fun with humans." Further, P12 commented, "Human is more fun as the game is much quicker and competitive." Consequently, some participants also commented that AI game lack a certain level of intensity. For example, P8 commented, "Playing with human is more fun. Rock, paper, scissors should be more intense, and I didn't feel the intensity. Good work on the app though!" Here, further research should be conducted on why users like the highly intense AI games and to what level the game should be intensified.

Further, out of 26 participants, 7 expressed that they had challenges regarding gesture detection when using the app, which caused frustration. For example P4 commented, "The app didn't recognize my hand gesture in a high percentage of cases. I wasn't sure if I should hold my hand vertically or horizontally." P13 stated, "The app recognizes rock all the time. It never recognizes scissors!" P3 noted, "Out of 10 rounds only two rounds were recognized properly … makes me feel frustrated." Regarding enriching the user experience, the need for interactive visual feedback in the form of either graphics or emojis was stated. For example, P12 said, "The app should give some graphical feedback or emoji signs to make it more interactive."

2.5   DISCUSSIONS

In light of concerns about why users behave differently with AI-infused apps/systems, the current study performed two tests. From a theoretical point of view, the results indicate that DSEs are important, and AI-based apps should have an ability to evoke and include emotions such as happiness, excitements, and relaxation at the same level as they have during dyadic interaction to develop mutual emotional attachments. The finding is in line with the proposal presented by LaGrandeur (2015), who states that inducing emotions is important to create safer and more attractive AIs, allowing both humans and AI to develop mutual emotional attachments. Brave and Nass (2002) also state that "any interface that ignores a user's emotional state or fails to manifest the appropriate emotion can dramatically impede performance and risks being perceived as cold, socially inept, untrustworthy, and incompetent" (p. 82). Further, supporting previous research (Thüring and Mahlke 2007), this study further reveals that poor app experience triggers the anger experience among users that leads them to behave differently. Based on the current study's findings, we suggest the below informal guidelines for application developers seeking to design better user experiences and minimize unpredictable behaviours, such as anger experience. These guidelines may not adequately address all AI-infused apps; however, they can be taken into consideration.

Improving the detection rate of gesture: although the model performed well during development, the participants reported that they were frustrated while playing because sometimes the app failed to detect their hand gestures. Thus, we recommend developers to (i) use a public data set to train the model offline and update the applications more frequently to improve gesture detection n, (ii) use images that are taken in different lighting conditions and with different backgrounds, (iii) to quantify the detection performance, use Detection Error Trade-off (DET) curves and miss rate versus False Positive Per Window (FPPW) as proposed by Lahiani and Neji (2018), and (iv) apply the gesture recognition method based on a convolutional neural network (CNN) and deep convolution generative adversarial networks (DCGAN) proposed by Fang et al. (2019), which can train the model using fewer samples and achieve better gesture classification and detection effects.

Considering interference time, response time, and progress indication with the AI-infused app to improve the user interaction: some participants reported that they were impatient because of the response time when interacting with the app. AI-infused apps are built with (i) a set of powerful algorithms within the app itself that can be used to collect the data sets, train the model, and display the results or (ii) can use different algorithms on the data set on the server side to analyze patterns in data and make predictions, displaying the results on the app by synchronizing them between the app and the server. In the latter scenario, there might be delays in the response because of mobile computational power and server response time because of slow database queries, libraries, and resource central processing unit (CPU) starvation, and slow application logic, which may increase users' cognitive load and affect their performance (Alnanih and Ormandjieva 2016). Therefore, we recommend application developers (i) consider running the longer running operation as the background task; (ii) test the interference time; and (iii) test the response time on apps utilizing different time frames, as mentioned by Nielsen (1993). Furthermore, in order to reduce uncertainty among users, we recommend developers to provide an accurate estimation of the waiting times using either a determinate or an indeterminate progress indicator (Luo 2017).

Automatic recognition of emotional states and visualizing in the form of graphical feedback or digital pictograms/emojis to develop human–AI emotional attachments: some participants were more interested in having graphics or having digital pictograms such as emoji as a way to receive feedback. This may be because apps that use emojis as a feedback method communicate a positive effect, specifically joy (Riordan 2017), or the graphical feedback makes users less frustrated and more engaged compared with textual feedback (Rieber et al. 1996). Stark and Crawford (2015) point out, "Emojis can act as an emotional coping strategy and a novel form of creative expression, even if, in both cases, working within real limits" (p. 1). Therefore, the form of feedback should manifest the appropriate DSEs with a new algorithm that describes the relationship between cognition and emotion better based on behavioural science and neuroscience (Wang et al. 2016). Aiming at inducing and evoking the emotions on apps through graphical feedback or digital pictograms/emojis, developers could refer to the work of Gao et al. (2012) on how tactile behaviour on the screen can reflect users' emotional state.

2.6 LIMITATIONS AND FUTURE WORK

We describe below several limitations. These limitations illustrate improvements that can be considered in future research. First, we only investigated self-reported DSEs with small samples and a case study on AI mobile games, and the results may not be generalized and inapplicable to other AI-infused apps. Therefore, the results from the current study can only be taken into consideration as informal guidance. Furthermore, because AI could be used in various app categories in different contexts, researchers may want to conduct further research with larger samples to identify which DSEs appear in which context, hence improving the AI–human relationship. Second, demographics such as age, gender, and culture could be added to future work to identify if DSEs vary during AI–human interactions.

2.7 CONCLUSION

In recent years, AI agency has been applied in forefront applications to simulate the intelligent behaviour and critical thinking to perform a range of activities that require human intelligence, such as decision-making, learning, sensing, and challenging humans through games. However, previous studies have shown that there is a need to understand the constraints of a user's interaction behaviours with AI-infused apps. We formulated and tested a hypothesis using the perspective of DSEs. The contributions of the current chapter are twofold. First, this research provides new knowledge about the influence of DSEs on users' interaction behaviours with AI-infused apps. Second, it provides an informal guideline on how to minimize unpredictable user interaction behaviours and improve their acceptance of these services or technologies in the future. In conclusion, the present study suggests DSEs should be an indispensable part of AI-infused apps (i) for individuals to react the same way as they act during human–human interactions and (ii) to minimize their adverse effects to enrich human–AI relationships.

Appendix A: https://doi.org/10.5281/zenodo.3361476

Acknowledgements. Thank you to Antti Knutas for your contribution during the data analysis phase and all the reviewers for their valuable comments and considerable time and effort. The first author would like to thank INVEST Research Flagship funded by the Academy of Finland Flagship Programme (decision number: 320162).

REFERENCES

Abdi H (2007) The Bonferonni and Šidák Corrections for Multiple Comparisons. In: Salkind N (ed) *Encyclopedia of Measurement and Statistics*. SAGE Publications Inc., Thousand Oaks, CA.

Allen M (2017) *The sage encyclopedia of communication research methods (Vols. 1–4).* SAGE Publications, Inc, Thousand Oaks, CA. doi: 10.4135/9781483381411

Alnanih R, Ormandjieva O (2016) Mapping HCI Principles to Design Quality of Mobile User Interfaces in Healthcare Applications. *Procedia Comput Sci* 94:75–82. ISSN 1877-0509, https://doi.org/10.1016/j.procs.2016.08.014

Amershi S, Weld D, Vorvoreanu M, et al (2019) Guidelines for Human-AI Interaction. In: *CHI Conference on Human Factors in Computing Systems Proceedings* (*CHI 2019*).

Angie AD, Connelly S, Waples EP, Kligyte V (2011) The Influence of Discrete Emotions on Judgement and Decision-Making: A Meta-Analytic Review. *Cogn Emot* 25:1393–1422. https://doi.org/10.1080/02699931.2010.550751

Baym NK, Zhang YB, Lin MC (2004) Social Interactions Across Media: Interpersonal Communication on the Internet, Telephone and Face-To-Face. *New Media Soc.* https://doi.org/10.1177/1461444804041438

Benson H, Beary JF, Carol MP (1974) *The Relaxation Response. Psychiatry.* https://doi.org/10.1080/00332747.1974.11023785

Bergmann R, Ludbrook J (2000) Different Outcomes of the Wilcoxon—Mann—Whitney Test from Different Statistics Packages. *Am Stat* 54:72–77. https://doi.org/10.1080/00031305.2000.10474513

Brave S, Nass C (2002) Emotion in Human–Computer Interaction. In: Julie A. J. L. (ed) *The Human-Computer Interaction Handbook.* Erlbaum Associates Inc, Hillsdale, NJ, 81–96. doi: 10.1201/9781410615862.ch4

Ceaparu I, Lazar J, Bessiere K, et al (2004) Determining Causes and Severity of End-User Frustration. *Int J Hum Comput Interact* 17:333–356. https://doi.org/10.1207/s15327590ijhc1703_3

Charlton JP (2009) The Determinants and Expression of Computer-Related Anger. *Comput Human Behav* 25:1213–1221. https://doi.org/10.1016/j.chb.2009.07.001

Cohen S (1994) Perceived Stress Scale. *Psychology* 1–3. http://www.mindgarden.com/products/pss.htm.

Corener Paul (2013) *The Fine Points of Feelings : Why Discrete Emotions Matter and How to Reveal Them.* Quirks. https://emotiveanalytics.com/the-fine-points-of-feeling-why-discrete-emotions-matter-and-how-to-reveal-them/.

de Freitas JS, Gudwin R, Queiroz J (2005) Emotion in Artificial Intelligence and Artificial Life Research: Facing Problems. In: Panayiotopoulos T, Gratch J, Aylett R, Ballin D, Olivier P, Rist T (eds) *Intelligent Virtual Agents.* IVA 2005. Lecture Notes in Computer Science, vol 3661. Heidelberg, Springer, Berlin. https://doi.org/10.1007/11550617_52.

Dfarhud D, Malmir M, Khanahmadi M (2014) Happiness & Health: The Biological Factors-Systematic Review Article. *Iran J Public Health* 43:1468–1477.

Fang W, Ding Y, Zhang F, Sheng J (2019) Gesture Recognition Based on CNN and DCGAN for Calculation and Text Output. *IEEE Access* 7:28230–28237. https://doi.org/10.1109/ACCESS.2019.2901930

Feng W, Tu R, Lu T, Zhou Z (2019) Understanding Forced Adoption of Self-Service Technology: The Impacts of Users' Psychological Reactance. *Behav Inf Technol* 38(8):820–832. doi: 10.1080/0144929X.2018.1557745

Gao Y, Bianchi-Berthouze N, Meng H (2012) What Does Touch Tell Us about Emotions in Touchscreen-Based Gameplay? *ACM Trans Comput Interact.* https://doi.org/10.1145/2395131.2395138

Griffiths MD (2008) Diagnosis and Management of Video Game Addiction. *New Dir Addict Treat Prev* 12:27–41.

Hadlington L, Scase MO (2018) End-user Frustrations and Failures in Digital Technology: Exploring the Role of Fear of Missing Out, Internet Addiction and Personality. *Heliyon.* https://doi.org/10.1016/j.heliyon.2018.e00872

Harmon-Jones C, Bastian B, Harmon-Jones E (2016) The Discrete Emotions Questionnaire: A New Tool for Measuring State Self-Reported Emotions. *PLoS One* 11:e0159915. https://doi.org/10.1371/journal.pone.0159915

Harmon-Jones E, Harmon-Jones C, Summerell E (2017) On the Importance of Both Dimensional and Discrete Models of Emotion. *Behav Sci (Basel)* 7:66. https://doi.org/10.3390/bs7040066

Harris JA, Isaacowitz D (2015) Emotion in Cognition. In: *International Encyclopedia of the Social & Behavioral Sciences*. Elsevier, pp. 461–466. doi: 10.1016/B978-0-08-097086-8.25003-4

Hilbert M, Aravindakshan A (2018) What Characterizes the Polymodal Media of the Mobile Phone? The Multiple Media within the World's Most Popular Medium. *Multimodal Technol Interact* 2:37. https://doi.org/10.3390/mti2030037

Hull DC, Williams GA, Griffiths MD (2013) Video Game Characteristics, Happiness and Flow as Predictors of Addiction Among Video Game Players: A Pilot Study. *J Behav Addict* 2:145–152. https://doi.org/10.1556/JBA.2.2013.005

Inkpen K, Choudhury M, Chancellor S, et al (2019) Where is the Human? Bridging the Gap Between AI and HCI. https://michae.lv/ai-hci-workshop/. Accessed 2 Apr 2019

Khakurel J (2018) *Enhancing the Adoption of Quantified Self-Tracking Devices*. LUT University. http://urn.fi/URN:ISBN:978-952-335-319-0

Khakurel J, Penzenstadler B, Porras J, et al (2018) The Rise of Artificial Intelligence under the Lens of Sustainability. *Technologies* 6:100. https://doi.org/10.3390/technologies6040100

LaGrandeur K (2015) Emotion, Artificial Intelligence, and Ethics. In: Romportl J, Zackova E, Kelemen J (eds) *Beyond Artificial Intelligence*. Topics in Intelligent Engineering and Informatics, vol 9. Springer, Cham. https://doi.org/10.1007/978-3-319-09668-1_7

Lahiani H, Neji M (2018) Hand Gesture Recognition Method Based on HOG-LBP Features for Mobile Devices. *Procedia Comput Sci* 126:254–263. ISSN 1877-0509, https://doi.org/10.1016/j.procs.2018.07.259

Luo G (2017) Toward a Progress Indicator for Machine Learning Model Building and Data Mining Algorithm Execution. *ACM SIGKDD Explor Newsl* 19:13–24. https://doi.org/10.1145/3166054.3166057

Lyubomirsky S, King L, Diener E (2005) The Benefits of Frequent Positive Affect: Does Happiness Lead to Success? *Psychol Bull* 131:803–855. https://doi.org/10.1037/0033-2909.131.6.803

Macefield R (2009) How To Specify the Participant Group Size for Usability Studies: A Practitioner's Guide. *J Usability Stud* 5:34–45.

Mehl MR, Vazire S, Holleran SE, Clark CS (2010) Eavesdropping on Happiness: Well-Being Is Related to Having Less Small Talk and More Substantive Conversations. *Psychol Sci*. https://doi.org/10.1177/0956797610362675

Messick DM, Thorngate WB (1967) Relative Gain Maximization in Experimental Games. *J Exp Soc Psychol*. https://doi.org/10.1016/0022-1031(67)90039-X

Meyer-Parlapanis D, Siefert S, Weierstall R (2017) More Than the Win: The Relation between Appetitive Competition Motivation, Socialization, and Gender Role Orientation in Women's Football. *Front Psychol*. https://doi.org/10.3389/fpsyg.2017.00547

Mill A, Kööts-Ausmees L, Allik J, Realo A (2018) The Role of Co-Occurring Emotions and Personality Traits in Anger Expression. *Front Psychol*. https://doi.org/10.3389/fpsyg.2018.00123

Nass C, Brave S (2010) *Emotion in Human-Computer Interaction*. doi: 10.1201/9781410615862.ch4

Nielsen J (1993) *Usability Engineering*. Morgan Kaufmann Publishers Inc. doi: 10.1145/1508044.1508050

R Core Team (2017) *R: A Language and Environment for Statistical Computing.* https://www.r-project.org/

Revelle W (2017) *Psych: Procedures for Psychological, Psychometric, and Personality Research.* https://cran.r-project.org/package=psych

Rieber LP, Smith M, Al-Ghafry S, et al (1996) The Role of Meaning in Interpreting Graphical and Textual Feedback During a Computer-Based Simulation. *Comput Educ* 27:45–58. https://doi.org/10.1016/0360-1315(96)00005-X

Riordan MA (2017) Emojis as Tools for Emotion Work: Communicating Affect in Text Messages. *J Lang Soc Psychol* 36:549–567. https://doi.org/10.1177/0261927X17704238

Russell S, Moskowitz IS, Raglin A (2017) Human Information Interaction, Artificial Intelligence, and Errors. In: *Autonomy and Artificial Intelligence: A Threat or Savior?* Springer International Publishing, Cham, pp. 71–101. doi: 10.1007/978-3-319-59719-5_4

Rzepka C, Berger B (2018) User Interaction with AI-enabled Systems: A Systematic Review of IS Research. *ICIS Proc* no. December:1–17.

Schroeder T (2017) Desire. In: Edward N Zalta (ed) *The Stanford Encyclopedia of Philosophy,* Summer 201. Metaphysics Research Lab, Stanford University. https://plato.stanford.edu/archives/sum2017/entries/desire/

Smith JC (2007) *The New Psychology of Relaxation and Renewal.* Biofeedback.

Stark L, Crawford K (2015) The Conservatism of Emoji: Work, Affect, and Communication. *Soc Media Soc.* https://doi.org/10.1177/2056305115604853

Stratton GM (1928) The Function of Emotion as Shown Particularly in Excitement. *Psychol Rev* 35:351–366. https://doi.org/10.1037/h0071406

Thüring M, Mahlke S (2007) Usability, Aesthetics and Emotions in Human-Technology Interaction. *Int J Psychol* 42:253–264. https://doi.org/10.1080/00207590701396674

Tofan D, Galster M, Lytra I, et al (2016) Empirical Evaluation of a Process to Increase Consensus in Group Architectural Decision Making. *Inf Softw Technol* 72:31–47. https://doi.org/https://doi.org/10.1016/j.infsof.2015.12.002

University of Stirling (2017) Can a Computer Learn Game Strategy as it Plays? *Dep Comput Sci Maths* http://www.cs.stir.ac.uk/~kms/schools/rps/index.php. Accessed 28 Nov. 2018.

Väänänen K, Pohjola H, Ahtinen A (2019) Exploring the User Experience of Artificial Intelligence Applications : User Survey and Human-AI Relationship Model. In: *CHI'19 Workshop on Where is the Human? Bridging the Gap Between AI and HCI.* Glasgow, p. 5.

Wang Z, Xie L, Lu T (2016) Research Progress of Artificial Psychology and Artificial Emotion in China. *CAAI Trans Intell Technol.* https://doi.org/10.1016/j.trit.2016.11.003

Williams R (2017) Anger as a Basic Emotion and Its Role in Personality Building and Pathological Growth: The Neuroscientific, Developmental and Clinical Perspectives. *Front Psychol* 8. https://doi.org/10.3389/fpsyg.2017.01950

Wohlin C, Runeson P, Höst M, et al (2012) *Experimentation in Software Engineering.* Springer Berlin Heidelberg, Berlin.

Yannakakis GN, Togelius J (2018) *Artificial Intelligence and Games.* Springer International Publishing, Cham.

3 AI vs. Machine Learning vs. Deep Learning

R. Lalitha

CONTENTS

3.1 INTRODUCTION: BACKGROUND AND DRIVING FORCES

While knowing about the futuristic issues of artificial intelligence (AI) and its applications, it is mandatory to understand about the correlation and differences between artificial intelligence, machine learning (ML), and deep learning (DL) as they are closely associated with one another. In a nutshell, machine learning forms the subset of artificial intelligence, and deep learning forms the subset of machine learning. Hence, any futuristic applications and ideas that emerge out of artificial intelligence algorithms will certainly contain machine learning algorithms and deep learning techniques. Therefore, it is essential to know where, when, how, and why machine learning or deep learning algorithms must be applied for developing innovative applications with them.

This chapter provides an overview of these domains and narrates their merits and demerits. The reader will be able to understand the features of each of these domains. The reader will also be able to know when and where these techniques are suitable and not suitable for developing applications.

3.2 OVERVIEW OF ARTIFICIAL INTELLIGENCE

Artificial intelligence is a branch of computer science that deals with creating intelligence artificially to a system. The system can be prepared to make decisions like human brains. The system can plan, move from one place to another, and recognize objects and sounds. The father of artificial intelligence, John McCarthy, says that "Artificial Intelligence is the science and engineering of making intelligent machines especially intelligent computer programs." Artificial intelligence systems are created by studying the process of thinking, learning, observing, and decision-making by the human brain.

The main objectives of creating AI systems are to create an expert system and to implement human intelligence in machines. The best programming languages to develop AI applications are Python, LISP, C++, Java, and Prolog. The humanoids, Robots, are examples of physical devices that are upgraded with artificial intelligence. They act as a substitute for human beings with efficiency in work, high speed in performance, and enormous memory.

3.3 STEPS TO IMPLEMENT ARTIFICIAL INTELLIGENCE ALGORITHMS

1. Study and analyze the real intelligence in a human brain.
2. Understand the problem/scenario.
3. Apply heuristic techniques to solve the problem or else select the appropriate AI technique.
4. Implement the technique to simulate the real intelligence.

3.4 WHEN/WHERE/HOW/WHY TO USE ARTIFICIAL INTELLIGENCE?

AI can be used in any domain where it is essential to exhibit intelligence in the system where it is implemented. It can be used in any sector like transport, healthcare, education, etc. When there is a need to simulate human intelligence and to make any system to behave like a human, AI can be implemented.

Human intelligence is not permanent and may vanish as years pass by. Hence, to retain expert knowledge for many decades, it is necessary to create expert systems by implementing artificial intelligence. AI is also needed to enhance the speed in prediction, forecasting, and decision-making by human efforts. The heuristic approach can be used to implement artificial intelligence, which includes learning techniques and experiences of the human expert. A large amount of data is processed and used with intelligent algorithms, iterative procedures to make a system to work with intelligence.

As technology has become an important and integral part of daily life, artificial intelligence is needed to automate the processes, reduce the errors, and provide accurate results at a greater speed.

3.5 EXAMPLES FOR ARTIFICIAL INTELLIGENCE APPLICATIONS

- Alexa is a well-known personal assistant with artificial intelligence that can voice commands and can execute actions based on that. It is used to set

reminders for the users to complete a task in time, to answer questions given by the user, and to order items online.
- With the help of AI, smart home devices can be controlled and used easily. Smart voice assistant and smart assistant for old age people are a few examples.
- AI can be used to diagnose disease in human bodies. For example, earlier cancer detection with AI is possible; radiology assistant with AI is also possible.
- Process automation in factories with AI.

3.6   OVERVIEW OF MACHINE LEARNING

Machine learning is the most promising and most relevant domain to apply artificial intelligence in systems. It is the most common way to process big data. Machine learning algorithms are designed in such a way that they are self-adaptive and are able to get new patterns to itself through experience. It is a way of learning from data. Tom Mitchell defines machine learning as follows: "A computer program is said to learn from experience (E), with respect to some class of tasks (T), and performance measures (P), if its performance at tasks T_i as measured by P_i improves with experience E." A target is called label in machine learning. A variable is called feature. The transformation which occurs for the variables is called feature creation. It combines computer science, mathematics, and statistics. Computer science is needed for implementing the algorithms. Mathematics is needed for developing machine learning models, and statistics is needed for generating inferences from the data. The best programming languages for machine learning are Python, R Programming, LISP, Prolog, and JavaScript.

3.7   STEPS TO IMPLEMENT MACHINE LEARNING ALGORITHMS

1. Identify the data set and prepare it for analysis.
2. Select the appropriate machine learning algorithm.
3. Develop an analytical model based on the selected algorithm.
4. Train the model with the test data.
5. Run and execute the model.

3.8   WHEN/WHERE/HOW/WHY TO USE MACHINE LEARNING?

Machine learning algorithms are used in situations where it is difficult to write code for the rules and also in cases where huge volumes of data have to be processed for prediction. It is used in places where process automation is needed, and time consumption has to be minimized. Machine learning is used to execute certain repetitive tasks, identify patterns in the input, and predict outcomes.

Machine learning is needed and important because as the data size grows exponentially large, computation and prediction become complex. Machine learning can be used to adapt to the changes in data by itself and can predict the desired result easily. As the data size grows enormously in any domain, computational algorithms are needed to derive meaningful insights from the data. Hence, machine learning algorithms are needed and essential to provide precise information to the user.

3.9   EXAMPLES FOR MACHINE LEARNING APPLICATIONS

- Speech recognition
- Weather forecasting and prediction
- Traffic prediction and sending alerts
- Filtering spam emails
- Product recommendation in online shopping
- Sentiment analysis
- Auto-driven cars

3.10   OVERVIEW OF DEEP LEARNING

Deep learning is a subset of machine learning. It has an artificial neural network to carry out the tasks of machine learning. It enables the system to process the data in a non-linear fashion. Deep learning can be defined as a class of machine learning algorithms which are capable of extracting more features from raw input data using multiple layers. It filters the input through many layers, and it will learn how to classify and predict the data. Deep means the many number of layers that are used to transform data. To implement deep learning techniques, many computational nodes will be created. Each node is trained to analyze the given information and to make decisions like human brains. It is exactly similar to how the human brain filters any information into deep layers to understand in depth. The best programming languages for deep learning are Python, R Programming, and LISP.

3.11   STEPS TO IMPLEMENT DEEP LEARNING ALGORITHMS

1. Provide the input to the system.
2. Classify the input.
3. Extract the features.
4. Generate the output.

3.12   WHEN/WHERE/HOW/WHY TO USE DEEP LEARNING?

Deep Learning is used when the data size is large and there is a lack of domain understanding for feature extraction. It is widely used to solve complex problems.

3.13 EXAMPLES FOR DEEP LEARNING APPLICATIONS

- Automatic translation of text
- Instant visual translation
- Classifying objects in photos
- Identifying and detecting objects in images
- Automatic generation of captions for images
- Game play automation

3.14 COMPARISONS OF ARTIFICIAL INTELLIGENCE, DEEP LEARNING, AND MACHINE LEARNING

Though artificial intelligence, machine learning, and deep learning are the subsets of one another, there are few differences among them based on certain features. Tables 3.1– show the comparison and differences between them.

Table 3.1 shows the comparisons based on classification.

Table 3.2 shows the comparison between AI, ML, and DL based on characteristics.

Table 3.3 shows the differences between AI, ML, and DL based on performance measures.

TABLE 3.1

AI vs. ML vs. DL: Based on Classification

S. No.	Artificial Intelligence	Machine Learning	Deep Learning
1	Meaning: It makes the software think intelligently. It is done by studying how human brains think, learn, and decide to solve a problem. AI systems behave like humans without any fatigue, emotion, and limitations	Meaning: It is a subset of artificial intelligence. It relies on patterns. The system uses statistical models to perform a specific task without any explicit instructions. It is widely used to make predictions and decisions	Meaning: It is also called deep structured learning or hierarchical learning. It is based on artificial neural networks. It is widely used for feature detection. It can be referred as the procedure to implement machine learning
2	Types: • Weak AI • Strong AI • Super-intelligence AI	Types: • Supervised learning • Unsupervised learning • Reinforcement • Learning	Types: • Unsupervised pre-trained networks • Convolutional neural networks • Recurrent neural networks • Recursive neural networks

TABLE 3.2
AI vs. ML vs. DL: Based on Characteristics

S. No.	Artificial Intelligence	Machine Learning	Deep Learning
1	Key components: The AI algorithms mainly depend on heuristics and the study of human intelligence	Key components: The algorithms depend on structured data. Based on the structured data, the algorithms classify and predict the required information	Key components: The deep learning networks rely on the layers of artificial neural networks
2	Benefits: • Error reduction is easy • Easy-to-handle repetitive tasks • Availability at all time • Applicable to all fields • Automation is easier • Permanent memory is available	Benefits: • Enormous volumes of data can be processed • Easy to identify patterns in data • No human intervention • Easy to handle multiple varieties of data	Benefits: • Possible to extract more features in an incremental manner • Domain expertise is not needed • As the scalability of data increases, the performance will become better

TABLE 3.3
AI vs. ML vs. DL: Based on Performance Measures

S. No.	Artificial Intelligence	Machine Learning	Deep Learning
1	Performance: The performance of AI algorithms is measured by comparing the observed values and inferred values in the data input. The intelligent agents are the autonomous entities which direct their activities to achieve the goal. The performance measure in AI is done through the intelligent agents. In AI, the different types of intelligent agents are as follows: 1. Simple reflex agent 2. Model-based reflex agent 3. Goal-based agent 4. Utility-based agent 5. Learning agent The performance measure is the criterion that measures the success of an agent	Performance: In machine learning, performance measures are used to evaluate the learning algorithms. The performance metrics in machine learning are as follows: 1. Classification accuracy 2. Logarithmic loss 3. Confusion matrix 4. Area under a curve 5. Mean absolute error 6. Mean squared error	Performance: The performance measures are used to evaluate the deep learning methods and models. The performance of deep learning algorithms can be measured through the following parameters: 1. Programmability 2. Latency 3. Accuracy 4. Size of the model 5. Throughput 6. Efficiency 7. Rate of learning

TABLE 3.4

AI vs. ML vs. DL: Based on Its Workflow

S. No.	Artificial Intelligence	Machine Learning	Deep Learning
1	Workflow: 1. Transform the real intelligence into the system 2. Process and train the system 3. Deploy and execute the system	Workflow: • Provide input data • Analyze the input • Identify the patterns • Make the future prediction • Generate feedback	Workflow: • Provide the training data • Identify the neural network model • Configure the model with the learning process • Train the model • Interpret the results

TABLE 3.5

AI vs. ML vs. DL: Based on Objectives and Limitations

1	Objective: The main objective is to simulate human intelligence in a system and to make the system to work smart and to solve complex problems	Objective: The main objective is to design the system for learning by itself and for making predictions	Objective: Deep learning has been introduced as a new area of machine learning to bring machine learning algorithms closer to artificial intelligence
2	Limitations: • Issues and challenges in integration • Implementation is time-consuming • Challenges in interoperability with cross-platforms • Difficult to interpret the results	Limitations: • Flaws in input data may lead to erroneous output • Bad input may bring down the reliability of the output	Limitations: • Need for a very large amount of data • Training the system is complex and time-consuming • Requires more graphical processing units and machines

Table 3.4 shows the comparisons between AI, ML, and DL based on workflow.

Table 3.5 shows the objectives and limitations of AI, ML, and DL.

Table 3.6 shows the futuristic issues in AI, ML, and DL.

Table 3.7 shows the differences between AI, ML, and DL based on future scope and tools.

3.15 SUMMARY

This chapter has provided an overview of artificial intelligence, machine learning, and deep learning and has provided solutions for when, where, how, and why

TABLE 3.6
AI vs. ML vs. DL: Based on Futuristic Issues

1 Futuristic issues: • When it gives imperfect solutions, it leads to a huge loss in business environment • Imperfect solutions lead to chaos and will have a great social impact • More human intervention and computation is needed to produce a perfect AI solution • Availability of skilled resources for developing solution with AI • End-user training cost and time • Data protection • Compatibility with devices and systems • Upgrading the algorithm periodically and its implementation in AI-based system is needed to produce reliable results	Futuristic issues: • Ethical issues • End-user satisfaction • Variation in norms and rules in different regions and countries • False correlation leads to misinterpretation • Time-consuming to build correct structured data • Reusability and integration	Futuristic issues: • Availability of very large data set for training • Time consumption in deep learning networks • Data overfitting occurs, when the number of parameters exceeds the number of observations • Information privacy • Latency occurs when retraining is needed for new data and information • Leads to unstable conditions when there is variation in input data

TABLE 3.7
AI vs. ML vs. DL: Based on Future Scope, Tools, and Areas of Application

1 Areas of application: Examples: • Medical diagnosis • Aviation • Robots • Workspace communication	Areas of application: Examples: • Social media services • Email spam and malware filtering • Virtual personal assistant	Areas of application: Examples: • Speech recognition • Image recognition • Natural language processing • Self-driving cars
2 Future scope: • It may spread across all spans of daily life • It will be applied more for language translations • It will be applied in all branches of engineering and for automation • Development of expert systems	Future scope: • It will be widely used in digital marketing • It has more scope in the field of education • It will have a greater impact on social media, search engines, and predictions	Future scope: • Wider scope for development of more deep learning tools and standards • Development of more simplified programming frameworks • Deployment of transfer learning concept through reusable components
3 Sample tools needed for development: • TensorFlow • Keras • PyTorch • Theano	Sample tools needed for development: • Weka • PyTorch • TensorFlow • KNIME	Sample tools needed for development: • Caffe • Torch • DeepLearning4J • Cuda

these techniques can be used. Artificial intelligence vs. machine learning vs. deep learning is compared based on their meaning, types, examples, benefits, limitations, futuristic issues, tools, future scope, areas of application, etc. Basic features of artificial intelligence, machine learning, and deep learning are summarized in this chapter.

4 AI and Big Data
Ethical Reasoning and Responsibility

Sweta Saraff

CONTENTS

4.1 INTRODUCTION

Do we have the capabilities to understand the infinite potentialities of Nature? This question was contemplated years ago by Sri Aurobindo (1993).

> "[O]ur science itself is a construction, a mass, of formulas and devices; masterful in the knowledge of processes and in the creation of apt machinery, but ignorant of the foundations of the being and of World-being, it cannot perfect our nature and therefore cannot perfect our life."

He further doubted the capabilities of science in reaching an ultimate truth or as a matter of fact, any ultimate truth:

> "One might ask whether science itself has arrived at any ultimate truth; on the contrary, ultimate truth even on the physical plane seems to recede as science advances. Science started on the assumption that the ultimate truth must be physical and objective – and the objective ultimate (or even less than that) would explain all subjective phenomena." Sri Aurobindo (1972)

The overwhelming speed at which the age of objectivity is celebrated by science – technology – data and artificial intelligence (AI) has elevated humanity at a material plane. This current state of mindset to achieve and compete relentlessly for success and the pleasures of life is trying to replace harmony with desire. Today everyone is in an unnerving race of achieving material success in life. What is real success? What we have achieved and we have lost? Can we still recover our losses? Is the damage repairable?

With the advancement of technology, there is an unease regarding the increasing rate at which the human race is snapping its ties with Nature, humanity, and their conscience. In the path of seeking objective pleasure, there is decline in subjective well-being. An attitude of liberated technology from the grasp of ethics is an insignia of regressive society. The presumption of progress is rested on hollow pillars of momentary achievement with no consideration of what is right or wrong unless it is relatable to self.

We all measure our success or happiness in terms of material possessions which are machines, clothes, hybrid food, luxury housing, and more dependability on others. At the same time, we are forgoing pleasure from staying close to Nature, family, and human values. This raises a question on measurement which was discussed by David Bohm, the renowned physicist. He succinctly differentiates between "measurable" and "immeasurable"; or it may be understood as a difference between finite and infinite. Bohm has postulated that the ultimate reality of Nature lies not in its fragmentation but in its unification. Bohm (1993) opines that this immeasurable universe encompasses all its parts and unites them in an irrefutable totality. The fragmented objectivity is far from "ethical and moral basis of human peace and happiness" (Chakraborty, 1998).

Professor S.K. Chakraborty expresses his concern that objective pathways may lead to the compromise of "ethical-moral sentiments" which unites mankind with each other and with our surroundings. Earlier philosophers like Gandhi and Tagore were proponents of self-sustained living, which was simpler, closer to Nature, similar feelings for self and others, and an unwavering sense of values.

4.2 ETHICS REASONING IN ARTIFICIAL INTELLIGENCE

Understanding human cognition and reasoning involved in taking any decision has always drawn attention. This field has been evolving with contributions from different disciplines like social sciences, humanities, neurosciences, mathematics, and artificial intelligence. Kay (2002) believed that it is imperative to decode the inception and root cause of human intuition to comprehend the complexities of the decision-making process. Cognition, action, and emotion are essential components of decision-making phenomena. Belief, thought about an event or another person, influences the behaviour and attitude, and these are compounded by psychological affect and physiological reaction within a person.

People most often analyse the opportunity cost in taking any decision. The decision-making process also incorporates one's assumption about others' expected behaviour in a context, as per their past experiences. It operates on a prototypical schema. There can be multiple outcomes and a chain of reactions following them. Then arises the dilemma of selecting the best possible option or making a choice and having confidence in the decision or analysis. A rational decision-maker always analyses several possibilities from different perspectives to make a specific choice. Some theorists proposed that humans used specific algorithms or normative reasoning (Goodwin & Wright, 1998), though Hoch et al. (2001) were of the opinion that there may be no specific rules followed by humans and they generally use heuristics to solve the problem if they are unique or novel. Rubinstein(1998) opined that people choose the best alternative by integrating desirability with feasibility.

Economists believe that people try to make decisions based on their satisfaction of wants or pursue a maximum satisfaction model. This model analyses the value of choice for its efficacy in a particular situation and also in accordance with their personal preferences, past experiences, etc. According to Oliveira (2007), one can analyze the "Expected Utility Theory" either analytically or synthetically. When one adopts the analytical view, they will observe the available choices and then decide their outcomes, whereas in a synthetic method, they will ascertain the utility first and then find out means to achieve it.

Based on these contemplations, the following pertinent questions arise – Is rationality overpowering ethics? Are we reasoning ethically? What strategies are we utilizing in differentiating right from wrong? Does science have the capacity to reason ethically? Can machines or robots reason like humans? Ethics, morality, and values are both subjective and culture based. It does not follow the "one size fits all" rule. Do these advanced algorithms follow policies of fairness above all? Are we careful about the specificity and sensitivity measures taken by the predictive tools? It evokes different opinions from utilitarian (consequential) as well as deontological (rule) school of ethics. People are held accountable for their doings, and similar responsibility needs to be fixed for decisions taken by automated machines. The engineers and the software experts need to take the ethical responsibility for the outcomes.

It is quite expected that automated vehicles (AV) such as self-driven (SD) cars would be facing such a dilemma in real-life situations where they must be programmed to take a quick decision. Which ethical rule will they follow? Would they harm an innocent pedestrian to save the car speeding from the other side? To come to an acceptable answer is quite difficult for AI engineers and scientists. We can also assume that if we present this question to people from different geographical locations, ethnicity, culture, education, work types and levels, value orientation, age, and gender, there will be disparities in responses.

Traditionalists may differ from pluralists in their outlook about ethical beliefs. Their reasoning style is more consequential and dependent upon the probable outcome. Some may consider saving more people over one due to higher order reasoning and future orientation, whereas people with emotional attitude may be overwhelmed by the thought of killing an innocent man. This dilemma still remains unanswered. First we must understand that the philosophy of ethics is based on inductive reasoning. Here, the order and straightforward rules of logic do not apply. There is hardly an agreement between nations regarding following the same laws and moral code of conduct for delivering justice. What ethical principles or design will then be imbibed by an automatic vehicle? Would one nation allow such self-driven cars which do not fully adhere to their legal systems?

The requirement of ethical reasoning is not limited to self-driven cars. Artificial intelligence today can predict stock prices, regulate capital markets, admissions, and tax evasions, and prepare data on a range of requirements like weather forecasts, cyclones, pollution levels, and epidemics. They are useful in maintaining applications in educational institutions, jobs, etc. AI has supported advancements and ease of living at most levels to the extent of pushing us to dependency and thinking like machines without emotions. Based on social media, AI can predict an individual's preferences, expected behaviour, or even reactions, intentions, beliefs, friends, etc.

Are these predictions on expected human opinion or behaviour reliable? Humans by nature are dynamic, subjective thinkers who act according to situations. They present different behaviours based on mood, situations, current wants, and expected utility. The argument remains – Can a machine accurately predict the behavioural patterns and emotional sensitivity of humans?

Another ethical challenge is taking informed consent before collection of data or any information. Data collected from naïve internet users without their knowledge or prior permission creates an environment of distrust and violation of personal space. The predictive technology used by Google and Facebook provides us with segregated information based on our search history, creating suggestiveness, which may be misleading. They sell this information to different companies which are dealing with products ranging from apparels, processed food, and travel agencies, making people vulnerable to different frauds.

They can predict locations travelled, home, school, and office addresses, shopping behaviour, suggestive friends list, chances of your partner's infidelity, investments, choice of candidates in an election, etc. All personal data are accessible easily over the internet. Just like a detective, AI can draw up a conclusion about the future behaviour, interests, or probable tendencies of an individual. This tool combines various factors to produce a single score for a person, and this prediction is used as a guide to take a future course of action. Does it follow the nuances of ethical reasoning? Predictions can be a brilliant tool, but are they adaptive? Does this intelligence conform to the principles of ethical sensitivity and awareness? Does it reason ethically?

Descriptive ethics, normative ethics, and applied ethics are recognized as major domains of ethical philosophy. Descriptive ethics take an empirical view of the beliefs of the people. This domain defines the principles behind the moral reasoning for differentiating right actions from wrong actions. The theorists working on descriptive ethics look into the values in which people repose their faith and the basic features of virtuous conduct. The importance of descriptive ethics lies in the fact that they are different from normative ethics. It believes that the ideals and values of various cultures are dynamic in nature. They are always evolving with development of societies and mankind. The theorists opine that ethical requirements of the current generation may be different from their predecessors. So descriptive theorists try to systematically inquire by collecting information and observations from different fields like basic sciences, applied sciences, and social sciences. Colby and Kohlberg's (2011) work on cognitive moral development is an example of descriptive ethics. This field of ethics deals largely with describing and predicting human behaviour (Donaldson & Dunfee, 1994). A framework on ethics may include three steps or major components (Figure 4.1): identifying the sensitivity or awareness, making a moral reasoning, and engaging in moral conduct.

Normative ethics studies how people should act or what course of action can be considered ethical (Hoffe, 1989). Normative ethical theories can be observed from

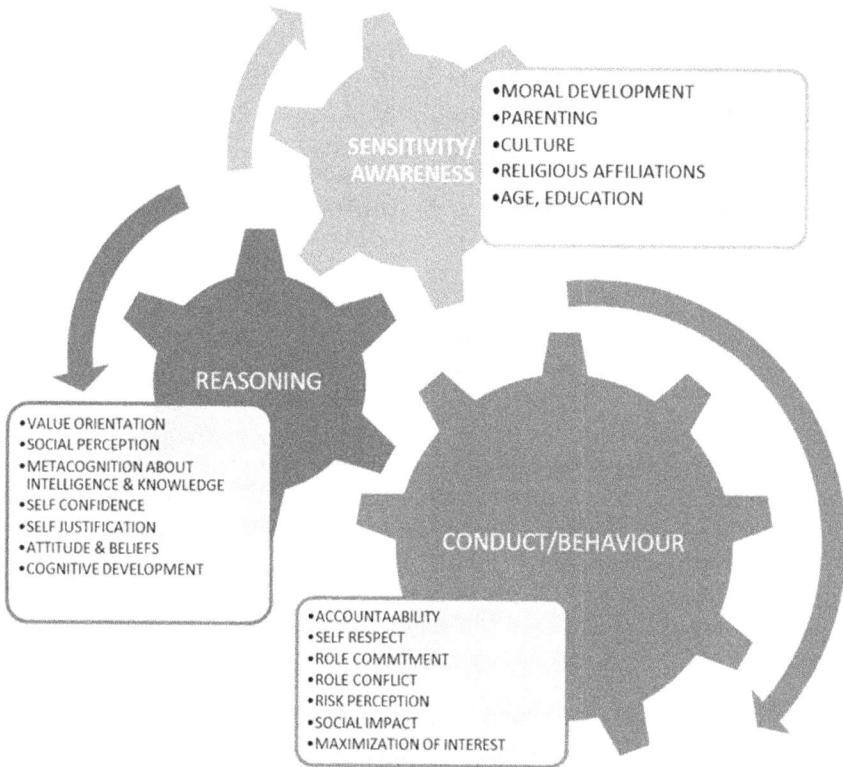

•MORAL DEVELOPMENT
•PARENTING
•CULTURE
•RELIGIOUS AFFILIATIONS
•AGE, EDUCATION

SENSITIVITY/
AWARENESS

REASONING

•VALUE ORIENTATION
•SOCIAL PERCEPTION
•METACOGNITION ABOUT
 INTELLIGENCE & KNOWLEDGE
•SELF CONFIDENCE
•SELF JUSTIFICATION
•ATTITUDE & BELIEFS
•COGNITIVE DEVELOPMENT

CONDUCT/BEHAVIOUR

•ACCOUNTAABILITY
•SELF RESPECT
•ROLE COMMTMENT
•ROLE CONFLICT
•RISK PERCEPTION
•SOCIAL IMPACT
•MAXIMIZATION OF INTEREST

FIGURE 4.1 Components of ethical reasoning.

three perspectives: virtue ethics, deontology, and utilitarianism. Virtue ethics lays emphasis on individual character, which includes how they are acquired, nurtured, and applied in real-life scenarios. It is based on a set of stable dispositions where decisions are not taken based on suitability in a particular context. Traits like honesty, truthfulness, and care are attributes of the personality and not just part of a habit. It includes choices, interests, values, attitudes, courage, and temperaments. Virtue ethics differs from deontology as it focuses on the inherent disposition of an individual rather than his or her adherence to an established set of rules. A virtuous action is representative of an inner sense of morality and "a way of being" leading to consistency in action and sensibilities.

Deontology ethics are well-defined or rule-based ethics (Waller, 2005). The main tenet of the theory lies in engaging in a behaviour when the action is morally justifiable rather than considering its consequences. Deontology is different from consequentialism as it believes in moral action rather than its effects (Flew, 1979). It deals with personal and professional duties and legal rights. Such ethics may contrast with personal or cultural values, but people have a moral obligation to follow a certain code of rules. The premise of utilitarian philosophy of ethics is grounded in selecting an action based on its perceived consequences. The main thought is to maximize utility in a conduct for general or optimal well-being of the humankind. In institutions or people engaging in utilitarian or consequential ethical practices, the main hindrance lies in limited perception or ability in understanding long-term consequences or effects on a new theory, practice, or invention.

Even with the best motive, a person of high integrity and good will may still make mistakes in justifying an action due to limitations of human perceptual ability and foresightedness. Current advancements in AI have created a deep interest among computer users, data scientists, psychologists, corporates, and the general public. The idea of gaming has transitioned from single user to multi-users with chat-bots, avatars, levels, badges, trophies, etc., creating a stimulating and challenging environment for players. This has shifted the focus of research in understanding the ethicality of implicit and explicit stimulations used in such technologies. Do they improve learning and development or they are creating blocks in the moral and cognitive development of millennials?

The most important issue related to sustainability in ethical reasoning lies in machine medical ethics: training ethics to machines as to how they should behave adhering to values and rules in different situations with pragmatism and care. Attempting to create algorithms to train an ethical AI is a very complex task, but still efforts are being made in this respect (Churchland, 2011; Wallach & Allen, 2010). A comprehensive and consistent moral theory which can balance social, cultural, and demographic differences and guide the actions of an ethical, logical, and intelligent robot is an urgent requirement in keeping pace with the innovations.

It must be programmed succinctly to deal with real-life situations. Since there are no available solutions to manifold ethical dilemmas and mutually agreed-upon rules and standards to follow, development of an acceptable theory of ethical machine looks a herculean task.

Tatjana Kochetkova (2015) suggests that use of AI in health and medicine should be limited to instances, where clear guidelines for ethical conduct already exist. A hybrid approach (mixed top-down–bottom-up) is advised where advantages of both top-down (use of moral principles in selection and control of action) and bottom-up (supporting with theoretical background and experimental learning) can be utilized (Allen et al., 2005). An AI system must be capable of understanding the legal, moral, and cultural impact of their decisions.

4.3 ETHICAL RESPONSIBILITY IN AI

With power comes great responsibility, and systems dependent on machine learning and artificial intelligence are no exception (Dignum, 2017). According to Dignum (2018),

> Whatever their level of autonomy and social awareness and their ability to learn, AI systems are artefacts, constructed by people to fulfil some goals. Theories, methods, algorithms are needed to integrate societal, legal and moral values into technological developments in AI, at all stages of development (analysis, design, construction, deployment and evaluation).

Both ethical reasoning and responsibility form the edifice on which future moral robots must be built. The primary goal is to train robots to behave according to the programmed code of conduct. However, can they explain the reasons behind their conduct? So, this responsibility of training and mentoring ethical robots lies on the software programmer. The next question is whether the programmer is equipped with moral reasoning? Does he understand the nuances of virtue, justice, and legality? How to solve the ambiguities of an ethical dilemma and make the most fair and honest choice?

A framework of ethical responsibility envisaged on the teachings of Vedanta is relevant for most of the cultures, societies, and overall development of humanity.

1. Delivering one's duty without any doubt and lethargy is expected from all. The world we live in will be better if each individual performs even the smallest task assigned to them. There is no use talking about reforming the world without caring for the pressing problems of the moment. The immediate crisis, the task that lies ready to hand, must engage the individual's attention, and if he does it well, he will indirectly help in the betterment of the world. Every man has a certain place to occupy and a certain function to fulfil in the social economy.

2. Belief in respecting dignity of all mankind above self-interest and personal desires is of paramount importance. Justice and benevolence are the primary attributes required to fulfil the ideals of social service. The laws made for the benefit of society, if given more importance than the existence of humanity, will lead to extremism by creating conflicts between duties. The laws serve as the guiding path to deliver justice without any biasness. But man has no call to act in blind obedience to them. The need is not conformity to rules, but conformity to the law of reason. When the rules come into conflict, we must fall back on the supreme commandment, and ask ourselves which course is most conducive to the realization of reason in the world. We must serve as moral rational beings, with acts expressive of the pivotal purpose.

3. Radhakrishnan (1914) said: "The inner spirit is more important than outward conformity to law. An action is good, not because of its external consequences, but on account of its inner will. Virtue is a mode of being and not of doing. It is not something to be found, but a function or an exercise of the will."

4. Reasoning credits us with accountability towards ideals of justice, humanity, and righteousness which must be delivered without any prejudices. If each individual is given the power to play according to their conviction, the universal goal of social balance and harmony would be destroyed. Logical judgement with moral intent must not be sacrificed to fulfil the norms of social convention. Myopic traditions and values are regressive and must not be favoured at the cost of knowledge and growth. We must build institutions which deliver harmony and progress with equal veracity.

5. No act should be done with a selfish interest or for the gratification of inclinations. An individual, if engaged in moral reasoning, serves humanity

with sensibilities and empathy. The highest ideal in life must be to serve society by exercising logical judgement even in the moment of crisis.

6. Character, which is the habit of will, is not determined but self-created. The rational self, if consciously agree to the immoral attributes or accept them as its stable disposition, then such characteristics are selfish in nature. Human beings cannot justify a mistake by blaming the genetic lineage. It is his wilful desire to accept or reject. He must control the acquired behaviour and reason with honesty and integrity.

7. Visionary idealism which lacks focus and intelligent direction is ineffective. If private interests are permitted, then there are chances of disruption in the society. A man cannot stop and debate within himself at every point as to what his duty is. He must fulfil his duty with enthusiasm and eagerness.

8. "Know the self to be sitting in the chariot, the body to be the chariot, the intellect (buddhi) the charioteer, and the mind the reins" – Radhakrishnan (1914). One must have control over senses for a steady and sharp intellect.

9. Knowledge should not be acquired for the purpose of completing a course but must enable us to take a right perspective and a pragmatic frame of reference about other things and our place in the world.

10. We must do the right whether the right is done or not. "Devoted each in his own work, man attains perfection" Radhakrishnan (1914). Every man is required to contribute to the national strength his quota of earnest work. It is by the endorsement of this philosophy of work that the world as a whole will flourish.

Bonnemains, Saurel, and Tessier (2018) in their paper on "Embedded ethics: some technical and ethical challenges" propose development of "formal tools" that are descriptive of different circumstances and "models" of ethical philosophies which have the potential to reason not only automatically but also ethically, giving proper justification. Today's millennials nurture the myth of a utopian world maybe, but such a world looks imperfect, even if it is the mightiest. An imperfect world, which is harmonious and hopeful of future growth, is more blissful and happier.

REFERENCES

Allen, C, Smit, I, Wallach, W. Artificial morality: Top-down, bottom-up, and hybrid approaches. *Ethics and Information Technology*, 7 (2005): 149–155.

Aurobindo Sri. *The Life Divine*. First Edn. Pondicherry: Sri Aurobindo Ashram; 1993. The Divine Life; p. 1034. 1939-40, Tenth impression 1993.

Aurobindo Sri, Mother. *On Science*. Pondicherry: Sri Aurobindo Ashram; 1972. Drawbacks and Limitations; p. 12.

Bohm, D. Science, spirituality, and the present world crisis. *ReVision*, 15(4) (1993): 147–152.

Bonnemains, V, Saurel, C, Tessier, C. Embedded ethics: Some technical and ethical challenges. *Ethics and Information Technology*, 20(1) (2018): 41–58.

Chakraborty, S.K. *Values and Ethics for Organizations: Theory and Practice*. New Delhi: Oxford University Press, 1998, 153–171.

Churchland, P. *Braintrust: What Neuroscience Tells Us about Morality?* Princeton: Princeton University Press, 2011, pp. 23–26.

Colby, A, Kohlberg, L. *The Measurement of Moral Judgment*, Vol. 1. Cambridge: Cambridge University Press, 2011.

Dignum, V. Responsible autonomy. In *Proceedings of the Twenty-Sixth International Joint Conference on Artificial Intelligence (IJCAI'2017)*, 2017, pp. 4698–4704.

Dignum, Virginia. Ethics in artificial intelligence: Introduction to the special issue. *Ethics and Information Technology*, 20(1) (2018): 1–3.

Donaldson, T, Dunfee, TW. Toward a unified conception of business ethics: Integrative social contracts theory. *Academy of Management Review*, 19(2) (1994): 252–284.

Flew, A. Consequentialism. In *A Dictionary of Philosophy*, 2nd Ed. New York: St Martins, 1979, p. 73.

Goodwin, P, Wright, G. *Decision Analysis for Management Judgment*. Chichester: John Wiley & Sons Ltd, 1998.

Hoch, SJ, Kunreuther, HC, Gunther, RE. *Wharton on Making Decisions*. New York: John Wiley & Sons, Inc, 2001.

Höffe, O. Kant's principle of justice as categorical imperative of law. In *Kant's Practical Philosophy Reconsidered*. Dordrecht: Springer, 1989, pp. 149–167.

Kay, J. Beware the pitfalls of over-reliance on rationality: Attempting to shoehorn complex decisions into the framework of classical theory can be a mistake. *The Financial Times*, 2002, p. 9.

Kochetkova, T. An overview of machine medical ethics. In *Machine Medical Ethics*. Cham: Springer, 2015, pp. 3–15.

Oliveira, A. A discussion of rational and psychological decision-making theories and models: The search for a cultural-ethical decision-making model. *Electronic Journal of Business Ethics and Organization Studies*, 12(2) (2007): 12–13.

Radhakrishnan, S. The ethics of the Vedanta. *The International Journal of Ethics*, 24(2) (1914): 168–183.

Rubinstein, A. *Modeling Bounded Rationality*. Cambridge, MA: MIT Press, (1998).

Wallach W, Allen C. *Moral Machines: Teaching Robots Right from Wrong*. Cambridge: The MIT Press, 2010.

Waller, BN. *Consider Ethics: Theory, Readings, and Contemporary Issues*. New York: Pearson Longman, 2005, p. 23.

5 Online Liquid Level Estimation in Dynamic Environments Using Artificial Neural Network

Thulasi M. Santhi and S. Sathiya

CONTENTS

5.1 INTRODUCTION

A liquid level is an essential parameter to measure in almost every process system such as the food and beverage industry, petrochemical plants, water reservoirs, and automotive systems. Level measurements based on the principle of differential pressure, vibrating wire, capacitive magnetic floating, and ultrasonic are conventionally used in the industries. In automotive, mainly variable resistive, capacitive, float, and ultrasonic sensors are used for measuring the level of the fuel tank. Resistive float-type sensors are widely used as automotive fuel indicators more than other conventional sensors. Many conventional level measurement techniques are accurate only when the liquid contained in the fuel tank is stationary. With the presence of dynamic environments, such as variation in temperature, an inclination of the fuel tank, and liquid slosh, most of the methods fail to produce accurate level measurements which are very essential [1]. The conventional sensors are incapable of compensating the effects of temperature variations, inclination, or slosh while measuring the liquid level. The level measurement in such environments requires the best effective system with high sensitivity, which will result in highly accurate level measurement.

The proposed method of liquid measurement integrates the neural network approach along with the measurement sensor for improving the system accuracy. As the neural network is a computing system that predicts the output accurately by considering the past input data, the ability of a neural network is employed to improve the accuracy of the proposed system in the dynamic environments [1, 2]. Another level measurement technique with optical fibre has been tremendously increasing for the past few years due to its advantages such as being unaffected from electromagnetic interference, having low signal loss and distortion, and being lightweight. Thus, designing a system with fibre Bragg grating (FBG), a type of optical fibre, will further enhance the performance of the measurement system. The proposed level measurement sensor based on FBG-embedded cantilever beam and a float combination [3], integrated with the machine learning approach, specifically wavelet neural network, contributes a highly accurate and sensitive system apt for level measurements in dynamic environments [4]. The proposed design of the liquid level measurement system is specifically formulated for the automobile fuel tank which is affected by the dynamic environment when the vehicle undergoes different acceleration.

5.2 LIQUID LEVEL MEASUREMENT IN DYNAMIC ENVIRONMENTS

The liquid level measurements in static environments are easy to detect and accurate, while it becomes very much erroneous in dynamic environments. The level measurement of the fuel tank present in a moving automotive system is a common example of a dynamic system. For the effective usage of fuel in an automobile, the measurement of fuel quantity is essential which will help the drivers to determine the total time a vehicle can drive without refuelling. The main disturbances which affect the measurement accuracy of the fuel level during the movement of vehicles are temperature variations, the inclination of the tank, and sloshes. All these disturbances and the variations must be compensated to determine the exact level of the fuel left in the automobile fuel tank.

5.2.1 INFLUENCE OF TEMPERATURE

The temperature has a great impact on the variation of liquid properties, notably density and viscosity. The movement of molecules in the liquid varies according to the temperature variations. The volume of the liquid in the fuel tank seems to be increased or decreased as a result of low or high densities, respectively (density is the ratio of mass and volume). Viscosity is based on the cohesive forces between the molecules of the

TABLE 5.1

Dynamic Viscosity, Kinematic Viscosity, and Density Variations Due to Temperature Change of Engine Oil SAE 15W-40 [5]

S. No.	Temperature (°C)	Dynamic Viscosity (mPa·s)	Kinematic Viscosity (mm²/s)	Density (g/cm³)
1	0	1328.0	1489.4	0.8916
2	10	582.95	658.60	0.8851
3	20	287.23	326.87	0.8787
4	30	155.31	178.01	0.8725
5	40	91.057	105.10	0.8663
6	50	57.172	66.464	0.8602
7	60	38.071	44.585	0.8539
8	70	26.576	31.350	0.8477
9	80	19.358	23.006	0.8414
10	90	14.588	17.467	0.8352
11	100	11.316	13.648	0.8291

liquid. As temperature increases, the cohesive force between the molecules decreases and results in a reduction of viscosity; in the same way, a decrease in temperature causes an increase in cohesive force between molecules and viscosity increases. In short, both density and viscosity have an inverse relation to temperature.

As the fuel level measurement is considered, the temperature dependency of the engine oil SAE 15W-40 is given in Table 5.1 [5]. The variation in dynamic viscosity, kinematic viscosity, and density over temperature from 0 to 100°C is mentioned. It is clear from the table that these three parameters are decreasing concerning the increase in temperature of the oil. In Figures 5.1 and 5.2, the three parameters are plotted to analyze the effect of temperature on them.

The viscosity values have a remarkably high dependency on temperature as it reduces rapidly to very low values up to 50°C, and for further increase in temperature, the rate of decrease in viscosity is found to be lesser. The density decreases linearly as the temperature of the fuel increases. From the graphs, it is clear that the consideration of temperature variation of liquid in the level measurement system is significant. Due to the presence of environmental temperatures, the defects in the coolant circulation system, and excessive loads, the engine temperature will increase which results in the variation of liquid density and viscosity. As the properties of liquid vary because of the temperature, the liquid volume in the tank will also vary accordingly. Thus, the fuel level measurement will give errors in level readings unless the disturbances are compensated properly.

5.2.2 INFLUENCE OF INCLINATION

The automobile is movable systems that pass through different geographical variations such as elevations, slopes, and irregularities in the road. In such cases, the

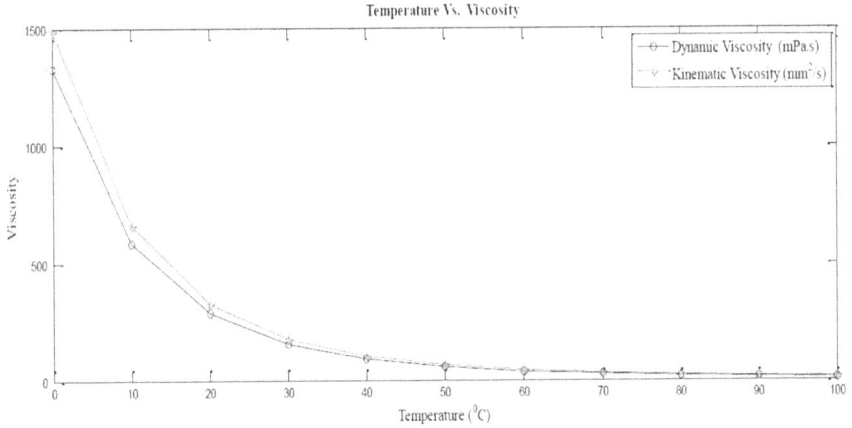

FIGURE 5.1 Dynamic and kinematic viscosity variations of engine oil SAE 15W-40 over temperature.

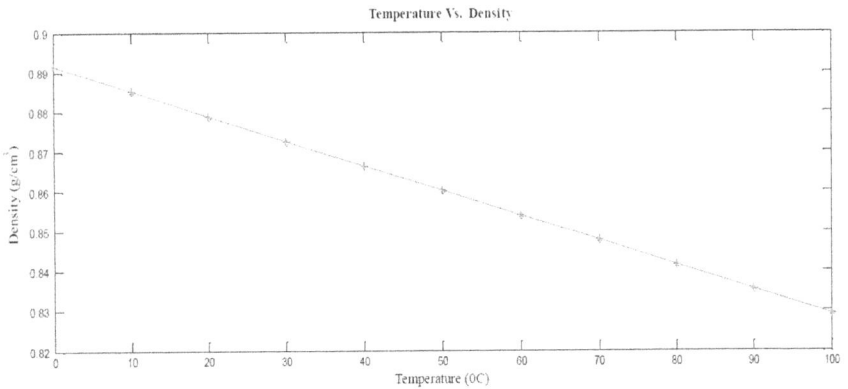

FIGURE 5.2 Density variations of engine oil SAE 15W-40 over temperature.

baseline of the sensor will be in inclination with the normal horizontal line and the liquid level will also get displaced in such a way that one side has the highest-level reading while the other has the lowest. If in case the position of the sensor is at the centre of the fuel tank, the level reading is at 45° inclination with the horizontal line and 15° inclination with the horizontal line, thus considering the inclination is significant in the level measurement system of a fuel tank.

5.2.3 INFLUENCE OF SLOSHES

Another main disturbance to be considered is the slosh of the liquid produced inside the automobile fuel tank during the movement of the vehicle with variable accelerations [6]. It is a highly unpredicted and undesired disturbance which seriously affects the accuracy of the liquid level measurement system. The sloshes of

the liquid vary the liquid level dynamically with respect to the different acceleration of the vehicle, which is difficult to compensate with the conventional liquid
measurement system. Therefore, the size and shape of the fuel tank, acceleration
of the vehicle, and inclination of the tank position are the parameters that vary the
slosh intensity.

5.3 SENSOR DESIGN

5.3.1 Fibre Bragg Grating Sensor

An optical fibre can be transformed into an FBG by exposing the core to intensive
laser light for introducing a periodic modulation in the refractive index. The broadband light signal passing through the FBG fibre gets refracted and reflected due
to the gratings. Because of the reflection and transmission of different wavelength
signals, interference occurs. The Bragg condition as given in Equation 5.1 must be
satisfied by the wavelength for the reflection of the optical signal:

$$\lambda_B = 2n\Lambda \qquad (5.1)$$

where λ_B is the Bragg wavelength, n is the refractive index of the optical fibre,
and Λ is the grating period. By changing the parameters n and Λ, the Bragg
wavelength can be varied. The grating length is also one factor that affects the
Bragg wavelength. Figure 5.3 shows the reflected signal from FBG looks like only
a single wavelength is reflecting, so it resembles peak signals. FBGs are passive,
robust, have less size, high sensitivity, and precision, which make the sensor suitable in many optical applications such as telecommunication fields and sensor
technologies, where wavelength selection is required [7]. The Bragg wavelength
change is described as

$$\frac{\Delta\lambda_B}{\lambda_B} = C_1\epsilon + C_2\Delta T \qquad (5.2)$$

where λ_B represents the Bragg wavelength under the unstrained condition of the FBG,
$\Delta\lambda_B$ is the Bragg wavelength shift due to the presence of the strain ε and temperature
variation ΔT, $C_1 = 0.78 \times 10^{-6}$ and $C_2 = 6.67 \times 10^{-6}$°C. For longer Bragg wavelengths,
the changes in wavelength occur due to the applied strain and temperature [3, 7, 8].

The axial strain and temperature are the physical parameters that the FBG sensor can measure directly. To measure displacement by the FBG, the displacement
must be converted to the axial strain experienced on FBG by any transducer. The
basic block diagram presenting the working of the FBG sensor for measurement of
displacement is shown in Figure 5.4. The cantilever structure of the sensing element
gets deformed and induces a strain according to the external displacement. This
strain is experienced by the FBG, which converts the displacement into variations in
wavelength, bandwidth, light intensity, and finally to voltage using an optical spectrum analyser (OSA) connected to FBG circuitry. Calibrations can be done with this
information concerning the displacement for the measurement.

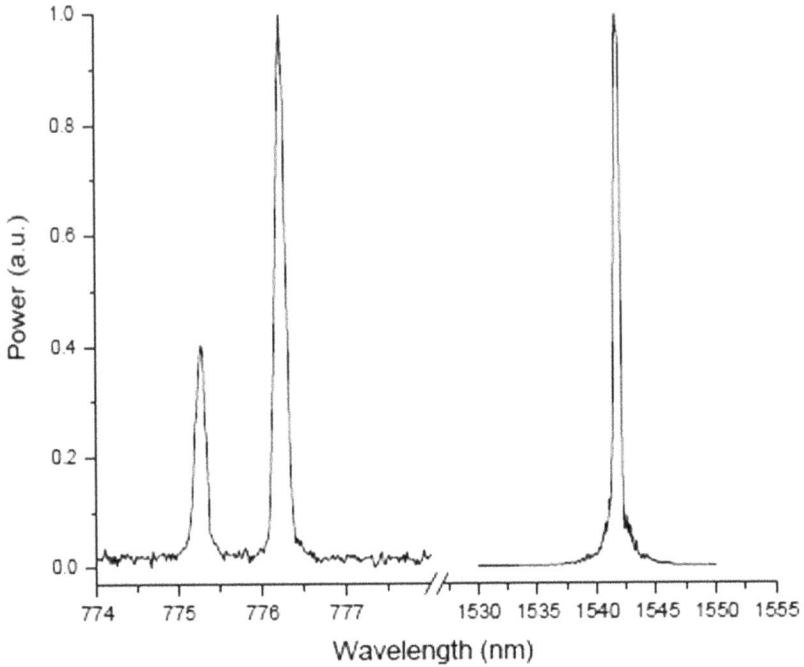

FIGURE 5.3 Example of reflection peaks from an FBG [7].

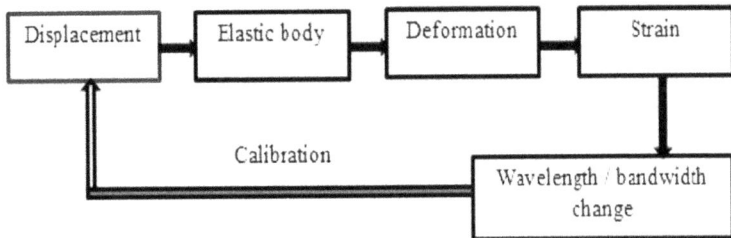

FIGURE 5.4 Schematic diagram of displacement measurement using FBG sensor.

5.3.2 Cantilever Beam

A cantilever beam is a beam that is fastened at one end and left free at the other end; it means one end is fixed to support and connected, while the other end is left without any support [9, 10]. When pressure or force is exerted on the free end of the beam, the beam carries this load to the support, i.e., fixed end, from where the moment of force and the shear stress can be managed. The tendency of an applied force to turn or twist the beam is the moment of force, while the stress applied parallel to the beam is the shear stress.

Due to the support on the cantilever beam's fixed end, the free end can carry a specific weight which effects in bending of the beam instead of breaking down due to the shear stress. Without external bracing or support, cantilever construction allows for overhanging structure, a stainless-steel cantilever beam is used in the described methodology of design. The cantilever beam deflection for the applied external force in the free end is illustrated in Figures 5.5 and 5.6.

5.3.3 FLOAT SENSOR

Float is a structure that is suspended or freely placed in the liquid surface in such a way that it will float fully or partially over the liquid surface. To follow the variations in liquid level, floats and float switches use the buoyancy principle which depends on

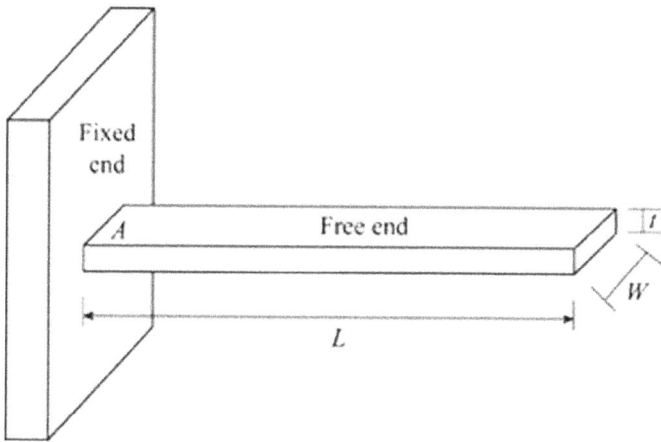

FIGURE 5.5   Typical cantilever beam.

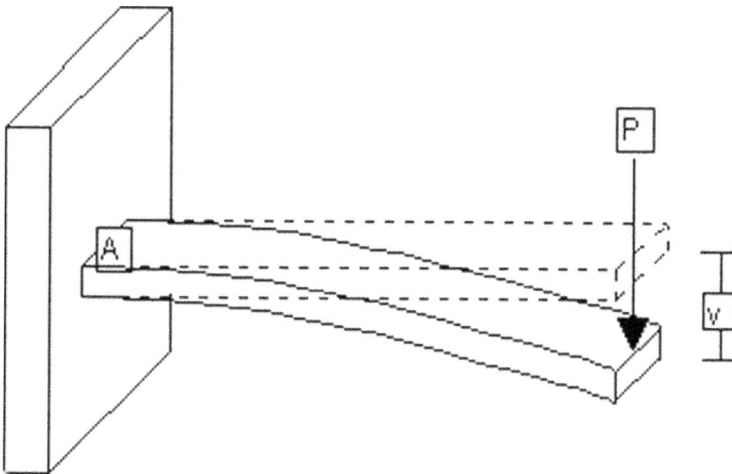

FIGURE 5.6   Cantilever beam deflection under load at the fixed end.

the liquid level and its density. Float structure is made up of plastics or PVC material having lesser density than the liquid whose level must be measured so that they can float over the liquid surface. Most of the floats are hollow spherical or capsule-shaped filled with air with lesser density than the fluid used in the applications.

For reed switch activation, commonly stainless-steel magnetic floats are used, and they are tubed magnetic floats with a hollow tubed connection for suspending from the top. In the measurement of the liquid level, with the consideration of strength, corrosion resistance, and buoyancy, the magnetic floats are suitable in a wide range of industries. The floats are made up of welding two symmetrical half-shells together in an airtight manner and the perfection in the process decides the strength and durability of the float. The welding should be done uniformly in such a way that it produces a smoothly finished seam, which is like the surface of the rest of the float.

5.3.4 System and Working Principle

The proposed sensor system design for measuring the fuel level of an automobile fuel tank is illustrated in Figure 5.7. This set-up is composed of an appropriate fuel-filled horizontal cylindrical automobile fuel tank, a cantilever beam made up of stainless steel, two FBGs mounted along with the optical fibre on the opposite sides of a cantilever beam, a float suspended from the free end of the beam, an optical source, and an optical spectrum analyser (OSA). The float suspended from the cantilever beam is submerged in the fuel. For increasing the buoyant force, a partially immersed float having a large cross-sectional area is used, and this arrangement results in the buoyancy force exerted by the liquid making vertical deflection according to the changes in level. Thus, the FBGs attached to the deflected beam get strained and the refractive index of the optical fibre core differs further shifting the wavelength of reflected light [8].

A broadband source is used as the optical source of the proposed FBG-based system which emits a wide range of wavelengths to the optical fibre. According to the refractive index of the optical fibre, the FBG sensors reflect a particular wavelength.

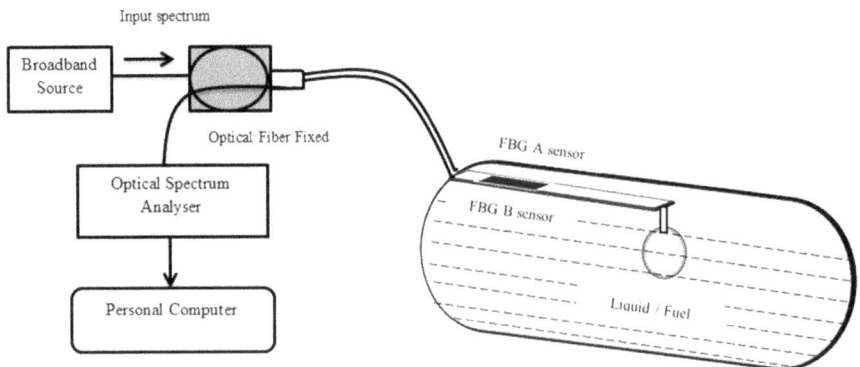

FIGURE 5.7 Sensor architecture.

In normal cases, i.e., when FBGs are not strained or no change in the liquid level, the refractive index does not experience any change. The increase or decrease in the liquid level forces the cantilever beam to deflect up or down, respectively, resulting in a change in the refractive index due to strain variations experienced by the optical fibre. This forms a wavelength shift in the back-reflected light, and the wavelength shift is detected by an optical spectrum analyser connected with a computer. The OSA results are stored for further processing, analyzing, and calculations of the liquid level [11]. This proposed design of the measurement system is used to improve the sensitivity and accuracy of the liquid level measurement system. In the proposed setup, the two FBGs are welded to both the surfaces of an elongated cantilever beam, in a way to provide temperature compensation [12, 13]. The FBG sensors provide high measurement accuracy and resolution and can measure strain experienced by the cantilever beam at the point where it is embedded.

5.4 INTRODUCING NEURAL NETWORKS FOR ACCURATE LEVEL PREDICTION

The accuracy improvement of any fluid level measurement system is mainly dependent on signal processing and signal classification, particularly in dynamic environments. The signal flow of the designed sensor is shown in Figure 5.8. The acquired signal from the sensor has to be processed by using suitable signal processing methods after sampling. Then the signal can be given to the neural network for the classification purpose.

5.4.1 SAMPLING OF SENSOR OUTPUT

The output voltage from any liquid level sensor continuously varies with time due to dynamic variations in the fuel level. The stages of the measurement system are mentioned in Figure 5.9. Discretizing the continuous signal at some constant sampling frequency fs by using sampling circuitry is necessary for digital signal processing

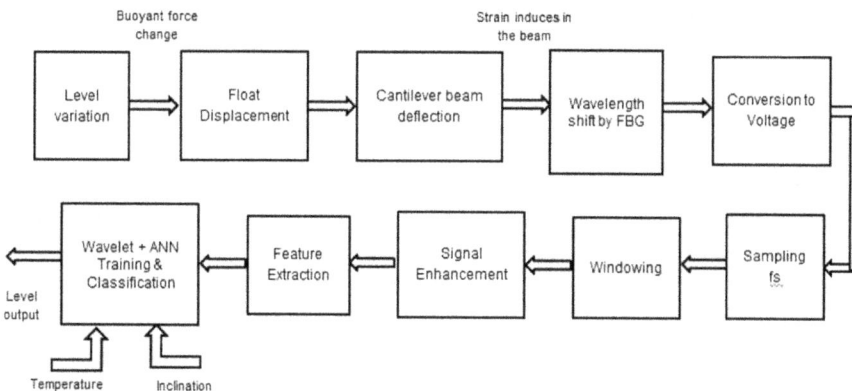

FIGURE 5.8 Signal flow of the designed level sensor system.

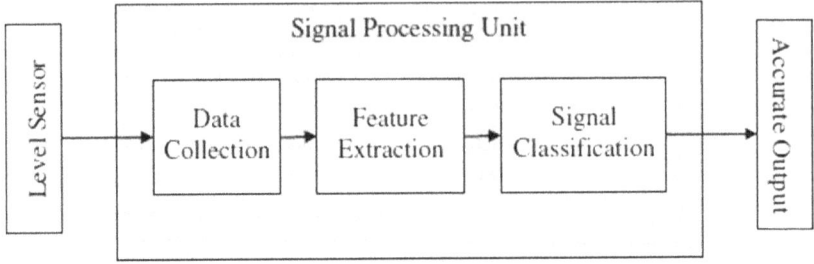

FIGURE 5.9 Overview of measurement system.

[14]. *Ts* is the sampling interval showing the time between two sample points of the signal, which is equal to the reciprocal of sampling frequency:

$$Ts = 1/fs$$

5.4.2 ARTIFICIAL NEURAL NETWORKS

The artificial neural network (ANN) model is a data processing and classifying technique developed by the inspiration from the way biological neurons process information. The neural networks have several neurons, interconnected in a manner to solve complex problems. The ANN learns by examples similar to human beings. For applications such as pattern recognition or data classification, the properly configured ANN using a suitable learning process can be used. The ANN learns by adjustments to the weights of interconnections that exist between the neurons which is similar to the biological systems [15]. ANN has a great ability to find relations from complex or imprecise data. Too complex relationships and patterns can be extracted, and it can detect the trend of the complex system by using neural networks. A trained neural network categorizes the information given to it for analyzing as an expert [16]. By introducing enough number of hidden neurons to the network, ANN can be trained effectively to form any multivariable function with very high precision level [17].

A typical neural network configuration is given in Figure 5.10. The neural networks are trained in such a way to obtain the specific target output from particular inputs. The

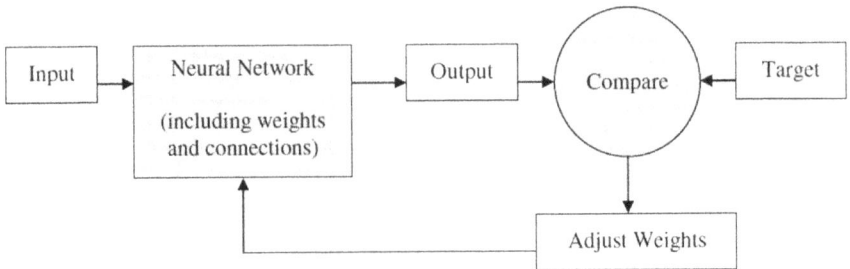

FIGURE 5.10 The typical configuration of an ANN.

network weights are adjusted and updated by comparing the output with the target to produce errors, till the error minimizes or network output becomes equivalent to the target. For high accuracy, more input and target pairs are required for training the network [18].

5.4.3 ACTIVATION FUNCTION

The activation function used by the neuron decides the output produced by the neural network. To deliver the output signal, the activation function considers both the inputs and the adjusted values of the weights [19]. This transfer function may come under one of the categories mentioned below:

- *Linear (ramp):* The output signal is proportional to all inputs and corresponding weights of the neuron.
- *Threshold:* The output will always be any one of the mentioned levels, comparing whether the total input given to a neuron is greater or less than the specified threshold value.
- *Sigmoid:* The output has no linear relationship between the inputs but varies continuously. This function resembles real biological neurons compared to linear and threshold transfer functions.
- *Wavelet:* The output has no linear relationship between the inputs but varies continuously. Different wavelets can be used as the activation function.

5.5   WAVELET NEURAL NETWORK

A trending technique used for signal classification is the machine learning approach. By combining the concept of artificial neural network and wavelet, a derived neural network is formed and named as wavelet neural network (WNN). WNN is a simple feed-forward neural network that uses different wavelets as activation functions instead of the binary, bipolar, or sigmoid functions as in conventional networks. For introducing non-linear function approximation, accuracy, and precision in outputs and fault tolerance, a relationship between neural network weights and wavelet transforms is defined [20].

The WNN considered is a three-layer network and the hidden layer activation function utilizes the Mexican hat wavelet, and the structure is shown in Figure 5.11. The Mexican hat wavelet function is defined as follows:

$$\phi\left(t\right) = \left(1 - x^2\right)\, e^{\left(-x^2/2\right)} \tag{5.3}$$

In WNN, the signal is passed in a forward direction (forward signal propagation), while the error is transmitted in the backward direction (backward error propagation). The left to right travel of the signal implies the forward propagation, i.e., through the input layer, then through the hidden layer, and finally through the output layer to the output end. Some intermediate results are produced by each layer which will be given as the input to the succeeding layer, and the final output is computed and given by the output layer. The backward propagation ensures the update of

$$\Phi(t)$$

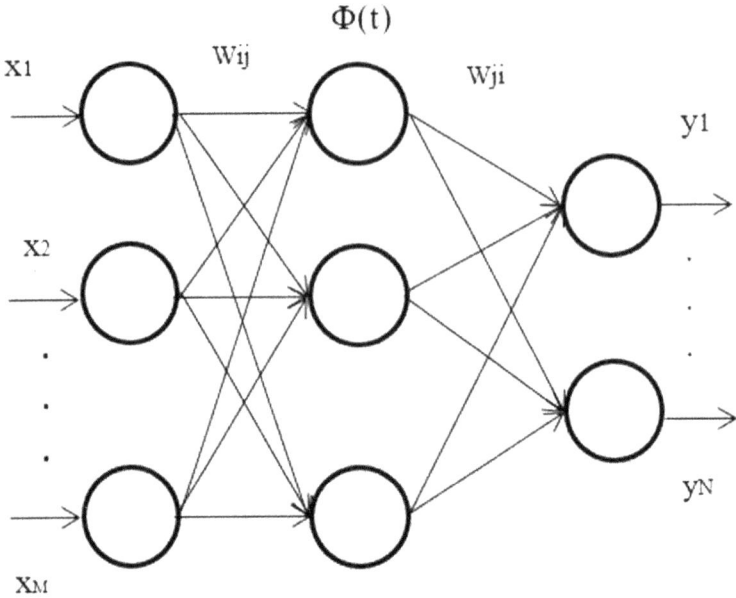

FIGURE 5.11   Structure of WNN.

weights and bias values. The non-linear and complex relationships can be modelled by the ANNs and it is applicable in many real-life practical systems. The response of unseen data can be inferred by the ANN by learning from the initial input and output relationships, i.e., an accurate prediction of output is possible from random inputs. The neural networks can learn even the hidden complex relationship between input and output by proper training to provide better results. All the mentioned features introduced the neural network-based machine learning approach in industrial, technical, image processing, banking, finance, and many other fields as a very powerful tool for the prediction of unknown, unexpected, and future outcomes.

Among the machine learning approaches, the WNN is faster and accurate, giving better results compared to the back propagation network (BPN) and support vector machine (SVM) techniques [4]. Training the programmed WNN algorithm with real-time data provides true, faultless level readings. Thus, the WNN is useful in different areas where exact true outputs are necessary, particularly for fuel level measurement in the dynamic environments where the disturbance behaviour is unknown.

5.5.1   TRAINING OF WNN

Neural networks can undergo training to perform a particular task. Many engineering tools are available for training neural networks. One of the powerful tools for training, analyzing, and simulating the neural network is the MATLAB software. The training procedure modifies the weights and bias weights of a network by following a training algorithm learning rule. Supervised learning and unsupervised learning are two broad classifications of learning rules for data clustering. Supervised learning always needs the support of a teacher for supervising and guiding the output according to what desired response has to be given for each input signal. A training set is provided with the learning rule for gaining proper network behaviour. While the inputs are given to the neural network, comparisons are made between the network outputs and the target values. To get the network outputs close to the target values, the weights and biases of the network are updated using the learning rules [15].

Unsupervised learning gets trained by the available information only, and this method never needs an external teacher. It is a self-organization technique because it self-organizes the data given to the neural network and finds out their common properties. According to the network inputs, the weights and biases are adjusted and modified. In unsupervised learning, there is no need for target outputs and a majority of the unsupervised learning algorithms are based on clustering operations. In the case of clustering, the algorithm categorizes the input patterns into several finite classes. For applications such as vector quantization, unsupervised learning is useful [15].

The Training Algorithm for WNN (Forward Calculations) [21]

$X(n)$ External input vector
$Wjk(n)$ Weight between input layer (k) and hidden layer (j)
$Wij(n)$ Weight between output layer (i) and hidden layer (j)
$aj(n)$ Dilation coefficient in the hidden layer at time n
$bj(n)$ Translation coefficient in the hidden layer at time n

At time n, the net internal activity of neuron j is

$$Vj(n) = \sum_{k=0}^{m} Wjk(n) * Xk(n) \tag{5.4}$$

$Vj(n)$ Sum of inputs to the jth hidden neuron
$Xk(n)$ kth input at time n

The output of jth neuron is computed by passing $Vj(n)$ through the wavelet $\phi(t)$ where

$$\phi a,b(Vj(n)) = \frac{\phi(Vj(n) - bj(n))}{aj(n)} \tag{5.5}$$

Sum of input neurons to output neuron is

$$V(n) = \sum_{j=0}^{N} Wij(n) * \phi a,b(Vj(n)) \tag{5.6}$$

The output of the neuron is calculated by passing $V(n)$ through the non-linear activation function, obtaining

$$Y(n) = \sigma(V(n)) \tag{5.7}$$

Learning Algorithm

At time n, the instantaneous sum of squared error is

$$E(n) = \frac{1}{2}e^2(n)$$

$$= \frac{1}{2}\left[y(n) - d(n)\right]^2 \tag{5.8}$$

$d(n)$ The desired response at time n

η Learning rate

The method of steepest descent used to minimize the above cost function. Weight between the hidden layer neuron j and input layer neuron k can be updated according to

$$\Delta Wjk(n+1) = -\eta * \frac{\partial E(n)}{\partial Wjk(n)} + \mu * \Delta Wjk(n)$$

$$= \eta * e(n) * \sigma(V(n) * Wij(n) * \phi a, b(Vj(n)) * \frac{Xk(n)}{ai(n)} + \mu * \Delta Wjk(n) \tag{5.9}$$

Weight between output layer neuron i and hidden layer neuron j can be updated according to

$$\Delta Wij(n+1) = -\eta * \frac{\partial E(n)}{\partial Wij(n)} + \mu * \Delta Wij(n)$$

$$= \eta * e(n) * \delta(V(n) * Wij(n) * \phi a, b(Vj(n)) + \mu * \Delta Wjk(n) \tag{5.10}$$

Translation coefficient can be updated according to

$$\Delta bj(n+1) = -\eta * \frac{\partial E(n)}{\partial bj(n)} + \mu * \Delta bj(n)$$

$$= -\eta * e(n) * \sigma(V(n) * Wij(n) * \phi a, b(Vj(n)) * \frac{1}{aj(n)} + \mu * \Delta bj(n) \tag{5.11}$$

Dilation coefficient can be updated according to

$$\Delta aj(n+1) = \eta * \frac{\partial E(n)}{\partial aj(n)} + \mu * \Delta aj(n)$$

$$= -\eta * e(n) * \delta(V(n) * Wij(n) * \phi a, b(Vj(n)) * \frac{Vj(n) - bj(n)}{(aj(n))^2} + \mu * \Delta aj(n) \tag{5.12}$$

```
┌─────────────────────────────────────────┐
│        Create and Initialize WNN         │
└─────────────────────────────────────────┘
                    ↓
┌─────────────────────────────────────────┐
│  Loading strain, temperature, inclination & │
│    level data for training and testing    │
└─────────────────────────────────────────┘
                    ↓
┌─────────────────────────────────────────┐
│             Pre-processing               │
└─────────────────────────────────────────┘
                    ↓
              ◇ All signals
                loaded?
                    ↓
┌─────────────────────────────────────────┐
│   Prepare training and target (level) data │
└─────────────────────────────────────────┘
                    ↓
┌─────────────────────────────────────────┐
│ Train the WNN for accurate liquid level prediction │
└─────────────────────────────────────────┘
                    ↓
┌─────────────────────────────────────────┐
│      Validate the WNN with test data     │
└─────────────────────────────────────────┘
                    ↓
┌─────────────────────────────────────────┐
│              Show results                │
└─────────────────────────────────────────┘
```

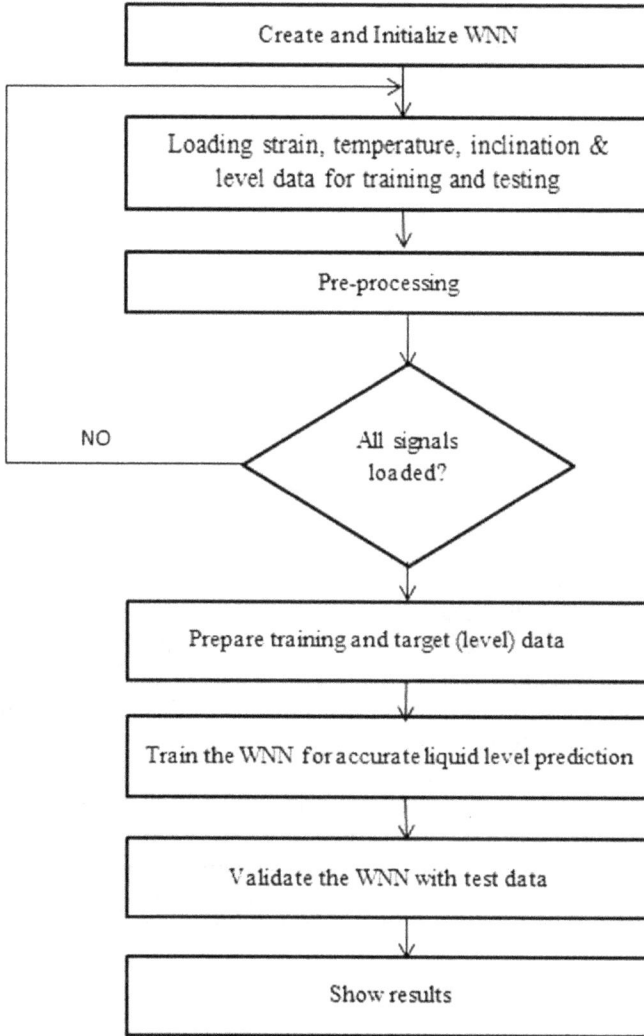

FIGURE 5.12   Flowchart of wavelet neural network training and validation.

The training and target data have been loaded to the network model once the training is done. The neural network parameters such as learning rate, the maximum number of epochs, and training function are fixed beforehand to call the training function. The trains are nothing but the training function having training vectors, network objects, and target vectors as parameters. The training steps involved are shown as a flowchart in Figure 5.12.

5.6   RESULTS

After calculating the samples from the mathematical model, the training data in disturbances such as variations in temperature and inclination of the vehicles are saved

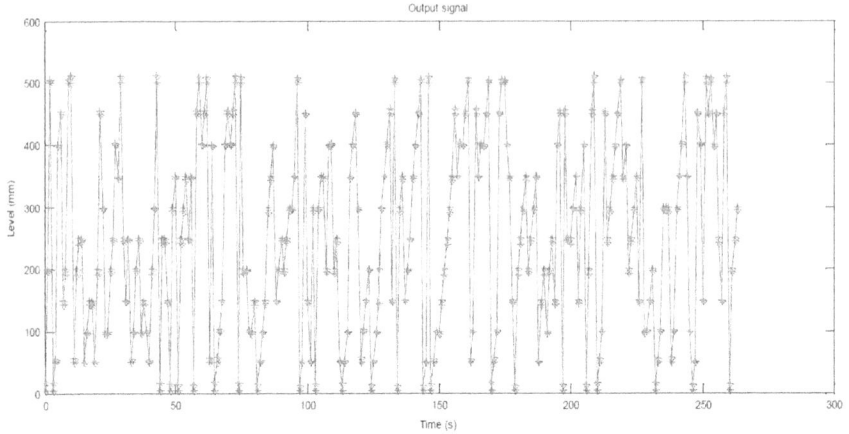

FIGURE 5.13 Final output: the red line shows the actual level and the blue line shows the output of WNN.

as individual data. The saved data are loaded to the wavelet neural network and classified efficiently. After training and validation of the wavelet neural network model, the proposed system is tested for its accuracy even in the dynamic environments. The testing inputs are given randomly with the disturbances and dynamic variations with respect to time. The neural network is tested to predict the level output and it is compared with the known true values of fuel level of the tank.

The output of the neural network is given in Figure 5.13. The red line shows the true level and the blue line shows the predicted level output, which are matching with trained output a at highest accuracy. The MATLAB simulation and validation results carried out based on the given algorithm shows that the wavelet neural network predicted accurate results, which are exactly matching with the true values of the level of the fuel tank at dynamic variations due to temperature and inclinations.

5.7 CONCLUSION

WNN-based signal classification and signal processing method integrated with FBG-embedded cantilever sensor was used for accurately predicting the level of the fuel tank in the automotive system under dynamic conditions. Extensive training was done to ascertain an effective configuration for the WNN-based liquid level measurement system. The WNN parameters selection and the signal pre-processing configurations were all based on training with the values from the model. While comparing with the existing level measurement methods, the results obtained from the WNN-based FBG-embedded cantilever beam sensor system were having higher sensitivity and higher accuracy in the dynamic environments.

Based on the accuracy, the WNN-based system is suitable to use in vehicles used for racing competitions where vehicles are exposed to high manoeuvres. The drivers of the automotive systems can monitor the dashboard for predicting the fuel level accurately without any errors due to variations in accelerations, temperature, and inclination. In conclusion, the proposed measurement system along with wavelet neural network method for signal processing and classification is effective and highly accurate in the determination of fuel level using FBG-embedded cantilever sensor in dynamic environments.

REFERENCES

1. Edin Terzic, Romesh Nagarajah and Muhammad Alamgir, "A neural network approach to fluid quantity measurement in dynamic environments", *Mechatronics*, vol. 21, no. 1, pp. 145–155, Feb. 2011.
2. Edin Terzic, Jenny Terzic, Romesh Nagarajah and Muhammad Alamgir, *A Neural Network Approach to Fluid Quantity Measurement in Dynamic Environments.* London: Springer-Verlag, 2012.
3. Dong-Sheng Xu, Hua-Bei Liu and Wei-Li Luo, "Development of a novel settlement monitoring system using fibre-optic liquid-level transducers with automatic temperature compensation", *IEEE Transactions on Instrumentation and Measurement*, vol. 67, no. 9, pp. 2214–2222, Sept. 2018.
4. Bashir Shokouh Saljoughi and Ardesir Hezarkhani, "A comparative analysis of artificial neural network (ANN), wavelet neural network (WNN) and support vector machine (SVM) data-driven models to mineral potential mapping for copper mineralizations in the Shahr-e-Babak region, Kerman, Iran", *Applied Geomatics*, vol. 10, no. 3, pp. 229–256, Sept. 2018.
5. https://wiki.anton-paar.com/in-en/engine-oil/
6. Mengmeng Han, Jian Dai and Kok Keng Ang, "Hydrodynamic aspects of moving vehicle with sloshing tanks", *MATEC Web of Conferences*, 2018. https://doi.org/10.1051/matec conf/201821115002
7. Ilian Haggmark and Michael Fokine, "Fiber Bragg gratings in temperature and strain sensors", Laser Physics Group, May 2014.
8. Thulasi M Santhi and S Sathiya, "Performance enhanced liquid level sensing system for dynamic environments", *IEEE 5th International Conference for Convergence in Technology (I2CT)*, March 2019.
9. B Vasuki, S Sathiya and K Suresh, "A new piezoelectric laminated cantilever resonance based hydraulic pump", *2013 IEEE Sensors Applications Symposium Proceedings*, vol. 978, no.1, pp. 4673–4637, April 2013.
10. S Sathiya and B Vasuki, "Higher resonant mode effect on the performance of piezo actuated 2-DOF rectangular cantilever shaped resonators (2-DOF RCR) for liquid viscosity and density sensing", *Microsystem Technologies*, vol. 23, no. 7, pp. 2431–2445, July 2017.
11. Hua-Fu Pei, Jian-Hua Yin, Hong-Hu Zhu, Cheng-Yu Hong, Wei Jin and Dong-Sheng Xu, "Monitoring of lateral displacements of a slope using a series of special fibre Bragg grating-based in-place inclinometers", *Measurement Science and Technology*, vol. 23, Issue 2, article id. 025007, 8pp, 2012.
12. Wenlong Liu, Yongxing Guo, Li Xiong and Yi Kuang, "Fiber Bragg grating based displacement sensors: state of the art and trends", *Emerald Insight*, vol. 39, no. 1, pp. 87–98, 2019.
13. T Guo et al. "Temperature-insensitive fiber Bragg grating liquid-level sensor based on bending cantilever beam," *IEEE Photonics Technology Letters*, vol. 17, no. 11, pp. 2400–2402, Nov. 2005.

14. M Pharr and G Humphreys. *Sampling and Reconstruction. Physically Based Rendering: From Theory to Implementation.* Amsterdam: Elsevier/Morgan Kaufmann, 2004, pp. 279–367.

15. Neural Networks. http://www.doc.ic.ac.uk/^nd/surprise_96/journal /vol4/cs11/report .html

16. BD Ripley, "Statistical aspects of neural networks", In OE Barndorff-Nielsen, JL Jensen and WS Kendall (Eds.), *Networks and Chaos—Statistical and Probabilistic Aspects.* London: Chapman and Hall, 1993, pp. 40–123.

17. H Demuth, M Beale and M Hagan, "Neural network toolbox 5 users guide", *MathWorks*, 2007.

18. R Rojas, *Neural Networks—A Systematic Introduction.* New York: Springer, 1996.

19. U von Luxburg and G Rätsch (Eds.), *Advanced Lectures on Machine Learning: ML Summer Schools, Canberra, Australia, February 2–14, 2003.* Tübingen, August 4–16, 2003, (Rev. lectures/Olivier Bousquet). Berlin, New York: Springer.

20. Ping Zhou, Chenyu Wang, Mingjie Li, Hong Wang, Yongjian Wu and Tianyou Chai, "Modeling error PDF optimization based wavelet neural network modeling of dynamic system and its application in blast furnace iron making", *Neurocomputing*, vol. 285, pp. 167–175, April 2018.

21. MR Mosavi, "Wavelet neural network for corrections prediction in single-frequency GPS users", *Neural Processing Letters*, vol. 33, no. 2, pp. 137–150, April 2011.

6 Computer Vision Concepts and Applications

Bettina O'Brien and V. Uma

CONTENTS

6.1   INTRODUCTION

Artificial intelligence (AI), which is the source of subfields like machine learning (ML), deep learning, computer vision, etc., is a technology that defines intelligence as knowledge acquisition and representation, where knowledge is acquired and applied to a problem. The objective of any AI problem [1] would be to elevate the success rate and not the accuracy, whereas an ML problem would aim at utilizing the model to increase its accuracy but does not concentrate on its success. In general, artificial intelligence is the ability to engineer intelligent machines and programs, while machine learning is the ability to learn without being explicitly programmed. AI pertains to decision-making where the goal is to simulate natural intelligence in machines. But machine learning learns from data to maximize the performance of the machine based on a specific task.

Deep learning [2, 3], a subset of machine learning, makes the computation of multilayered neural networks feasible, working with large amounts of data. Algorithms with single-layered neurons worked just fine, but the weights remained the same during the training phase, which didn't have a very good impact on the accuracy. So, algorithms were designed such that adjustment of weights made changes which reduced the error. This, in turn, concentrates on delivering high accuracy for task-specific applications such as object detection [1, 4], image segmentation [5, 6], speech recognition [7], feature extraction [8, 9], language translation, and other recent upgrades in technology. These image-related applications significantly co-relate with another domain of AI: computer vision.

Computer vision [10, 11] is a simple phenomenon that aims at making computers view and sense its surroundings. It has been the source for almost all computations which implicate visual content, especially in the form of digital images. Machines simply illuminate images as a series of pixels with their own set of colours [4] and values. Hence, it helps to replicate how a human brain reads an image. This is the basic way of how intelligence will be modelled in a machine. It ultimately aims at extracting information from pixels. This field of computer science, which initially worked based on statistics alone, is now switched to deep neural networks [7]. Here, both the objective and accuracy of achieving the goal are taken care of.

Image processing [2] is another term that is often thought to resemble computer vision, but it is not. The commonality is that both subfields of AI require images to work with, but the method in which these images are used is varied. Image processing, as the name implies, processes images by adding in functionalities like smoothing, sharpening, adding contrast, and stretching, or performs other transformations in images and the output would be an image. But computer vision aims to identify images and interpret information to provide an output in terms of the image size, colour intensity, etc. It also tries to classify images, identify instances from an image, etc. Trying to classify images or associate the information between images to differentiate between individuals are applications that involve deep learning techniques. Image processing uses probabilistic methods like the hidden Markov model, editing, restoration techniques, and filtering. These methods use neural networks to incorporate more learning and to model intelligence.

Computer vision has a non-complete list of applications that tends to grow over the years. It has a dramatic impact on almost all sectors like retail, healthcare, financial services, and so on. In general, computer vision [10, 11] is applied in automatic inspection, identification systems, detecting events [12], controlling processes, modelling or recreating objects or environments, navigation, etc. Section 6.2 gives a detailed explanation of feature extraction and the techniques used to select the necessary information from the image. In Section 6.3, methods used to identify and locate objects and instances in images are dealt with. In Section 6.4, some of the important hardware, software, and services in computer vision are discussed. In Section 6.5, the applications in the field of healthcare, self-driving cars, and automatic target detection are dealt with. A case study of how computer vision and deep learning methods influence autonomous robotic path planning around a dynamic environment with obstacles is explained.

6.1.1 EVOLUTION OF COMPUTER VISION

Computer vision, before deep learning came into existence, had a simple procedure in processing images. Initially, the captured images were stored in a database. For each image, data points regarding the properties and measurements of the image were defined. This procedure is called data annotation. The same annotation procedure was to be done for other new sets of images. After these processes, the objective of application-specific problems was carried out. This traditional approach requires much manual work to identify features during the pre-processing stage. Such a time-consuming process can now be handled by the concept of feature extraction.

6.2 FEATURE EXTRACTION

Feature extraction is one of the indispensable preconditions for processing images. In the field of machine learning, feature extraction [8] learns and identifies patterns from images. It can be interpreted as the transformation from a set of raw data into derived values called features. A feature [9] represents a function of characteristics or measurements specifying the assessable property of an image. These features must be both instructive and non-repetitive. Instead of using the entire data as a whole, relevant information from features is extracted to perform the expected task. The extracted features that are rich in the needful information are further used for feature selection and classification tasks. Alternatively, feature extraction seems to be a significant form of dimensionality reduction. In the problem of image classification, the extracted features should contain the relevant information required to differentiate between classes.

6.2.1   Types of Features

The outcome of feature extraction would be a set of features referred to as the feature vector, which is a representation of images. A feature is of good quality when it contains discriminative information that can differentiate between objects. The types of features can be classified as follows:

1. *Local features* are features derived from the results of the edge detection or image segmentation process. They are individual results or features gained from processing different partitions of an image.
2. *Global features* are features calculated from the overall image or a well-ordered sub-section of an image.
3. *Pixel-level features* are calculated at each pixel in terms of colour and texture. Colour holds features such as prominence, colour spread, and co-occurrence, whereas texture holds features like contrast, sharpness, and change in intensity.
4. *Object-level features* are calculated in terms of shape, size, texture, and spatial distribution. Shape and size hold features like perimeter, bending energy, Fourier shape descriptor, model dimensions, area, and solidity. The texture holds the same features as pixel-level features. Spatial distribution holds features like edge connectedness, length of edge, compactness, and distance among peer neighbours.
5. *Domain-specific features* are conceptual features that are application dependent; e.g., character recognition pertains to the formation of letters alone, whereas face identification deals with facial features alone.
6. *Semantic-level features* are based on a subset of low-level features that obtains a high-level concept. The inference or relation is identified between two or more low-level features.

6.2.2   Feature Extraction Methods

In general, feature extraction is classified into low-level and high-level features. Low-level features are the basic features of an image like shape, size, and texture. Such features can be directly identified and extracted. They do not rely on individual object description in the image. High-level features are those that concentrate on the object's shape, edge, or corners of the image.

6.2.2.1   I. Low-Level Features

6.2.2.1.1   a. Local Features

Local features generally give a representation of the image patches, which tends to identify unique patterns or structures from an image. Local feature detection is a preliminary procedure in a computer vision algorithm. It tries to find distinct features that differentiate its surrounding pixels in terms of edges, corners, texture, colour, or intensity. Such features extracted locally from an image are classified into texture, shape, and colour features. These local features contribute to influential factors for images. Despite image occlusion or any presence of clutter, local features identify image compatibility. So, it helps in differentiating between patches in an image, which leads to the application of image classification.

i. *Texture*

The texture of an image is a feature that identifies the region of interest (RoI) by performing image segmentation. Later, image classification is performed to classify between regions. Texture gains information from the spatial representation of colours in an image. Since the texture is identified in terms of regions, as a repeating pattern of local features, it cannot be defined for a point in an image. An image's texture can be seen in terms of fine, coarse, grainy, smooth, sharp, etc.

These features are described in the tone and structure of a texture. Tone defines the properties of texel's pixel intensity, and structure defines the spatial representation or relation between texels. Structurally, the texture is defined as a set of elementary *texels* in a regular pattern, and statistically, it can be defined as a measure of intensity arrangement in any particular region of an image. These measurement sets define the feature vector.

6.2.2.2 Texture Estimator

- The range is the simplest operator which calculates the difference between maximum and minimum intensity values of neighbours in an image. Range estimates texture by converting the original image into a brighter image. The texture is read as the brightness value.
- *Variance* calculates the sum of squares of the differences between the intensities of the central pixel and its neighbours, i.e. it simply calculates the variance between neighbour regions.

ii. *Colour*

Colour is the primary image perception in understanding how human brains perceive an image. It is a robust descriptor which simplifies object detection and image segmentation problems. The colours [4] that are perceived from an image are made up of light of different wavelengths. A colour space depicts the colour representing its intensity values. Colour space models can either be hardware-oriented based on a three-colour stimulus like RGB or user-oriented.

6.2.2.3 Colour Histogram

A colour can be described in various shades (e.g., green colour being in shades of dark or light green). But for the machine to interpret variations between shades of the same colour, the probability of pixel intensities needs to be defined. Colour histograms define a set of bins, where each bin holds the probability of pixels of a particular colour. A vector representation of a histogram is Equation 6.1:

$$H\{H[0], H[1], ..., H[i], ..., H[N]\} \tag{6.1}$$

Here, each $H[i]$ in Equation 6.1 represents the number of pixels in the colour i in an image and N represents the total number of bins. Histograms need to be normalized if two different sized images are to be compared. The normalized vector representation is given in Equation 6.2 and the reduced form is given in Equation 6.3:

$$H'\{H'[0], H'[1], ..., H'[i], ..., H'[N]\} \tag{6.2}$$

$$H'[i] = H[i]/P \tag{6.3}$$

P tends to the number of pixels in the image. The range of intensity values for a greyscale image can be given in the range $I(u,v)$, denoting the pixel intensities which belong to $[0, k-1]$, where k represents the number of colours. For example, for an 8-bit image, $k = 2^8 = 256$ colours. So, in general, colour histograms deal with problems like over- and underexposure, brightness, contrast, and dynamic range.

6.2.2.4   Colour Descriptor

Descriptors provide a way to relate pixels in an image. A colour descriptor is categorized under the general information descriptors. Since colour is the most significant feature of any visual content, it has to be described exclusively. For any image, the colour description can be performed using the following methods:

- Dominant colour descriptor (DCD)
- Scalable colour descriptor (SCD)
- Colour structure descriptor (CSD)

To trace the relation between colours among a group of images, the following methods are used:

- Colour layout descriptor (CLD)
- Group of frames (GOF) or group-of-pictures (GOP)
    a. **Global Features**
       Global features represent images in terms of a single vector. Mostly the texture and shape descriptors are considered while addressing the global features of an image.
       i. **Principal Component Analysis (PCA)**
          PCA is a dimensionality reduction tool that reduces a large set of feature variables into a smaller set but still retains the needed information from the larger set. PCA emphasizes variations in features and identifies strong patterns from data sets. The set of uncorrelated variables derived as the identified patterns from an image is known as principal components and these components are simply eigenvectors of a covariance matrix and are orthogonal. For example, if two dimensions, say, the height and weight in a data set are considered, PCA builds the coordinates (x,y) value for every point. Here,

the combination of height and weight is called principal components. The variance of a linear composite of variables i and j are given by Equation 6.4:

$$\sum_{i=1}^{P} a_i x_i = \sum \sum a_i a_j \sigma_{ij} \tag{6.4}$$

where σ denotes the covariance between the ith and jth variables. The variable which contributes less information along the axes would be dropped off. In terms of matrix representation, $a'Ca$, where C represents the covariance matrix and a represents the vector of variable weights, PCA finds the weight vector a that minimizes $a'Ca$ with a constraint defined in Equation 6.5:

$$\sum_{i}^{P} a_i^2 = a'a = 1 \tag{6.5}$$

Hence, PCA projects data in the directions of maximum variance and finds the most precise representation of data in a lower dimensional space.

ii. ***Linear Discriminant Analysis (LDA)***

LDA is another way that strives to reduce dimensionality while still maintaining the class-biased information. So, LDA best applies to classification problems. The basic objective of Fisher's LDA is to find a projection to a line such that the samples from various classes are segregated. Let x be a p-dimensional predictor with k class, where k represents linear discriminant functions, then Fisher's approach projects x to $(k - 1)$ dimensional spaces to extract new features in which the variance between classes is maximized and the variance within classes are minimized.

The objective is to acquire a scalar y by projecting the samples x onto a line $y = w^T x$. To define a good projection vector, a measure of separation between classes needs to be defined. So, the mean vector of each class w_i in x-space and y-space is given by

$$\mu_i = \frac{1}{N_i} \sum_{x \in w_i} x \tag{6.6}$$

$$\mu_i' = \frac{1}{N_i} \sum_{y \in w_i} y \tag{6.7}$$

Since y tends to $w^T x$, $y = w^T x$. Substituting y in Equation 6.7,

$$\mu_i' = \frac{1}{N_i} \sum_{y \in w_i} w^T x \tag{6.8}$$

Substituting Equation 6.6 in Equation 6.8,

$$\mu_i' = w^T \mu_i \tag{6.9}$$

The objective now can be framed to find the distance between the projected means, which in turn yields better class separability. Figure 6.1 shows a graph of randomly chosen means between classes and how their separation measure is calculated from the distance between means.

I. **High-Level Features**
High-level representations of an image are well-built features inclined to be seemingly hidden. But minimizing the bridge between low-level and high-level features is significant in emphasizing object detection problems. The low-level local features after detection can be combined to classify images. So, to learn in deep about these hidden features, one requires time and computational power. This is where the most important segment of deep learning is used. Deep neural networks use several layers to learn high-level representations of an image.

a. **Template Matching**
Template matching is a high-level method that tries to identify areas of an input image using a given image pattern or any patch. So given an image, this method works by sliding the template over the whole image to detect edges of instances. The matching approach completely

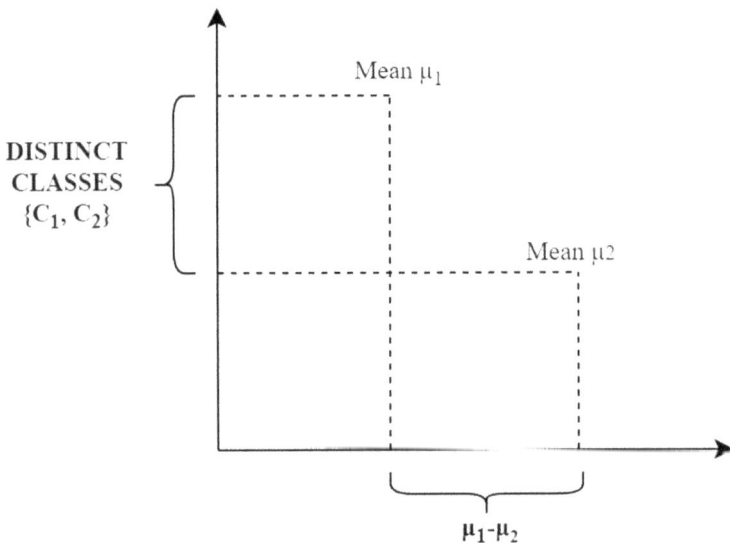

FIGURE 6.1   Distance between means and class separability of x and y spaces.

FIGURE 6.2 Correlation analysis.

depends on the type of image and objective chosen. Template matching can be performed based on features, area, template, or motion.

The matching process in Figure 6.2 is done pixel-by-pixel, where the template image moves overall viable positions in the given input image to compute an integer index which interprets how well a match is found. For a greyscale image, the exactness cannot be precise, so a difference in levels between the source image and template has to be known. Thus, the distance *dist* is calculated by Equation 6.10:

$$dist(I,t,a,b) = \sqrt{\sum_{i=1}^{n}\sum_{j=1}^{m}\left(I(a+i,b+j)-t(i,j)\right)^2} \qquad (6.10)$$

where

(*a,b*) denotes the top corner coordinates of the template *t*

I   denotes the greyscale image with a grey-value template *t* of size *n*m*

Correlation is the measure of degree where two variables correlate with each other in terms of general behaviour if not directly through values. Correlation is given by Equation 6.11:

$$C = \frac{\sum_{i=1}^{N}\left(u_i - u'\right)\left(v_i - v'\right)}{\sqrt{\sum_{i=1}^{N}\left(u_i - u'\right)^2 \sum_{i=1}^{N}\left(v_i - v'\right)^2}} \qquad (6.11)$$

where

u and *v* denote the template grey-level image and source image, respectively

u' and *v'* denote the average grey-image level in the template image and source image, respectively

N denotes the image template size in terms of pixels and is given by $N = columns * row$

C takes values that range between ±1, where greater values represent stronger relationships between images

   b. ***Hough Transform***

        Hough transform is a method for calculating global feature representation, where it gradually builds classes by working with the local features with the number of classes initially being unknown. Hough transform is, in particular, used to detect shapes such as lines, circles, or curves. It is a generalized localization problem aiming to find the location of shapes in images.

6.3 OBJECT DETECTION

Object detection [9] is the principal challenge in computer vision. It is known to rapidly grow to give improved results in terms of accuracy and testing time when used with machine learning [1, 13] and deep learning [14] approaches. One such latest software system that implements object detection algorithms using deep learning concepts is the *Detectron,* developed by Facebook. Nowadays, there are more deep learning pre-trained models [3, 7] used for object detection like Convolutional Neural Network (CNN), Region-based Convolutional Neural Network (RCNN), You Only Look Once (YOLO), Fast R-CNN, etc. The basic structure for computer vision is CNN implemented using deep learning. CNN combines image classification, object classification, and localization. For the past few years, CNN has proved to be a reliable method for object detection and classification due to its speed and accuracy.

6.3.1 IMAGE CLASSIFICATION

To classify images, the image given as input goes through a deep convolutional network, and this network gives a feature vector to the fully connected layers that provide different class scores. Finally, the output delivers the content of the entire image as a whole. This is the basic possible task in computer vision. There are various types of other computer vision tasks that detect spatial pixels inside the image. These are achieved through a convolutional neural network.

6.3.1.1 Classification and Localization

To know where and what the object detected is, object classification and localization [15] algorithms are used. In addition to predicting a category-labelled object, a boundary box is drawn around the region of the object in the image to show where the object is located in that image. The distinction between classification with localization and object detection is that in the localization scenario the number of objects is known ahead of time, while in the latter bounding boxes are generated exactly proportional to the identified quantity of objects found in the image.

6.3.2 IMAGE SEGMENTATION

6.3.2.1 Semantic Segmentation

In semantic segmentation [6], the output image is the decision of the category of every pixel in that image. It is similar to image classification. Rather than assigning a single category labelled to the entire image, the output of this segmentation [5] produces a category label for each pixel of the input image. Semantic segmentation does not differentiate instances, it only labels the category of that pixel. When two similar objects are close together, the semantic model does not distinguish between the objects. Instead, it labels the whole mass of pixels. This is a small drawback for semantic segmentation that can be fixed using instance segmentation. One potential approach to attack semantic segmentation is through classification. The sliding window method can be approached to segment an image semantically.

For a sliding window, the input image is cropped into many small images and each cropped image is treated as a classification problem. This will help to find the category of the central pixel of the cropped image. The sliding window is computationally expensive. To label a pixel, the full image needs to be cropped into smaller images and this would be expensive to run. A model that would be even more efficient than semantic segmentation is a fully convolutional network. Rather than extracting single patches from the image and classifying it separately, it can be imagined as a whole stack of convolutional layers.

6.3.2.2   Demerits of Sliding Window

- *Expensive*

  Using CNN by cropping multiple images for detecting objects from the image will be expensive. The method can be altered to pass the entire image instead of passing the cropped image to detect objects.
- *Inaccurate boxes*

  The algorithm can find and localize objects in an image, but the shape of the box is not accurate, sometimes it has a rectangular object and at some point, the object does not fit in the box at all.

6.3.2.3   Instance Segmentation

Given an input image, the objective is to identify and detect the location and identities of objects in the image, which is, perhaps, similar to detecting objects [16] but, instead of predicting a bounding box for each object, the whole segmentation mask has to be predicted so that a particular instance from the image would be captured.

6.3.3   REGION-BASED METHODS

6.3.3.1   Region Proposal

This method does not use deep learning as such, but it is mostly based on a traditional approach to computer vision. This region proposal network uses the basic signal processing technique to generate several proposals. For a given input image, the region proposal network gives an output of thousand boxes as to where the object might be present. These are relatively fast to run.

6.3.3.2   Region-based Convolutional Neural Network(R-CNN)

Given an input image, the model executes the region proposal network to get RoI. One of the problems in R-CNN is that the regions in the input image can be in different sizes. To run them all through the convolutional network for classification, the input image needs to be of the same size. Each of the region proposals is taken and wrapped to a fixed square size that is expected as input. This will run through a convolutional network, which in turn makes classification decisions for each of those images to predict categories. Besides, R-CNN also predicts a regression for each of the input region proposals. The main disadvantage of R-CNN is that it is computationally expensive, and the runtime is slow.

6.3.3.3 Fast Region-based Convolutional Neural Network

Problems faced with R-CNN can be fixed by Fast R-CNN. Both methods are almost the same but rather than each RoI being processed separately, the entire image is sent through the convolutional network all at once. It gives a feature map that effects high resolution that corresponds to the entire image. If there are fully connected layers downstream, those layers will be expecting fixed-size input. So, reshaping the cropped images from the convolutional feature map is done using the RoI pooling layer, which is similar to max pooling. Once the images are wrapped from a feature map, classification scores can be predicted by running those images through fully connected layers. In terms of speed, Fast R-CNN is ten times faster than the SPP net (which is in between R-CNN and Fast R-CNN) and R-CNN. In terms of test time, Fast R-CNN is superfast since the test-time is dominated by computing region proposals. So, Fast R-CNN ends up being bottle-necked by computing these region proposals. This problem is solved by faster R-CNN.

6.3.3.4 Faster Region-based Convolutional Neural Network

In Faster R-CNN, self-generated region proposals are created. It runs the entire input image through convolutional layers to get a feature map representing the entire image in high resolution. There is a separate region proposal network that works above those convolutional features, which in turn predicts its region proposals inside the network. It looks like Fast R-CNN when the region proposals are predicted. Now the image is cropped from those region proposals and passed to the network for classification.

6.3.4 ALTERNATIVE METHODS

6.3.4.1 HOG Features

The HOG is an important technique that can be used for object detection, mainly in self-driving cars. It is a feature descriptor, powerful when used with a support vector machine (SVM) that works for object detection. To determine whether the object is found or not, each computed HOG descriptor is fed to the SVM classifier. By changing the image scale, HOG solves the scaling issues. HOG represents a single vector which describes a segment of an image.

6.3.4.2 You Only Look Once (YOLO)

YOLO algorithm is a better solution to create accurate boxes around the images and is also faster than CNN. YOLO is just a modified version of CNN. In YOLO, the image is divided into multiple grids and for each grid, classification and localization algorithm is implemented. There are multiple versions for YOLO models. The latest version is YOLO9000 which is faster, stronger, and better. It is trained on 9000 classes.

6.3.4.3 Demerits of YOLO

- Detecting one object multiple times can be confusing and time-consuming to correct them. This can be solved by Non-Max suppression. This helps to remove boundary boxes that have a low probability that is close to high-probability boxes.

- No possibility to detect multiple objects in one grid. With the use of Anchor box, each object is assigned to each anchor box in a single grid to detect multiple objects. This can also be solved by selecting a small grid, but when objects are too close to each other this method can fail.

6.4 COMPUTER VISION HARDWARE, SOFTWARE, AND SERVICES

The advancement of tools and libraries in computer vision has evolved over the years. The evolution of hardware and software has grown such that services are now available and are offered by standard companies, as depicted in Figure 6.3. Various hardware products and services are available in the market, out of which the most significant ones are listed below.

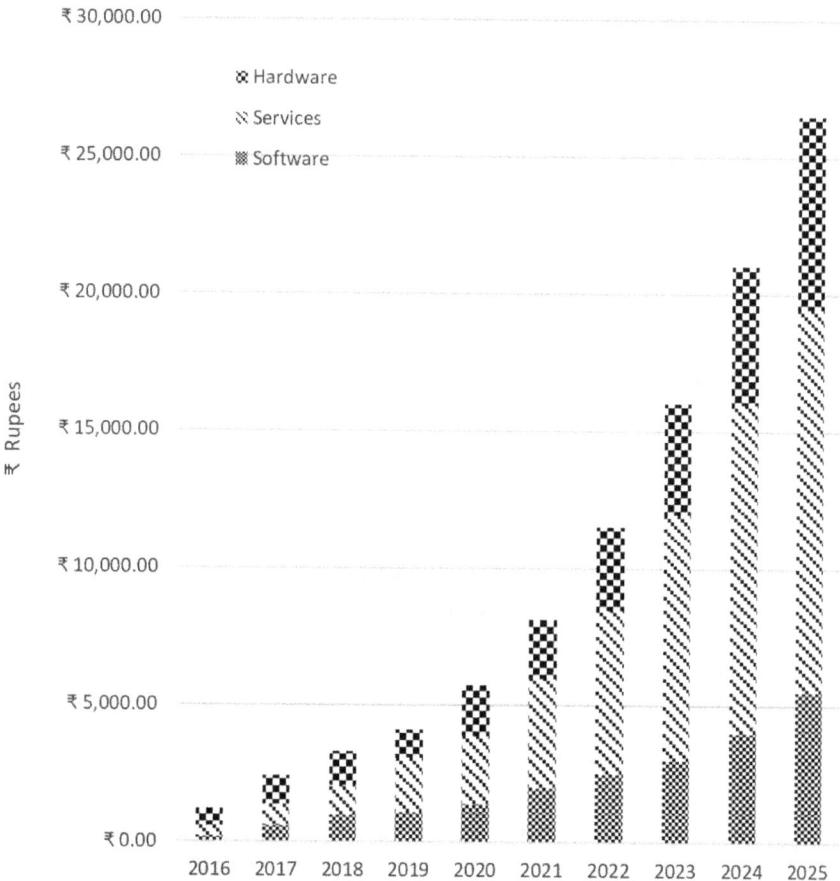

FIGURE 6.3   Computer vision revenue by world markets.

6.4.1 COMPUTER VISION HARDWARE

a. *Commercial Products*

Some of the most frequently used products are 3D Scanners, Bandit, CycloVision, DALSA (Digital Imaging), Eltec, Genex Technologies, IQinVision, NanoSystem, Optel Vision, Robocams, and Vista Imaging.

b. *Research Systems*

Some of the hardware products used in research systems are BiTEC, Binocular Camera Platforms, Homemade 3D Laser Scanner, and VLSI Vision Chips.

6.4.2 SOFTWARE LIBRARIES AND TOOLS

Recent advancements in GPUs, software tools, libraries, and frameworks have made it possible to build various image-processing applications. Out of the available ones, the basic and most widely used tools are as follows.

a. *OpenCV*

OpenCV is the most fundamental multi-platform open-source library that has in-built techniques and algorithms to implement image-related problems. It works well for a smaller image data set. But for larger data sets, since OpenCV does not support GPU, it uses CUDA.

b. *MATLAB*

Due to the ability of prototyping, MATLAB is the most used research tool for building computer vision applications. The ease of debugging using MATLAB is higher. But its demerit is that it is a pay-and-use platform.

c. *TensorFlow*

TensorFlow is an important tool that is gaining attention nowadays. It provides a way to integrate deep learning with computer vision. It also has APIs that can be used for other applications like classification, detection, and so on. One demerit of TensorFlow is that it consumes a lot of resources.

d. *CUDA*

CUDA is a platform developed by NVIDIA to enable increased computing performance in GPUs. The toolkit includes a primitive library that uses image and video functions. Memory distribution and power consumption is one downside to this platform.

6.4.3 COMPUTER VISION SERVICES

a. *Google Cloud Vision API*

Cloud and Mobile Vision service developed by Google allows users to call simple APIs to their application. The so-called APIs hold resourceful machine learning models and their functionalities. It also allows barcode scans and text detection using the OCR functionality.

b. *Amazon Rekognition*

Rekognition is a deep learning-based service developed by Amazon to analyse images and videos. The basic technique supported by this service is object detection and segmentation. This service provides the best analysis for facial recognition as well as sentimental analysis.

c. *Microsoft Azure Computer Vision API*

This service invented by Microsoft provides the same functionalities with images and videos. Besides, it provides handwriting recognition.

6.5   APPLICATIONS OF COMPUTER VISION

Through the years, the applications of computer vision have been exponentially expanding in almost all the fields. The phrase application here can be explained as a high-level function which tends to handle an objective with a high level of complexity. In general, computer vision is used in the fields of healthcare, manufacturing, security systems, OCR recognition, tracking systems, agriculture, autonomous vehicles, robotics, and much more. In this section, applications relating to healthcare [17], autonomous vehicle, automatic target detection, and a case study on how computer vision is used in robotic path planning are discussed.

6.5.1   HEALTHCARE

Computer vision combined with machine learning has made adequate progress in the field of healthcare [2]. The possible areas in healthcare where technology can directly intervene are medical imaging analysis, monitoring of patients using gadgets, predictive analysis of diseases, and much more. Medical imaging is the process of making visual representations of the inner body organs for analysis. Such analysis requires feature extraction techniques [17], which is one of the key concerns. Other disciplines like electroencephalography (EEG) and electrocardiography (ECG) provide graphical representations of the body organs, whereas on computer vision the concept of image co-registration is used. Image co-registration is the transformation of multiple images or photographs from different sources and time into one coordinate system. Co-registration implies the integration of information from different images into a single image for analysis. This concept has a significant impact in the field of healthcare as well as in satellites.

A fundamental machine learning algorithm for visual perception is the polyharmonic extreme learning machine (PELM) algorithm. This algorithm visually learns the image details after which it tries to classify images with a particular disease, based on the learned parameters. This is where machine learning and deep learning are used along with computer vision for disease prediction. One such important accomplishment in the healthcare field is the prediction of breast cancer [18] using deep learning.

6.5.2   AUGMENTED REALITY

Augmented reality can be discerned as adding digital information to a real-world environment. AR just creates additional graphics and sounds to an existing

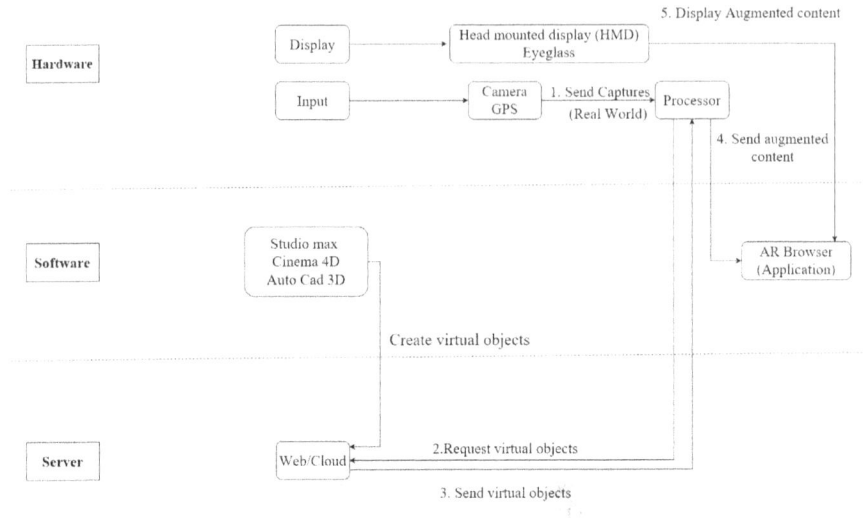

FIGURE 6.4   General AR-based system.

environment, whereas virtual reality (VR) virtually creates a wholly artificial environment of the real world. AR tends to change the perception of the reality observed from a physical space, which completely involves computer vision techniques. There are various types of AR [19], under which generic digital augmentation is one. This type visualizes multiple 3D objects in its environment. Figure 6.4 gives a general outline of how an AR-based system works and how the hardware, software, and servers integrate to visualize virtual objects in a captured real-world environment.

To randomly include objects into a physical space, a combination of three techniques is required.

- *Simultaneous localization and mapping (SLAM)* [15]: It involves a map construction of an unknown environment and based on the virtually included 3D object's location, a regular update of the map is done.
- *Depth tracking*: It measures the distance between the added virtual objects and the AR device.
- *Image projection*: The captured information from input sensors is processed and then projected onto the augmented setup.

6.5.3   VISION-BASED SELF-DRIVING CARS

Autonomous driving [13] has been a major application of computer vision along with deep learning techniques, where the computer itself perceives and learns from a non-deterministic environment. Object detection is the most significant strategy based on which vision-based self-driving cars work. The general procedure involves stereo vision, space estimation to drive, lane recognition, traffic light recognition, object detection and tracking, semantic scene understanding, mapping and localization, temporal and spatial recognition of the vision, and vision fusion from various sensors. Stereo vision [1, 7] involves the perception of the environment from sensors, which can be categorized as direct and mediated perception. Perception [20] can be taken directly from the working space regarding the texture, 3D shape, etc., without any knowledge of object categories. The basic methodology of how object detection and classification with localization are performed for a given image has been discussed in the previous section.

6.5.4 AUTOMATIC TARGET RECOGNITION AND DETECTION

Automatic target detection is the potential of any learning algorithm to identify, sense, and detect the target from an environment based on the input sensors. From a given environment, the spatial and temporal features are learned, which would be useful in recognizing the target. An application of robotic navigation is discussed as a case study in which the robot's path is semantically generated, given a target label.

6.5.4.1 Case Study: Robotic Path Planning Using Visual Percepts

The objective [8] is to navigate a robot to a target object in an undeterministic environment, i.e. through a previously unseen physical space. The environment is interpreted as representations that capture spatial and semantic information [12]. The workflow of the paper is depicted in Figure 6.5. The representations are in terms of detection (Det), segmentation (SSeg), depth of the image, and RGB. This problem is perceived using visual observations, handled by Markov's decision process. Semantic segmentation and detection are performed using state-of-the-art computer vision algorithms, where the features are identified and classified using a convolutional neural network (CNN) and a long short-term memory (LSTM) network to remember previous actions of the robot in the environment towards the goal: the target label. Thus, deep networks are used to learn navigation policies. During the training phase, an optimal path planning algorithm is used to calculate the progress towards the goal, which in turn helps in identifying what action to perform during the testing phase. The representations Det and SSeg, in particular, with no requirements of domain adaptation, identify layouts of scenes, location of the given target, and the obstacles along the path. For simulation, a renderer generates bounding boxes to identify objects. Thus, given a target label in an environment, the environment is interpreted into representations that identify and capture the objects and their location. Based on these representations and the memory about previous actions from

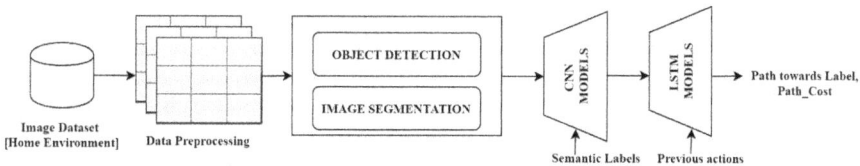

FIGURE 6.5 Path planning using visual percepts.

LSTM, the next action towards the goal will be learned. Finally, the model generates a path from the start to the target.

6.6 CONCLUSION AND FUTURE DIRECTIONS

The incorporation of deep learning and artificial neural networks in almost all fields of computer vision has made it possible to replicate human vision. Computer vision works basically with images and includes techniques such as image analysis, scene analysis, and image understanding. Computer vision without artificial intelligence works well but including deep learning methods achieves accurate results. Researchers have identified that computer vision is more effective in identifying image patterns than human cognitive systems. This chapter includes the most important aspects of computer vision like feature extraction, object detection, and image segmentation. The applications of computer vision have been growing exponentially, especially in terms of control systems, detecting events, modelling environments, target detection, automatic inspection, navigation, etc.

The future of computer vision has great scope because of its faster, reasonably economic, and more accurate result-providing capabilities. Computational time is one inconvenience that should be noted, because of the immense mathematical operations involved for each iteration. Further methods are being researched so that the computationally time and power can be gradually reduced.

BIBLIOGRAPHY

1. K. Grauman and B. Leibe, *Visual Object Recognition (Synthesis Lectures on Artificial Intelligence and Machine Learning)*. Morgan and Claypool, 2011.
2. Muhammad Imran Razzak, Saeeda Naz, and Ahmad Zaib, "Deep learning for medical image processing: Overview, challenges and the future," in *Classification in BioAapps*, pp. 323–350. Springer: Cham, 2018.
3. N. Buduma and N. Locascio, *Fundamentals of Deep Learning: Designing Next-Generation Machine Intelligence Algorithms*. O'Reilly Media Inc., 2017.
4. F.S. Khan, R.M. Anwer, J. Van De Weijer, A.D. Bagdanov, M. Vanrell, and A.M. Lopez, "Color attributes for object detection," in *2012 IEEE Conference on Computer Vision and Pattern Recognition*, pp. 3306–3313. IEEE, 2012.
5. Y.-J. Zhang, *Advances in Image and Video Segmentation*. IGI Global, 2006.
6. M. Seyedhosseini and T. Tasdizen, "Semantic image segmentation with contextual hierarchical models," *IEEE Transactions on Pattern Analysis and Machine Intelligence* 38(5): 951–964, 2015.
7. M. Alam, M.D. Samad, L. Vidyaratne, A. Glandon, and K.M. Iftekharuddin, "Survey on deep neural networks in speech and vision systems," *arXiv Preprint ArXiv:1908.07656*, 2019.
8. I. Guyon, S. Gunn, M. Nikravesh, and L.A. Zadeh, *Feature Extraction: Foundations and Applications*, vol. 207. Springer, 2008.
9. Y. Zhang, Y. Chen, C. Huang, and M. Gao, "Object detection network based on feature fusion and attention mechanism," *Future Internet* 11(1): 9, 2019.
10. R. Szeliski, *Computer Vision: Algorithms and Applications*. Springer Science & Business Media, 2010.

11. S.J. Prince, *Computer Vision: Models, Learning, and Inference.* Cambridge University Press, 2012.
12. A.T. Mousavian, M. Fišer, J. Košecká, A. Wahid, and J. Davidson, "Visual representations for a semantic target driven navigation," in *2019 International Conference on Robotics and Automation (ICRA)*, pp. 8846–8852. IEEE, 2019.
13. J. Stilgoe, "Machine learning, social learning and the governance of self-driving cars," *Social Studies of Science* 48(1): 25–56, 2018.
14. C.C. Aggarwal, *Neural Networks and Deep Learning.* Springer International Publishing, 2018.
15. T. Taketomi, H. Uchiyama, and S. Ikeda, "Visual SLAM algorithms: A survey from 2010 to 2016," *IPSJ Transactions on Computer Vision and Applications* 9(1): 16, 2017.
16. L. Xu, Y. Li, Y. Sun, L. Song, and S. Jin, "Leaf instance segmentation and counting based on deep object detection and segmentation networks," in *2018 Joint 10th International Conference on Soft Computing and Intelligent Systems (SCIS) and 19th International Symposium on Advanced Intelligent Systems (ISIS)*, pp. 180–185. IEEE, 2018.
17. J. Gao, Y. Yang, P. Lin, and D.S. Park, "Computer vision in healthcare applications," *Journal of Healthcare Engineering*, 2018, p. 5157020.
18. D. Wang, A. Khosla, R. Gargeya, H. Irshad, and A.H. Beck, "Deep learning for identifying metastatic breast cancer," *arXiv Preprint ArXiv:1606.05718*, 2016.
19. R. Silva, J.C. Oliveira, and G.A. Giraldi, "Introduction to augmented reality," *National Laboratory for Scientific Computation* 11, 2003.
20. O.I. Abiodun, A. Jantan, A.E. Omolara, K.V. Dada, N.A. Mohamed, H. Arshad, K.V. Dada, N.A. Mohamed, and H. Arshad, "State-of-the-art in artificial neural network applications: A survey," *Heliyon* 4(11): e00938, 2018.

7 Generative Adversarial Network

Concepts, Variants, and Applications

K. Rakesh and V. Uma

CONTENTS

7.1 INTRODUCTION

Humans always prefer automation of tasks. The invention of computers and computer programming has made it possible to automate some tasks by creating a series of instructions, referred to as algorithm. As we approach more and more complex problems, algorithms start to demand a certain degree of intelligence that is similar to human cognitive functions. This kind of intelligence that is induced to a machine preparing them to mimic our intelligence is termed as artificial intelligence (AI). A subfield of AI that is responsible for most of the recent advances of many fields is machine learning (ML). ML algorithms are statistical models built by finding patterns and inferences from the data that is provided. One class of ML algorithms that learns a significant amount of cognitive tasks is deep learning (DL) [1].

Deep learning comprises multiple layers of artificial neural network (ANN), a computing system that learns low-level to high-level features of the input data. DL is the technology behind many of the current awesome products and services like voice assistants, recommendation systems, self-driving cars, medical image analysis, computer vision systems, and many more [2]. The discussion on the basics of deep learning and deep neural network (DNN) is made in Section 7.2.1. Though we made significant progress in recognition and prediction tasks, the creative applications of AI are demanding more attention and in recent years more research momentum is noticed in this application. The research works on generative tasks have many applications in areas like content writing, film industry, game development, advertisement, etc.

The class of deep learning networks that is capable of generating data from the input is known as deep generative networks [3]. Over the years, many generative models have been proposed. All these different approaches for generative modelling have been extensively discussed in Section 7.2.2. Out of these models, generative adversarial networks (GAN) is the most popular model used in generation-based research works and applications. It provides state-of-the-art results and there prevails a debate on social and cultural aspects of its applications.

GAN can mimic the distribution of a data set, such that new data of the same distribution can be generated, which will be similar to the input data [4]. The GAN and its architecture are detailed in Section 7.3. Their architecture can be changed accordingly to our needs or use cases. This flexibility has provided a lot of variations while dealing with the distribution. Even the distribution of two different data sets has been utilized to produce some interesting results. The overview of the variations in GAN and the techniques used in these variations are discussed in Section 7.4.

The various research directions on GAN show several types of GAN that can be used for different architecture and functions. These diverse types of GAN have made it possible to use them in a variety of applications. Some of the interesting applications include generating photo-realistic images, converting an object in an image to another object, changing features of a person, generating missing areas in an image, image editing, converting a text to image, future scene prediction, creating 3D objects, medical image generation, and many more. Many of these applications and their impact on the industry have been discussed in Section 7.5. The future directions of GAN are discussed in Section 7.6.

7.2 OVERVIEW

7.2.1 Deep Learning

Machine learning algorithms have been sufficient for basic classification tasks like spam detection. But for solving complex tasks like image classification, a huge number of parameters have to be understood by the algorithm. This is not possible in classical machine learning algorithms. Then, there evolved a particular type of ML algorithm that was modelled after the human brain known as the artificial neural network. ANN comprises many connection units called neurons. Weight and bias values are set as parameters to connections that prevail between neurons in one layer to every other neuron of the adjacent layer. These parameter values are automatically figured out using algorithms like backpropagation as the network progresses along with the data samples. ANN shows significant results than classical ML algorithms. More and more complex problems have been solved by adding more layers and progressing deep into ANN. These deep structures are termed as deep learning.

Deep learning is capable of learning even from unsupervised data which makes it learn features automatically without much human effort. Deep learning has progressed over the years with the rise of graphical processing units (GPUs) and a huge increase in data and research over different ANN architectures and techniques. Many standard deep learning architectures have been proposed for various applications. DL architectures that are mostly applied in industries include deep neural networks, convolutional neural networks (CNN), recurrent neural networks (RNN), deep belief networks (DBN), AutoEncoders, and GAN.

Each network performs better in different applications. DNN is used for basic prediction and classification tasks of structured data like medical records. These data sets have rows of data and many parameters that are tuned to provide the required results. CNN performs better at image processing, object recognition, and image segmentation. RNN processes sequential data like time-series data and paragraphs of text for prediction of the upcoming data. Autoencoders are good at capturing the essential data from the input, and it is used in data compression and image processing. DBN is used in image clustering, recognition, and generation. At present, GAN is used for the generation of images and videos with high accuracy.

7.2.2 Deep Generative Models

Deep generative models work mainly by learning the distribution of the input data and exploring through the latent space. The model describes a probability density function (PDF) that represents the distribution of the input data. Mostly the generative model performs maximum likelihood from the PDF [2]. Each generative model differs from each other in the implementation of maximum likelihood. The representation of the density function can be explicitly defined or implicitly defined in a generative model with each one having its advantages and disadvantages.

In explicitly defined functions, a tractable PDF is crafted to provide the data distribution or approximates an intractable PDF. Some explicit PDF generative models that have tractable PDF include PixelRNN, NADE, MADE, and non-linear IOC.

Explicit PDF generative models that use approximation are Variational AutoEncoder (VAE) and Boltzmann machine.

In implicitly defined functions, the density function is not mentioned and the model will internally figure out a way to draw samples from the same distribution provided. Some implicit PDF generative models are generative stochastic networks (GSN) and GAN.

The explicitly defined model requires defining the density function which is hard to be established. Because the density function should be able to model the whole distribution, defining such a function is not possible. In the explicitly defined models that have been mentioned, a tractable density function is modelled or approximation of a density function is utilized. The tractable density function is a density function designed in such a way that it can model most of the distribution. But, for a complex distribution like distributions in Nature, the distribution of speech, and natural image, designing a parametric function that can capture the distribution perfectly is difficult. The problem is that it cannot handle the vast distribution of some data sets and this limits the usability of these models. The models that approximate the density function learn the distribution using two approaches. One by approximating an intractable density function by placing a lower bound in log-likelihood and maximizing the bound. Another approach uses the Markov chain to use an estimation of the density function or its gradient. Both of these approaches include approximation which does not guarantee to capture the complex distribution. These complications led to the design of implicit density function which is mostly automatically learned by the model itself. The GAN is the only implicit density function that shows promising results.

A lot of research has been done to track and improve these promising models. Most recent researches show results of GAN to be state-of-the-art in generative modelling. GAN has various applications when compared to other types of generative models.

7.2.3 GENERATIVE ADVERSARIAL NETWORKS

GAN was proposed by Ian Goodfellow in 2014 [5]. GAN contains two networks that compete against each other. One network helps the other to figure out the distribution of the data. There is no need for defining the PDF for the network. The architecture automatically figures it out internally which removes one huge problem. More than 12,000 citations have been made at present on the original research paper of GAN by Ian Goodfellow. It gained popularity for its robust architecture and ability to add many different kinds of models replacing the generator or discriminator in the architecture of GAN. Over the years, many different variations of GAN architecture made up of a combination of other neural network models have been proposed and used. The main concept in GAN that makes the whole architecture effective is adversarial training. Facebook's AI Research Director LeCun referred to adversarial training as "The most interesting idea in the last 10 years of ML." Apart from images and videos, GAN is also used for the generation of text, music, speech, and 3D objects.

7.3 GAN ARCHITECTURE

7.3.1 GENERAL STRUCTURE

GAN is a combination of two neural networks: the generator model and the discriminator model. The architecture of vanilla GAN is represented in Figure 7.1.

The generator model tries to generate data similar to the distribution of training data set and the discriminator network checks whether the generated data is agreeably realistic.

In the vanilla GAN which was first introduced by Ian Goodfellow in 2014 [5], a high-dimensional input noise is generated randomly and given as input to the generator. Any distribution can be used to generate these random noise values. The generator network has randomly assigned weights and biases as parameters. Both the weight and bias are numerical values that are subjected to mathematical operations along the process. The generator network processes the input noise with its parameters and produces output with the required dimension of the output image. These processes depend on the functions that are used in the neural network. The generated output will be mostly noisy and does not have any similarity with the distribution of our training data set. The data set can be an image, video, text, audio, or even 3D objects. Then, the generated output and the real data are passed into the discriminator. The discriminator classifies the real and fake images after processing it with its parameters, randomly assigned weights, and biases. From the output of the discriminator, the generator loss and discriminator loss are calculated and then backpropagation is performed. It helps the generator and discriminator to update their randomly assigned weights.

The generation, discrimination, and backpropagation are repeated many times with new samples of input data; this is referred to as training of GAN. Throughout

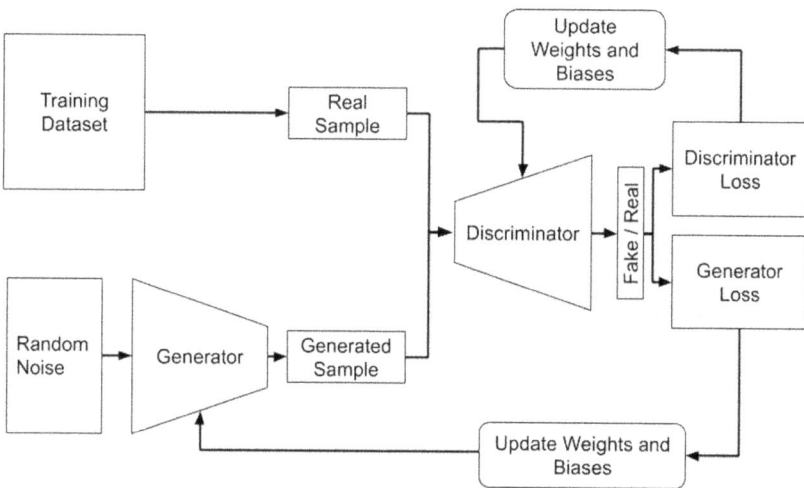

FIGURE 7.1 GAN architecture.

the training, the generator model gets better at generating more realistic data and the discriminator improves in classifying generated and real data [6, 7]. Both the generator model and the discriminator model converge at a point, after which both do not improve. Now, the discriminator model is discarded and the generator alone is used for generating the samples. The discriminator model was used as a tool for improving the generator and not used again in the generation of samples. However, the discriminator is a good classifier and sometimes used for validation of other models [4].

7.3.2 Adversarial Process

The generator and discriminator networks compete against each other to improve each other in their respective tasks. This adversarial technique was inspired by game theory, a mathematical modelling study of the interaction between two rational decision-makers. The interaction is performed based on the zero-sum game, in which the participant either wins or loses. Here, when the discriminator successfully identifies the real and fake data, the generator is penalized and vice versa when it fails to identify. GAN is mainly based on the Nash equilibrium to balance the generator and the discriminator. Nash equilibrium is a solution state when many players are playing a game and each knows the other player's strategies and makes decisions without changing his or her strategy. The generator and discriminator know the strategies of each other but make the best decisions by not changing their strategy, making the models to compete with each other and achieve the Nash equilibrium state and converge at a point.

Both the generator and discriminator models are modelled as differentiable functions. A value function is defined using the generator and discriminator functions in the form of a two-player minimax game and then optimized throughout the training. If the discriminator is completely optimized before starting to optimize the generator, then the generator will not get better. So, k-steps of discriminator optimizations are performed before one step of optimizing the generator. Then, the generator model tries to increase the chances of the discriminator making mistakes.

```
Algorithm
Step 1: Iterate through the number of iterations.
Step 2: Iterate through the discriminator for a fixed
number of times.
    Step 2a: Sample noise and minibatch of n examples from
    given examples.
    Step 2b: Calculate the error and gradients.
    Step 2c: Update the discriminator weights.
Step 3: Sample minibatch of n noise samples.
Step 4: The generator is provided with the noise sample
and output is generated.
Step 5: Output is sent to the discriminator and loss value
is calculated.
Step 5: The weights of the Generator are updated using the
loss value from Discriminator.
```

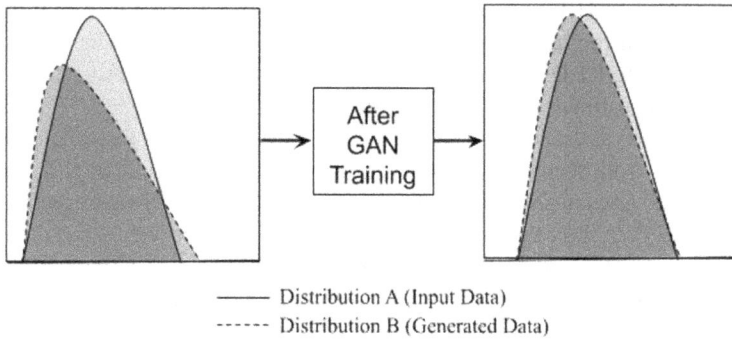

——— Distribution A (Input Data)
------ Distribution B (Generated Data)

FIGURE 7.2   Divergence of distribution.

7.3.3 BACKGROUND MATHEMATICS

Every data set has a distribution of data. Every model tries to understand the distribution of input data and captures relevant information from the distribution using probability estimation. The divergence of distribution between two data sets is provided in Figure 7.2. The generative models capture joint probability and discriminative models capture conditional probability. The generator and discriminator models are mathematically differentiable functions G and D, respectively. A value function is defined using these differentiable functions [5].

$$\min_G \max_D V(D,G)$$

$$V(D,G) = E_{x \sim pdata(x)}\Big[\log D(x)\Big] + E_{z \sim pz(z)}\Big[\log\big(1 - D\big(G(z)\big)\big)\Big]$$

The whole training process is designed in such a way that D tries to maximize the probability of labelling the generated and input data correctly. Log $D(x)$ is the logarithmic probability of discriminator predicting the real data label. The generator tries to minimize $\log\big(1 - D\big(G(z)\big)\big)$, the log probability of discriminator predicting generated images label correctly. E denotes the expectation of a random variable.

In Equation 7.1, there are two parts. The first part handles the real data sample and the latter part after the plus symbol deals with the generated samples. But, practically, the equation may not fare well as the discriminator can reject generated samples easily with high confidence. To change this in the equation, the generator function is changed. Rather than decreasing the log probability of discriminator making correct predictions, the generator increases the log-probability of discriminator making mistakes $\Big[\log D\big(G(z)\big)\Big]$. The value function is optimized by using backpropagation. This eliminates the use of a Markov chain which adds a huge advantage over other generative models.

The value function is based on Jensen–Shannon divergence (JSD), which measures the similarity between two distributions. JSD is based on Kullback–Leibler divergence (KLD). KLD measures the difference of one distribution to another one. The difference in the distance between the two distributions can be estimated by KLD. Minimizing KLD can minimize the distance between the distributions. Log-likelihood is the estimator that helps to minimize KLD.

The GAN internally finds the density function (PDF) of the input data distribution. A sample from the density function provides a relative likelihood. Likelihood expresses different plausible parametric values of the sample. The generator of GAN tries to model data that have a similar distribution of input data using the density function. At first, the generated distribution will be different from the real input distribution. This difference is then reduced with minimizing KLD by maximizing the log-likelihood of the distributions.

7.4 GAN VARIATIONS

7.4.1 OVERVIEW

The vanilla GAN invented in 2014 by Ian Goodfellow was popular mainly because of the adversarial process. Many different variations of GAN which improve its functionality, accuracy, and applications are proposed over the years. The variations may possess changes in architecture or formulations. Every variation provides a variety of applications for GAN. The number of GAN variant papers proposed has exponentially increased each year. The architectural variation includes changes by including concepts and components of other deep learning architectures. These changes can be an increase in the number of generators or discriminators, the way they are connected, and the introduction of other deep learning components in the discriminator or generator. The variations are mainly classified into architecture-based variants and formulation-based variants [8, 9].

7.4.2 TECHNIQUES

7.4.2.1 Architecture-based Variant Class

The architectural variant class has two subclasses: condition-based variants and structure-based variants.

I. **Condition-based Variants**

Condition-based variants involve conditional input that changes the output according to it. This class of variants started with the conditional GAN (CGAN) proposed in 2014 [10]. It is based on vanilla GAN with additional conditional input to generator and discriminator. This extracts features based on the modelling of the conditional input. The InfoGAN proposed in 2016 provides the condition as latent information to the generator. Auxiliary Classifier GAN (ACGAN) is an extension of CGAN that adds conditional input only to the generator. All these GANs include modifying the objective function according to the condition structure.

II. **Structure-based Variants**

The structure-based variants include changes in the architecture of GAN. The architectural change includes an increase in generators or discriminators, the addition of convolution layer (network), or encoder–decoder.

a. **Increasing Components**

The generators are increased in multiple generator GAN (MGAN) and discriminators are increased in multiple discriminator GAN (MD-GAN).

b. **Convolutional Components**

The addition of the convolutional network to the GAN architecture was first proposed in deep convolutional GAN (DCGAN) in 2015 [11]. Based on DCGAN, the Laplacian Pyramid GAN (LAPGAN) was introduced which has a series of DCGAN that processes images from coarse to fine. The 3D convolution was utilized in 3D-GAN for 3D object generation.

c. **Encoder–Decoder Components**

The encoder–decoder-based GAN variant includes an encoder, a decoder, or an auto-encoder network in the GAN architecture. Adversarial AutoEncoders (AAE) in 2015 was the first paper to combine encoders with GAN. In the same year, variational AutoEncoder was also combined with GAN. Then, Adversarial Learning and Inference (ALI) combined a generator and inference model, which is monitored by a discriminatory model. The energy-based GAN (EBGAN) which uses a different energy-based optimization function also utilizes the encoder–decoder network in the discriminator. Later, many architectures, including Adversarial Generator-Encode (AGE), Boundary Equilibrium GAN (BEGAN), Bidirectional GAN (BIGAN), and Decoder-Encoder GAN (DEGAN) were designed by combining encoder–decoder networks in different strategies [8].

d. **U-NET Components**

A convolutional network called U-NET which is in the form of encoder–decoder style with connections between the encoder and decoder layers (skip connections) has been used extensively in recent works. The GAN architectures like progressive GAN (ProgGAN) and Pixel-to-Pixel GAN (Pix2Pix) [12] provided profound results by utilizing U-NET in their models. ProgGAN uses U-NET in the generator and decoder network in discriminator. ProgGAN gradually increases the decoder in the generator model and discriminator model to improve the image quality of the generated image. Pix2Pix uses U-NET as generator and PatchGAN, a series of deconvolutional layers, as the discriminator.

7.4.2.2 Formulation-based Variant Class

The formulation-based variant class includes GAN architectures that have modified objective function, normalization method, regularization method, and loss function. The variant typically tries to solve a problem that GAN faces by mathematically modifying the formulation. The original GAN was unstable and hard to train. It had many issues in training and convergence. Many proposed variants use these alternative objective functions to solve these issues.

I. **Divergence-based Modification**

Original GAN used JSD for the objective function. f-GAN and Wasserstein GAN (WGAN) improved stability of GAN training by changing JSD to other divergence formulations. f-GAN utilizes f-divergence for the objective function. WGAN formulated the objective function using Wasserstein distance. WGAN claims to be more stable than GAN, uses JSD, and also performs weight clipping to avoid mode collapse, a condition where data generated by the generator are not diverse or even produces the same sample for different inputs.

II. **Loss Function-based Modifications**

In some GANs, loss function is tuned to achieve desired results. Least-square GAN (LSGAN) uses the least-square loss function and f-GAN uses gradient-based loss function. Discover cross-domain relationship GAN (DiscoGAN) [13] attempts to automatically discover relationships and features between different domains. This is achieved by modifying the objective function with reconstruction loss. Cycle consistency GAN (CycleGAN) [14] introduces an alternative loss function, named cycle-consistency loss. It helps CycleGAN to minimize the distribution of data between two domains. StarGAN aims at the multi-domain image-to-image translation. It uses domain classification loss along with the reconstruction loss.

III. **Technique-based Modifications**

Another technique to optimize the objective function was performed in energy-based GAN (EBGAN), where discriminator is formulated as an energy function. The mode collapse problem was also handled in unrolled GAN (UGAN) by using unrolled optimization in the objective function.

IV. **Regularization**

GAN has to converge properly to achieve good results. The generator and discriminator model training has to be stable to achieve convergence. The unstable training process could lead to mode collapse. The regularization technique is a trick that could lead to the stabilization of the training process. Many different kinds of regularization are used to solve this problem. Regularization techniques like mode regularization, Lipschitz regularization, Jacobian regularization, and orthogonal regularization are used in variants like MRGAN, LSGAN, JRGAN, and BigGAN, respectively [15]. WGAN-LP uses a penalty function that uses Lipschitz constraint.

V. **Normalization**

Normalization is another trick performed to scale the values in GAN. It also aids in the stabilization of GAN training and convergence. The normalization techniques like gradient penalty and spectral normalization are widely used [15]. The gradient penalty is used in WGAN-GP, DRAGAN, and BWGAN. WGAN-GP uses the gradient penalty in WGAN. Along with the gradient penalty, DRAGAN uses deep regret analytics and Banach Wasserstein GAN (BWGAN) uses dual normalization in Banach space. Spectral Normalization is used in Spectral normalization GAN (SNGAN) and Self-attention GAN (SAGAN) [16]. The attention method is also used along with spectral normalization in SAGAN.

7.5 APPLICATIONS

7.5.1 IMAGE GENERATION AND PREDICTION

The creation of visual content remains challenging for humans. The generation of new images can be useful in many areas where more visual data is needed. More data can provide more possibilities and opportunities. In the case of DL, huge data sets are required to train models. These data can be hard to collect and pre-process. Data augmentation is a process done to increase the data set by performing transformations like cropping, scaling, padding, and flipping to the existing images. The augmented images are extremely useful in creating machine learning models for medical and geological image classification, where collecting medical images is hard, and augmented images serve the purpose of training these models. These problems can be tackled by GAN. They have the potential to create many images similar to the input image.

All the GAN variants are capable of data augmentation, depending on the type of data they process. DCGAN [11] and StyleGAN [17] are famous for the generation of human faces from scratch. Figure 7.3 provides an example of the image generation. BigGAN [18] is capable of generating photo-realistic images. Images of objects are also generated which can be used in websites, clothing lines, media, and many more areas.

INPUT

OUTPUT

FIGURE 7.3   Image generation.

7.5.2 IMAGE TRANSLATION

Image translation is the process of converting an image into another modified image or other forms of data like text, audio, etc. Image-to-image translation leads to many fascinating applications. Image translation includes the conversion of objects from one domain to objects of another domain. For example, images of a horse can be converted into images of zebra and vice versa. Many neural networks like CycleGAN [14], Pix2Pix [12], and BigGAN [18] have demonstrated this using various data sets. The conversion between day and night images, seasons, colourization of black and white images, and many more have been demonstrated using Pix2Pix. Even altering human features like hair colour and facial expression has been translated using StyleGAN [17] and DiscoGAN [13]. An example of converting label image from CMP facades data set [19] to real-world buildings using Pix2Pix [12] is provided in Figure 7.4. Even the conversion of faces into cartoon or anime characters has been demonstrated. Image super-resolution is another application where GAN shines.

Image super-resolution is the problem where fewer dimension images are converted into images with higher resolution. Variants like Super Resolution GAN (SRGAN) [20] and Enhanced Super Resolution GAN (ESRGAN) [21] have achieved remarkable results in image super-resolution tasks. Style transfer is another problem where the style or feature of one image is transferred to the input image. The image which has to be referred for the features or style is processed as conditional input and the main input is the image where style has to be applied. Many variants under conditional GAN class are capable of solving this problem with good accuracy.

Image translation into textual data has been achieved in DCGAN, StackGAN [22], AttnGAN, GAN-INT-CLS, and OP-GAN. Image inpainting can also be performed in GAN. In some images, certain parts of the pictures are missing or hidden by overlapping objects. Image inpainting is a process where these images are translated into images that are filled with accurate details in missing regions. Image inpainting can also be used to remove watermarks in images.

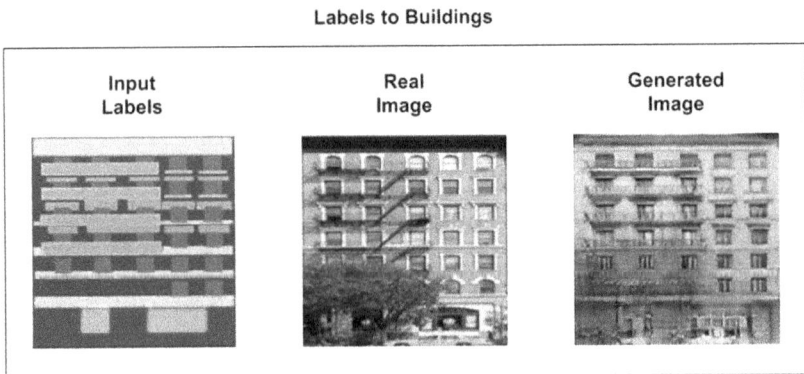

FIGURE 7.4   Image translation using Pix2Pix.

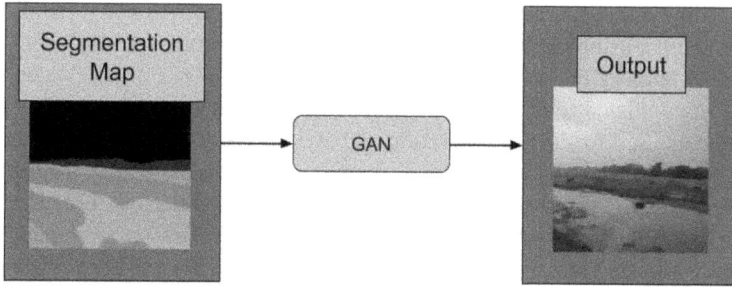

FIGURE 7.5   Image editing using GAN.

7.5.3   IMAGE EDITING

Image editing is the technique of manually providing input to the model that alters the image accordingly. This technique enables us to paint objects and scenes like trees, architectural structures, sea, and sky into the image. The variants like IGAN, GP-GAN, and GauGAN are capable of image editing. This is under development as the technique has a long way to follow, even though the results seem stunning and natural in some cases. Unlike other models, GauGAN [23] has achieved some remarkable quality in images. Figure 7.5 describes a made-up example of image editing using GAN.

7.5.4   3D OBJECT GENERATION

The virtual environment has been extensively created and experienced by humans. These environments are used in gaming industries, virtual reality, and many more areas. Virtual environments are filled with 3D objects that are hard to create, involving a lot of man-hours as well. Altering and recreating 3D objects will still delay such projects on a large scale.

Automating the creation of 3D objects and environments can increase the development process on a large scale. The automated 3D object creation can also pave the way for real-time object creation and simulations. 3D GAN [24] and 3D Scene-GAN [25] are two prominent 3D object generation architectures. 3D GAN can generate high-quality 3D objects with sample images and objects. 3D Scene-GAN tries to create the whole environment with the sample images. These variants laid the foundation for the whole application in these areas. There are still more fascinating applications in these areas that are yet to be discovered.

7.5.5   VIDEO MANIPULATION

Video generation has been an interesting area of work for humans. Creative video content creation is an important profession and many industries have been built upon this platform. Video content has to be created by using video capture devices or graphically creating each scene through software. In each case, the creator needs plenty of time and resources to complete the task. The generation and manipulation of video content using GAN can make it easy for the creators.

Video is nothing but a sequence of pictures, the same as images. Video can be created, manipulated, and predicted. Generation of video can be done using the GAN models used for image generation and style transfer. Video manipulation has created a great discussion in society, as there occurs a possibility of changing faces and expressions in video content. Face swapping GAN (FSGAN) [26] is a recent addition to the class of GANs that can manipulate videos. Few research papers have proposed works on video prediction. Video prediction guesses the next frame and it can evolve machines into a better content creator.

7.5.6 AUDIO GENERATION AND TRANSLATION

GANs are capable of generating audio with ease like other forms of data. The generation of audio initiates other operations with audio content. The adversarial models like MidiNet, WaveGAN [27], and MuseGAN [28] have received a good response from the research community. A representation of using GAN to generate music is provided in Figure 7.6. These models have guided audio manipulation and translation models. FusionGAN [29] can fuse different genres of music and create quality music. GANs can also perform voice impersonation of one input over the other. The translation of audio to text and vice versa has been performed in GAN-TTS.

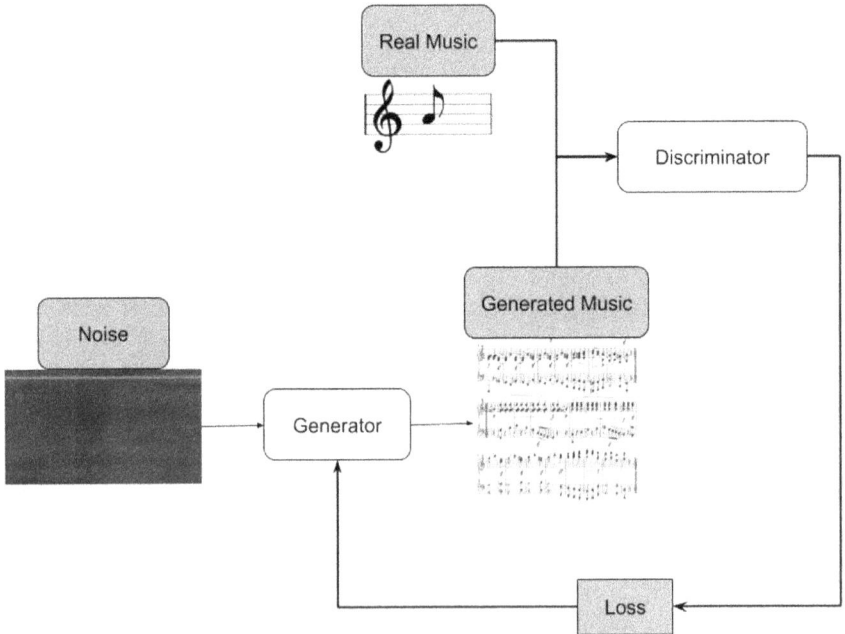

FIGURE 7.6 Audio generation.

The conversion of audio content to the image has been demonstrated in Wav2Pix architecture.

7.5.7 MEDICAL IMAGE PROCESSING

Image prediction is used in medical fields. The anomalies in medical reports are hard to detect and need medical expertise. The introduction of GAN can greatly help professionals automate prediction tasks. AnoGAN [30] is one such GAN that can predict tumours in medical images. Generating medical images also have been demonstrated in MedGAN [31]. The PET scans have been translated into CT scans using MedGAN. Medical reports like X-rays, CT scans, and MRI scans have been translated into clear, understandable images. Motion correction and positron-emission tomography (PET) have also been performed in MedGAN.

7.6 CONCLUSION AND FUTURE DIRECTIONS

Generative adversarial network has great potential and a lot of applications. They can be used in nearly every possible field and production sector. The rapid research developments had provided hope for developing a lot of creative applications using GAN. Each variation of GAN provides new features. These variations are enveloping various kinds of neural networks and techniques that help gather complex features and generate more relatable and realistic images as discussed. The vast collection of GAN implies the variety of its applications. The impact of GAN had been on every aspect of creative tasks. Every form of natural and synthetic data has been manipulated using GAN with proper research work. These works on GAN take us closer and closer to the possibility of creating automated machines that are creative.

There has been a vast production area eagerly waiting to utilize these ideas and research works. GANs would be widely used in medical fields to predict diseases and abnormalities. They can help generate augmented images, which in turn improves the quality of image classification. They would also be used in creating content for media and industries. The graphical rendering machines and game development teams can utilize GANs to create and modify 2D images and 3D objects on the fly. The 3D object generation quality of GAN is not up to mark at present. If the quality increases down the line, these rendering techniques with GAN can rapidly minimize the time taken for developing games. In the near future, many tasks that are considered impossible today can be demonstrated using GAN.

BIBLIOGRAPHY

1. Francois, Chollet, *Deep Learning with Python* Manning Publications, 2017.
2. Goodfellow, Ian, Yoshua Bengio, and Aaron Courville, *Deep Learning.* MIT Press, 2016.
3. Nielsen, Michael A., *Neural Networks and Deep Learning*, Vol. 25. Determination Press, 2015.
4. Langr, Jakub, and Vladimir Bok, *GANs in Action* Manning Publications, 2018.

5. Goodfellow, Ian, et al., "Generative adversarial nets," in *Advances in Neural Information Processing Systems*, Curran Associates, 2672–2680, 2014.

6. Wang, Kunfeng, et al., "Generative adversarial networks: Introduction and outlook," *IEEE/CAA Journal of Automatica Sinica*, IEEE, 4(4): 588–598, 2017.

7. Creswell, Antonia, et al., "Generative adversarial networks: An overview," *IEEE Signal Processing Magazine*, IEEE, 35(1): 53–65, 2018.

8. Pan, Zhaoqing, et al., "Recent progress on generative adversarial networks (GANs): A survey," *IEEE Access*, IEEE, 7: 36322–36333, 2019.

9. Wang, Zhengwei, Qi She, and Tomas E. Ward, "Generative adversarial networks: A survey and taxonomy," *arXiv Preprint ArXiv:1906.01529*, 2019.

10. Mirza, Mehdi, and Simon Osindero, "Conditional generative adversarial nets," *arXiv Preprint ArXiv:1411.1784*, 2014.

11. Radford, Alec, Luke Metz, and Soumith Chintala, "Unsupervised representation learning with deep convolutional generative adversarial networks," *arXiv Preprint ArXiv:1511.06434*, 2015.

12. Isola, Phillip, et al., "Image-to-image translation with conditional adversarial networks," in *Proceedings of the IEEE Conference on Computer Vision and Pattern Recognition*, IEEE, 1125–1134, 2017.

13. Kim, Taeksoo, et al., "Learning to discover cross-domain relations with generative adversarial networks," in *Proceedings of the 34th International Conference on Machine Learning*, Volume 70, JMLR, 1857–1865, 2017.

14. Zhu, Jun-Yan, et al., "Unpaired image-to-image translation using cycle-consistent adversarial networks," in *Proceedings of the IEEE International Conference on Computer Vision*, IEEE 2223–2232, 2017.

15. Kurach, Karol, et al., "A large-scale study on regularization and normalization in GANs," *in Proceedings of the 36th International Conference on Machine Learning*, JMLR, Volume 97, 3581–3590, 2019.

16. Zhang, Han, et al., "Self-attention generative adversarial networks," *arXiv Preprint ArXiv:1805.08318*, 2018.

17. Karras, Tero, Samuli Laine, and Timo Aila, "A style-based generator architecture for generative adversarial networks," in *Proceedings of the IEEE Conference on Computer Vision and Pattern Recognition*, IEEE, 4401–4410,2019.

18. Brock, Andrew, Jeff Donahue, and Karen Simonyan, "Large scale gan training for high fidelity natural image synthesis," *arXiv Preprint ArXiv:1809.11096*, 2018.

19. Tylecek, Radim, The cmp facade database. Tech. rep., CTU–CMP–2012–24, Czech Technical University, 8 p, 2012.

20. Ledig, Christian, et al., "Photo-realistic single image super-resolution using a generative adversarial network," in *Proceedings of the IEEE Conference on Computer Vision and Pattern Recognition*, Elsevier, 4681–4690, 2017.

21. Wang, Xintao, et al., "Esrgan: Enhanced super-resolution generative adversarial networks," in *Proceedings of the European Conference on Computer Vision (ECCV)*, Springer, Workshops, 0–0, 2018.

22. Zhang, Han, et al., "Stackgan: Text to photo-realistic image synthesis with stacked generative adversarial networks," in *Proceedings of the IEEE International Conference on Computer Vision*, IEEE, 5907–5915, 2017.

23. Park, Taesung, et al., "Semantic image synthesis with spatially-adaptive normalization," in *Proceedings of the IEEE Conference on Computer Vision and Pattern Recognition*, IEEE, 2337–2346, 2019.

24. Wu, Jiajun, et al., "Learning a probabilistic latent space of object shapes via 3d generative-adversarial modeling," in *Advances in Neural Information Processing Systems*, Curran Associates, 82–90, 2016.

25. Yu, Chong, and Young Wang, "3D-scene-GAN: Three-dimensional scene reconstruction with generative adversarial networks," in *ICLR Workshop*, ICLR, 2018.

26. Nirkin, Yuval, Yosi Keller, and Tal Hassner, "Fsgan: Subject agnostic face swapping and reenactment," in *Proceedings of the IEEE International Conference on Computer Vision*, IEEE, 7184–7193, 2019.

27. Donahue, Chris, Julian McAuley, and Miller Puckette, "Adversarial audio synthesis," *arXiv Preprint ArXiv:1802.04208*, 2018.

28. Dong, Hao-Wen, et al., "Musegan: Multi-track sequential generative adversarial networks for symbolic music generation and accompaniment," in *arXiv preprint arXiv:1709.06298*, 2018.

29. Ma, Jiayi, et al., "FusionGAN: A generative adversarial network for infrared and visible image fusion," *Information Fusion*, Elsevier, 48: 11–26, 2019.

30. Schlegl, Thomas, et al., "Unsupervised anomaly detection with generative adversarial networks to guide marker discovery," in *International Conference on Information Processing in Medical Imaging*. Springer, 146–157, 2017.

31. Armanious, Karim, et al., "Med GAN: Medical image translation using GANs," *Clinical Orthopaedics and Related Research (CoRR)*, Elsevier, 79, p.101684, 2018.

8 Detection and Classification of Power Quality Disturbances in Smart Grids Using Artificial Intelligence Methods

Gökay Bayrak and Alper Yılmaz

CONTENTS

8.1 INTRODUCTION

Integration of DG units to the grid improves the distribution system performance, reliability, stability, security, power quality (PQ), and voltage profile. Also, DG-based microgrid should have fewer losses, high-degree voltage support, and a reliable power quality disturbance (PQD) detection system [1]. The interconnection of DG with a grid may reduce the cost, losses, complexity, a load of lines, interdependence, and inefficiencies associated with low-carbon emission in microgrids [2].

The DG-based power plants must adapt the defined electrical standards to provide reliable grid operation and grid stability in cases of any fault or disturbances at the point of common coupling (PCC). The PQDs [3], islanding detection and protection, and electromagnetic interference are critical issues to adapt DG systems to the grid. According to Ref. [4], this fault and disturbance conditions costs in the European Union (EU) are predicted to be high.

DGs inclusion in the system has a lot of positive effects, but it creates technical difficulties facing the integration of DG. These difficulties are voltage regulation, increased distribution losses, harmonic control and harmonic injection problem, islanding condition control, and the sensitivity of existing protection schemes.

The conventional power system is designed to have a unidirectional flow of power from the high-voltage to the low-voltage side. Today, power distribution systems with bidirectional power flows have been created due to grid-integrated DG systems, which have increased in numbers rapidly. As a result, classical analysis, operation, and design methods cannot manage this complex grid structure. DG integration into the main grid is an essential issue if considering this aspect. The DG systems are expected not to operate out of the defined threshold values of the related standards' voltage and frequency. In a microgrid, unintentional islanding conditions cause several damages for both the system operators and the DG unit side. An incorrect detection in islanding conditions with conventional methods also leads to switching of DGs and economic losses. The voltage, frequency, and current balance of DG systems operating with the grid are also essential parameters for microgrids.

Figure 8.1 shows the general structure of a DG system. There are some technical electrical criteria to switch DG systems from the grid, and these conditions must be considered to provide the sustainable operation, reliability, and high-power quality of the grid. IEEE 929-2000 defines some technical electrical rules for microgrids to provide a reliable power system [5].

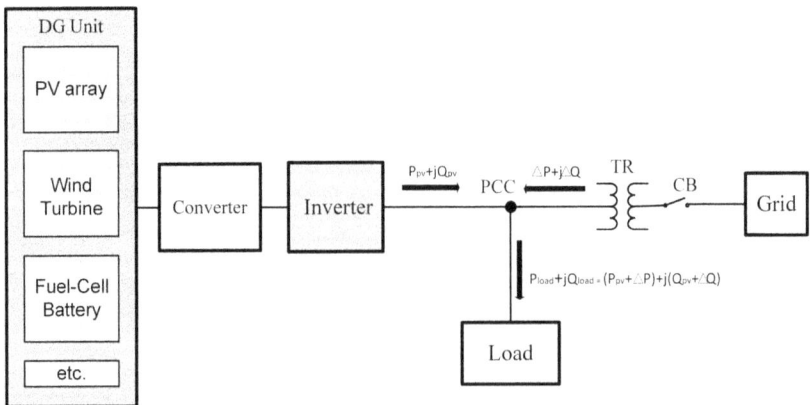

FIGURE 8.1   A general schematic for DG systems.

PQ parameters are limit values that allow an electrical device to operate as intended without any significant loss of performance and life expectancy. Any power system problem that leads to voltage, current, and frequency deviations can cause malfunction of hardware on the consumer side. Voltage sag/swell and interruption, current/voltage harmonics, transients, notches, spikes, flicker, and voltage imbalances are among the main factors affecting PQ in systems. Also, abnormal operating conditions of the grid affect the DG system, so the DG system has to be out of islanding condition, which is a security problem in a DG system. Several international standards have been established to regulate PQ. IEEE Std. 1547 [6] and IEEE Std. 519 [7] determine harmonic voltage and current harmonic limits. IEEE Std. 1159 [8] also offers some suggestions for monitoring PQDs in DG systems.

8.1.1 Signal Processing (SP)-based PQD Detection Methods

Many methods have been proposed to detect and classify PQDs in microgrids in the last decade [9]. Today, PQ analysers are used to detect short-term and long-term voltage disturbances using effective value conversion (true RMS method). These analysers performed the harmonic analysis of a signal using the fast Fourier transform (FFT) method. These methods are easily applicable but have many drawbacks. These disadvantages are difficulties in the selection of threshold values, noisy measurements, incorrect detection tendencies, etc. Thus, intelligent methods are more reliable to detect PQDs [10]. Signal processing-based methods use both time and frequency components of a signal. Thus, this characteristic is a useful property for detecting disturbances compared to conventional methods. SP-based methods used in fault detection can also be used to extract the features to be given to the classifier input [10]. Fourier transform (FT), wavelet transform (WT), Hilbert–Huang transform (HHT), s-transform, curvelet transform, Kalman filter, Gabor transform, etc. constitute SP-based methods for detecting PQDs. Figure 8.2 shows a classification of SP-based PQD detection methods used in smart grids.

PQD detection is proposed in Ref. [11], and this method used Dbn wavelets to detect events accurately. The discrete wavelet transform (DWT)-based method is investigated for wind turbines in Ref. [12], and UWT decomposition coefficients are used to obtain detailed coefficients. The WT method was proposed to detect fault conditions for the grid-connected PV-based microgrid investigated in Ref. [13]. WT method is also used to identify PQDs in a hydrogen energy-based microgrid in Ref. [14]. Threshold values were used in Ref. [15] for a DG system by determining harmonics, entropy, and energy by DWT decomposition. In another study, three-level decomposition is performed in a DG system using WT to the classification of PQDs [16]. The un-decimated WT (UWT)-based disturbance detection method was suggested for microgrids in Ref. [17] for the first time. This method was tested for real-time applications, and it overcame the limitations of WT-based methods.

8.1.2 Artificial Intelligent (AI) Methods for PQD Detection

Various artificial intelligence (AI)-based algorithms have been used in the literature to classify any faults and PQDs in microgrid applications. Some of the AI-based

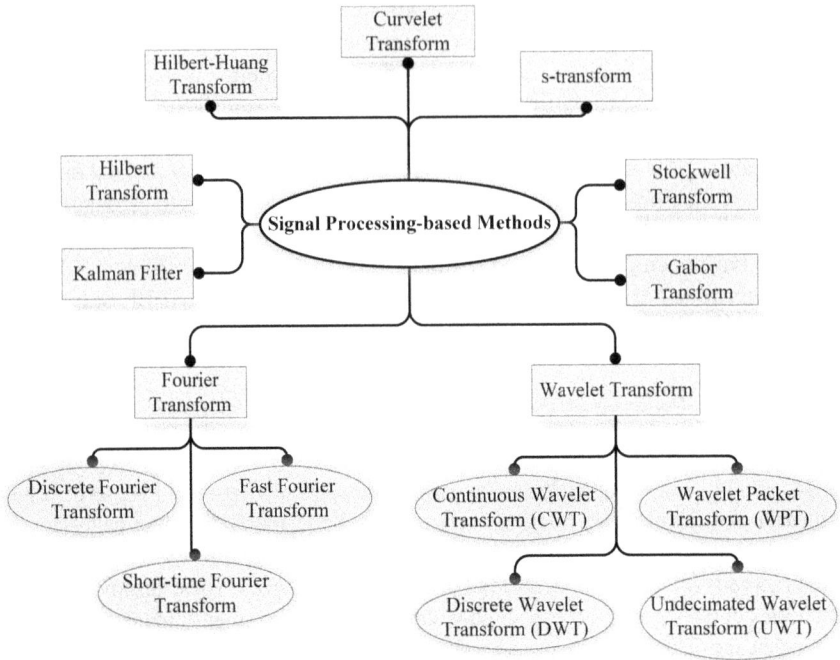

FIGURE 8.2   SP-based methods using in smart grids.

methods for detection of PQDs are fuzzy expert system-based classifiers, neural net-work (NN)-based classifiers, decision tree (DT) and random forest (RF) classifiers, support vector machine (SVM)-based classifiers, and deep learning (DL)-based clas-sifiers. Figure 8.3 shows a classification of AI-based PQD detection methods used in smart grids. The classifier is an algorithm that takes raw data or extracted futures data as an input and makes a decision about the normal operating state of the system. Artificial neural network (ANN) has been used for fault detection in PV systems and wind turbines in Refs. [18, 19]. A SVM is a classification method that is used suc-cessfully in many pattern recognition problems. It is also originally a proposed and defined classifier to distinguish two classes from each other [20]. A multiclass SVM method that performs multiple PQDs is proposed in Ref. [21]. A DT-based algorithm developed for the classification of complex PQDs is proposed in Ref. [22]. Adaptive probabilistic neural networks (APNN) and probabilistic neural networks used the classification of PQDs [23, 24].

Complex learning networks gradually replace the few layers used in regression and classification problems with developing information technologies. DL algorithms can be considered as multilayers of ANNs. In the study where the DL algorithm was used for the first time in islanding detection is Ref. [25]. In another study [26], a new approach based on single spectrum analysis, curvelet transform, and deep convolu-tional neural network (DCNN) has been performed to realize the classification of PQDs. A novel DL method proposed using raw data for the classification of PQD

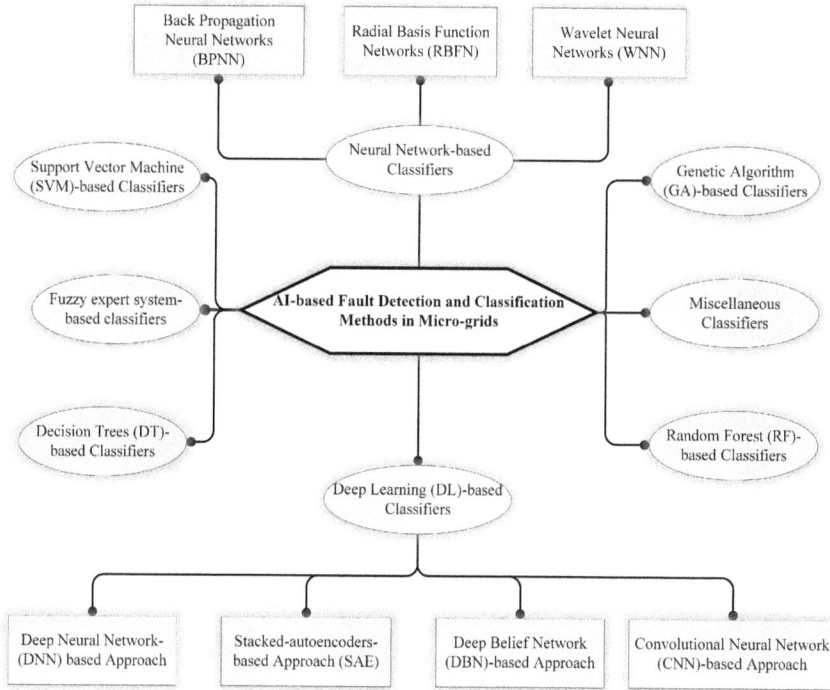

FIGURE 8.3 AI-based classification methods in smart grids [9].

[27]. This method recognizes automatic feature extraction and selection using any filter or signal processing method. The proposed algorithm is compared to other DL network structures – long short-term memory (LSTM) network, ResNet50, stacked auto-encoder (SAE), and gated recurrent unit network – in this study. Results show that the proposed DCNN-based method has higher accuracy and less training cost. In Refs. [28, 29], 1D time series PQD data converted 2D image data and then 2D convolutional neural network (CNN) performed the classification of PQDs. The data type for this method is not suitable because PQDs are 1D time series signals. While there is a significant correlation between the pixels on the horizontal and vertical axes in the image data, there is only one direction correlation in the PQD signals. Results are compared to other AI-based networks but noise independence, training time, and model size of the proposed network are unspecified.

8.2 WAVELET TRANSFORM (WT)-BASED PQD DETECTION METHODS

Fourier transform (FT) methods are the most commonly used SP-based methods founded by J. Fourier. PQDs generally have stationary and non-stationary signal characteristics [30]. These methods do not allow the local analysis of frequency components, so they cannot use non-stationary signals. This problem is solved by the windowing process using a short-time Fourier transform (STFT) [31]. STFT has a fixed window size, and it does not contain decomposed low- and high-frequency components simultaneously [32, 33]. WT-based methods have flexible time-frequency representation, locally decompose discontinuities, and sudden changes in signal in high-grade derivatives where other signal processing methods fail to detect PQDs.

8.2.1 WAVELET TRANSFORM (WT)

WT, which is often used in pattern recognition applications, can decompose signals at different scales and resolutions. WT-based methods used in DG systems can be examined under four headings as continuous WT (CWT), discrete WT (DWT), wavelet packet transform (WPT), and undecimated WT (UWT). WT uses the mother wavelet $(\psi(x))$ function as a scalable window concept instead of the windowing concept used in the STFT. The CWT is using the parameter b which is the shifting factor, and the scaling factor is s. The CW function $(\psi_{s,b})$ is expressed as in Equation 8.1. CWT is determined as in Equation 8.2 by using the wavelet function [34]:

$$\psi_{s,b} = \frac{1}{\sqrt{s}} \psi\left(\frac{t-b}{s}\right) \tag{8.1}$$

$$CWT(s,b) = \frac{1}{\sqrt{s}} \int_{-\infty}^{\infty} f(t)\psi\left(\frac{t-b}{s}\right) dt \tag{8.2}$$

CWT method is used for limited real-time processes because of intensive computation. DWT is designed by using filter banks to remove the disadvantages caused by the computational load in CWT. DWT is performed to signal as in Equation 8.4, using $g_{j,k}(n)$ main wavelet function (Equation 8.3):

$$DWT(j,k) = \sum_{n \in Z} \sum_{k \in Z} S(n) g_{j,k}(n), \qquad g_{j,k} \in Z, j \in N, k \in Z \tag{8.3}$$

$$g_{j,k}(n) = a_0^{\frac{-j}{2}} g\left(a_0^j n - kb_0\right) \tag{8.4}$$

The relation between main wavelet function and $S(n)$ can be expressed as in Equation 8.5:

$$S(n) = \sum_{n \in Z} \sum_{k \in Z} d_{j,k} g_{j,k}(n) \tag{8.5}$$

The DWT applied signal is observed approximation coefficients (cA_n) and detail coefficients (cD_n). Equation 8.6 gives the formula of the wavelet coefficients:

$$cA[n] = \sum_k S(k) g[2n-k]; \qquad cD[n] = \sum_k S(k) h[2n-k] \tag{8.6}$$

The UWT method uses filter banks like DWT, but the coefficient lengths are the same as the original signal [17]. Detail and approximation coefficients are upsampled using multi-resolution analysis (MRA) by the "á trous" algorithm in the UWT method [35]. UWT can reduce oscillations and noise, and it is useful for selecting threshold values rather than DWT [36]. UWT detail and approach coefficients can be calculated by using Equation 8.7.

$$c_j^U[k] = \left(h_0^{(j)} * c_{j-1}^U[k]\right) = \sum_n c_{j-1}^U\left[k + 2^j n\right]h_0(k)$$

$$w_j^U[k] = \left(h_0^{(j)} * c_{j-1}^U[k]\right) = \sum_n c_{j-1}^U\left[k + 2^j n\right]h_1(k) \qquad (8.7)$$

8.2.2 Proposed DWT-based PQD Detection Method

In this part of chapter, a DWT-based method using threshold values of WTs to detect PQDs is proposed for smart grids. The developed MATLAB/Simulink model of a developed electrical power distribution system is shown in Figure 8.4 [30]. This system consists of a grid-connected PV plant, transformers, inductive and resistive loads, an induction motor, a capacitor bank, and three-phase non-linear loads. Table 8.1 shows the generated PQD signals with a developed model following IEEE standards [5,6]. Single-phase information is used for PQD events in the proposed model indicated in Figure 8.4.

The flowchart of the proposed DWT-based PQD detection method is shown in Figure 8.5. Firstly, the voltage signal of a developed grid-connected PV system is acquired at PCC. The two-level decomposition process is performed using the high-pass and low-pass filters with daubachies4 (db4) mother wavelet, which reduces the number of samples at each filter output. The selected db4 type contains a minimum sample and is a short wavelet type. Thus, it reduces the computational load. In the last stage, the algorithm decides the nominal condition or PQDs according to the determined threshold value. Different experimental studies have been conducted. The most appropriate parameters have been identified by selecting the type of wavelet, sampling frequency, wavelet level, and threshold value to be used in the proposed DWT-based method for PQD detection. PQD is detected when the specified coefficients of the d2 threshold value are exceeded. After the PQD detection, the event duration time and the overvoltage/undervoltage condition amplitude are determined WT coefficients [17]. The PQD conditions are detected by the proposed method correctly and in a short time. The voltage signal, detail coefficients, and approximation coefficients for PQD signals are shown in Figure 8.6.

8.3 AI-BASED PQD CLASSIFICATION METHODS

Fuzzy expert system-based classifiers, neural network (NN)-based classifiers, decision tree (DT) and random forest (RF) classifiers, SVM-based classifiers, and DL-based classifiers are AI-based methods for detecting PQDs.

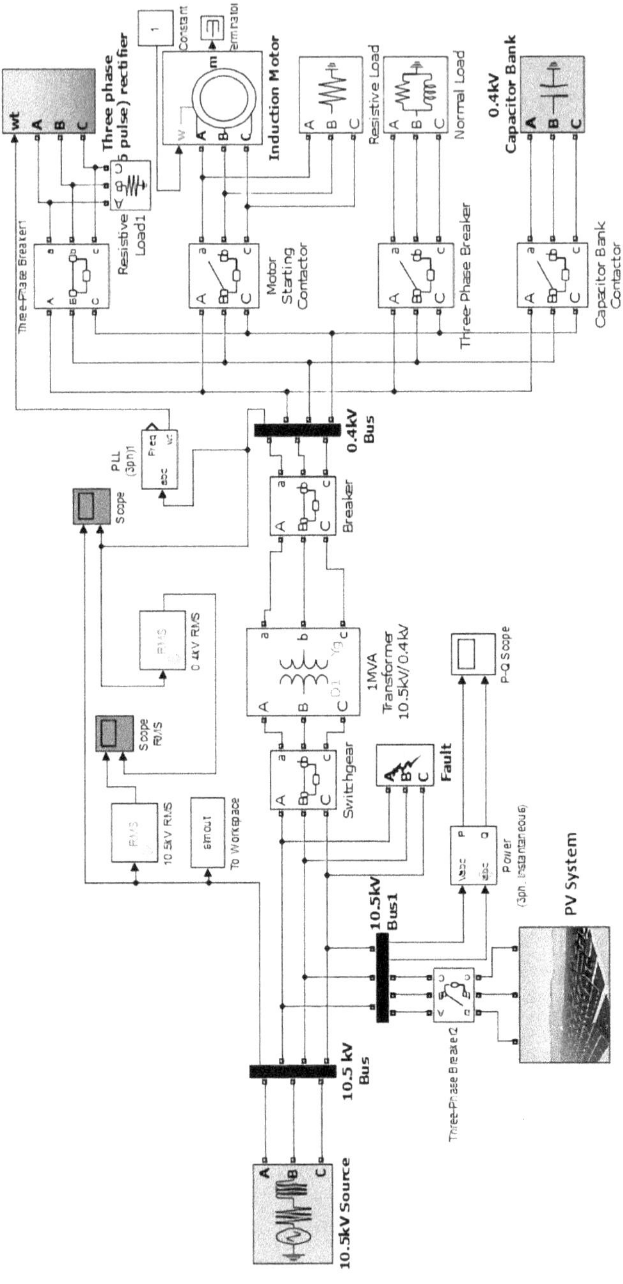

FIGURE 8.4 MATLAB/Simulink model of developed electrical power distribution system [30].

TABLE 8.1

PQD Events Obtained from Developed Model

PQD Event	Causes of Event
Voltage sag	Short-circuit faults, switching on a large load
Voltage swell	Short-circuit faults, switching of large capacitors
Voltage interruption	Short-circuit faults, component failures, CB tripping
Harmonics	Non-linear power electronic-based drivers, arc welders
Oscillatory transient	Switching of large capacitor bank and inductive loads
Notch	Non-linear power electronic-based drivers

In the training phase with ANN, PQD signals or extracted futures are given as input to the system, and the weight coefficients of each node in the model are determined. In the test phase, the unknown signal is provided as an input to the trained ANN system, and the class of test signal is decided according to the value calculated in the output node. As a result of each output layer, the probability of input signal belonging to that class is obtained.

SVM classifier, which is a classification method used successfully in many pattern recognition problems, was originally a proposed and defined classifier to distinguish the two classes. Multiple classifications are performed with the help of kernel function. The purpose of kernel functions is to move the features that cannot be separated linearly in the existing space into a higher dimensional space and to make the features linearly separable in this new high dimensional space. Decision tree (DT) is a supervised learning method, and it has tree-shaped architecture. DT establishes the relationship between the class and the attributes. Random forest (RF) is a further supervised learning method, and it has multiple tree-shaped architectures. RF classifier improves the generalization performance of DT.

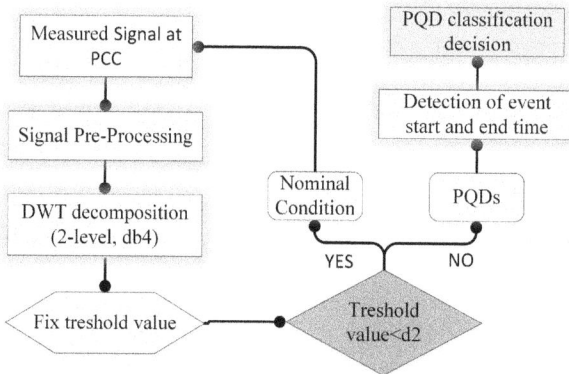

FIGURE 8.5 Proposed DWT-based PQD detection method.

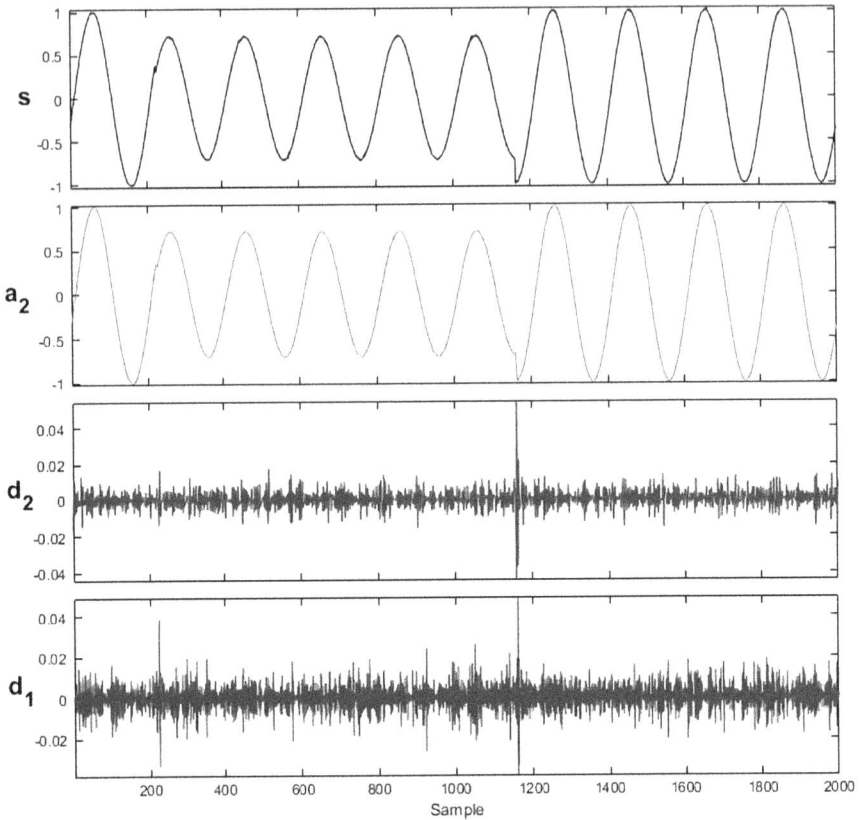

FIGURE 8.6 PQD detection using the proposed DWT-based method. (a) Voltage sag. (b) Voltage swell. (c) Voltage interruption. (d) Oscillatory transient.

8.3.1 Deep Learning Structures

Complex learning networks gradually replace the few layers used in regression and classification problems with developing information technologies. DL algorithms can be considered as multi-layers of ANNs.

A two-step procedure can be followed to train the deep neural network (DNN):

1. Initiation of unsupervised learning algorithms such as stacked auto-encoder (SAE) for determining weight values.
2. Fine-tuning of initiation weights using a supervised learning algorithm to provide better classification.

8.3.1.1 SAE-based Methods

AE structures provide better results than other AI methods in extracting features from data, removing unnecessary information, and compressing useful information. AE consists of an encoder and decoder network. Represented features are defined as

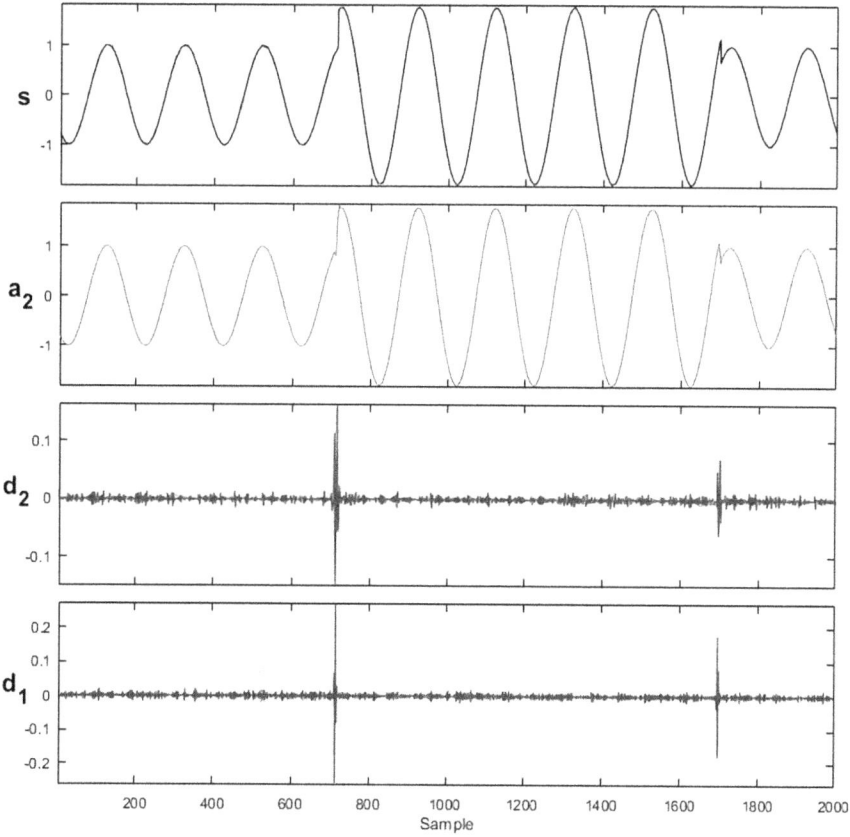

FIGURE 8.6 Continued.

Equation 8.8 using the SAE structure where σ is the activation function and θ is the training parameters:

$$h_i = f_0(x_i) = \sigma_f\left(w^T \cdot x_i + b\right) \tag{8.8}$$

The reconstructed sample $\hat{x}_i$ can be obtained by Equation 8.9:

$$\hat{x}_i = g_{\theta'}(h_i) = \sigma_g\left(w'^T \cdot h_i + b'\right) \tag{8.9}$$

A two-step procedure can be followed to train the DNN and SAE:

1. Initiation of unsupervised learning algorithms such as SAE for determining weight values.
2. Fine-tuning of initiation weights using a supervised learning algorithm to provide better classification.

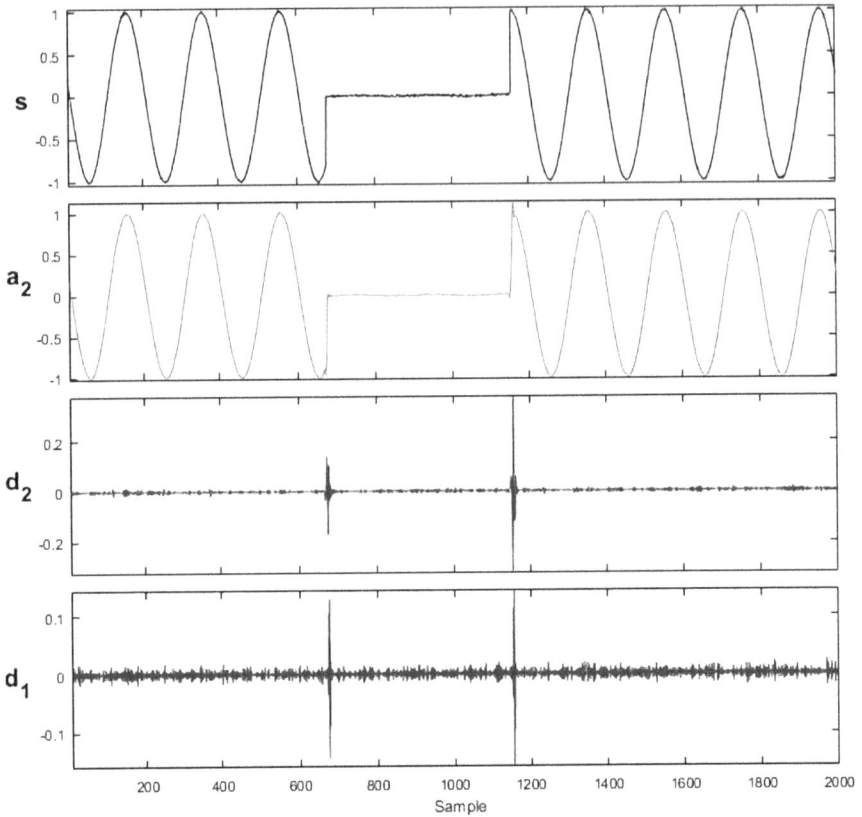

FIGURE 8.6 Continued.

8.3.1.2 DNN-based Methods

DNN algorithms can be considered as multilayers of ANNs.

8.3.1.3 DBN (Deep belief network)-based Methods

Restricted Boltzmann machine (RBM) is a type of generative stochastic neural network (GSNN) which has been widely used as greedy-layer wise pre-training strategy to train the DBN [37]. DBN-based methods could automatically learn features from the input data different from the SAE. DBN structures solve the vanishing problem using the backpropagation (BP) algorithm for fine-tuning DBNs. DBN maps the learned features into the label space by adding the classification layer to classify the PQDs or faults in microgrids.

8.3.1.4 CNN-based Methods

CNN is capable of capturing the shift-variant properties of input data [38]. CNN-based PQD classification methods are learned features from the raw data without using any signal processing methods compared to other DL structures. Different

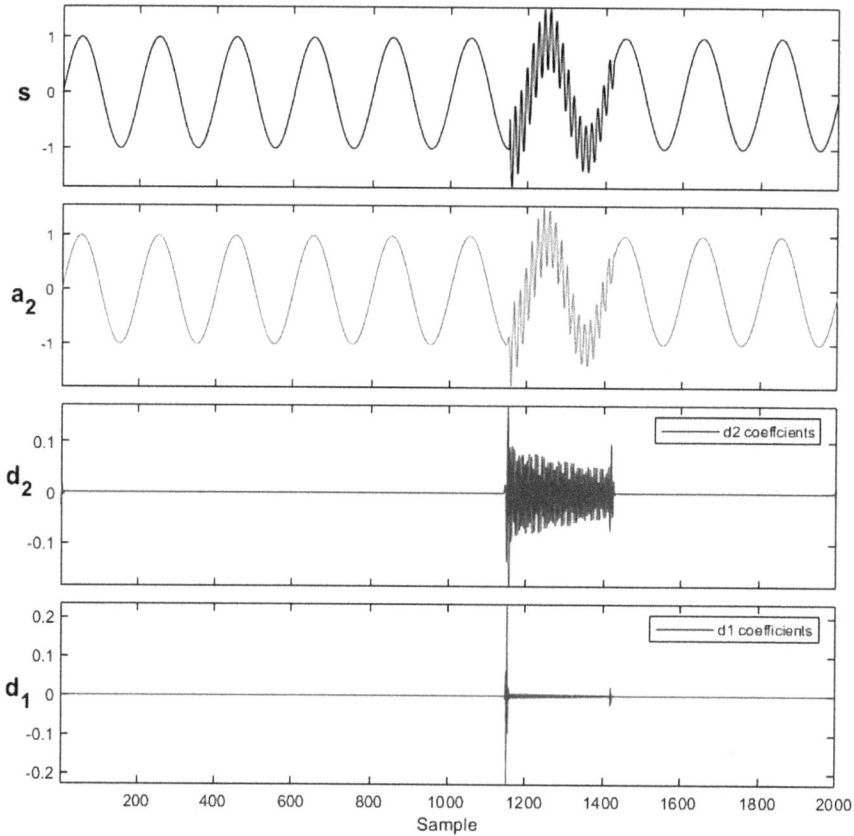

FIGURE 8.6 Continued.

from traditional methods, CNN methods were performed for detection and classification of PQDs without using an additional method of pre-processing, feature extraction, and classification. Besides, the number of training parameters in classification algorithms is reduced using CNN.

The signal processing and classification-based hybrid method and DCNN-based method are shown in Figure 8.7. DCNN has stacked units to extract features. Stacked units consist of pooling layers, convolutional layers, and batch-normalization layers.

8.3.2 Proposed Deep Learning and WT-based Hybrid PQD Classification Method

In this part of the chapter, a hybrid method based on DWT and feed-forward deep neural network (FDNN) is proposed for the classification of PQDs occurring in grid-connected PV systems. In the developed process, the nominal and PQD signals were subjected to five-level DWT decomposition, and then the energy values of the detail coefficients were calculated. The calculated energy values constitute the inputs of

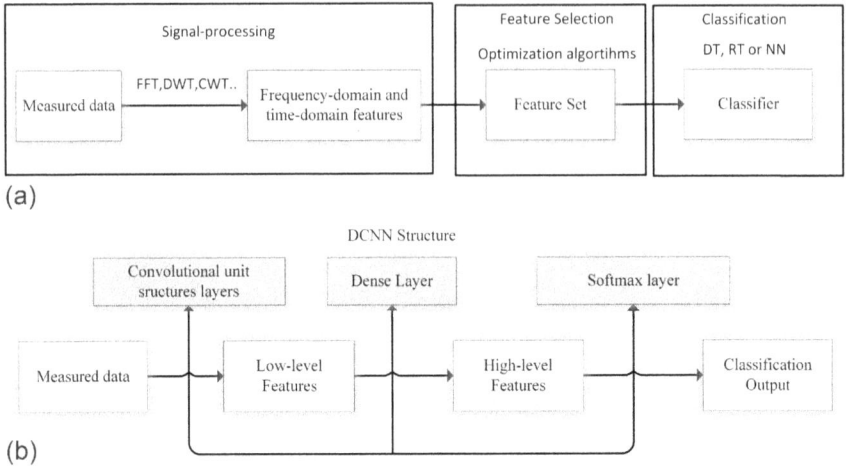

(a)

(b)

FIGURE 8.7 (a) Signal processing and classification-based hybrid method. (b) DCNN-based PQD classification method [27].

the multilayer FDNN. At the last stage, PQDs were classified using the softmax activation function. Figure 8.8 shows the flow diagram of the proposed method. The classified power quality events by proposed FDNN-based method are as follows:

- Class 1 – Nominal signal
- Class 2 – Voltage sag
- Class 3 – Voltage swell
- Class 4 – Voltage interruption
- Class 5 –Harmonics
- Class 6 – Flicker
- Class 7 – Oscillatory transient
- Class 8 – Impulsive transient
- Class 9 – Notch

Several parameters for PQD signals:

- The fundamental frequency value is 50 Hz, and the sampling frequency is 10 kHz.
- The number of cycles of fundamental frequency is ten cycles (0.2 s–2000 points).
- The number of samples for each class is 2000.
- The amplitude of the PQD signals is per unit (p.u.).

Most of the available literature has gone into using a mathematical model or simulation to build data sets due to the difficulty of obtaining real signals related to PQD. In this study, the integral-based mathematical model was used to generate PQD signals

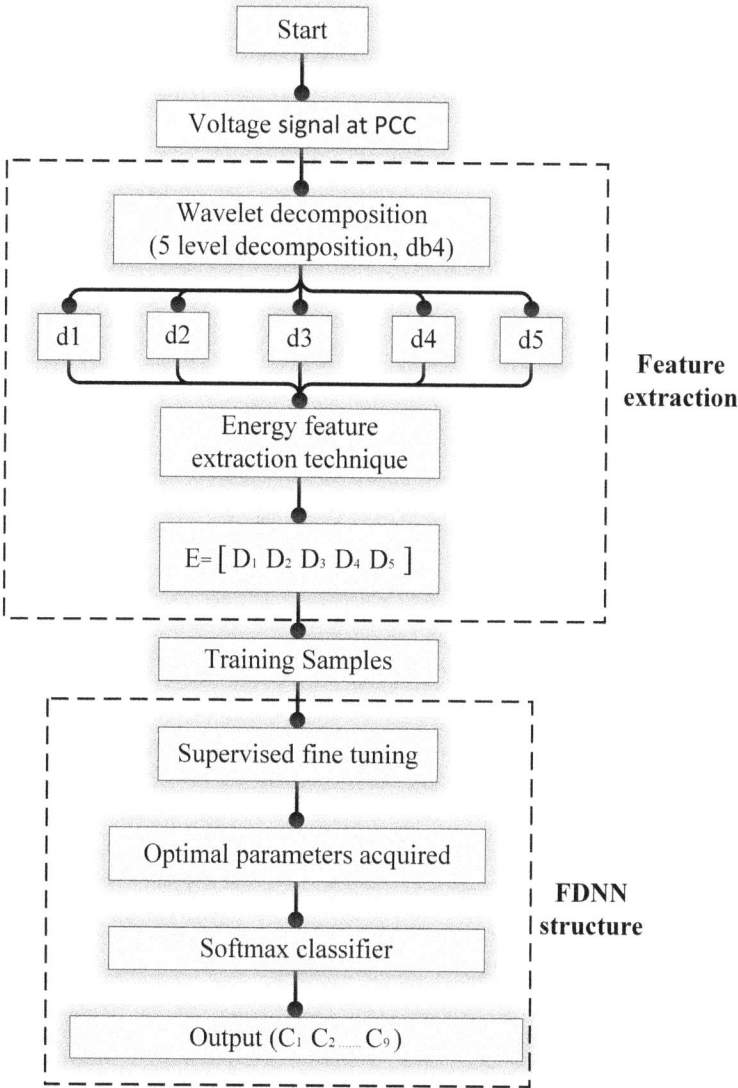

FIGURE 8.8 FDNN-based model flowchart.

[39, 40]. Train, validation, or test data sets can be generated randomly by the software created based on the equations specified in PQD parameters, possible values in Table 8.2. Mathematical definitions of five of the most common PQDs mentioned in IEEE standards are given in Table 8.2. Other PQDs were generated with the help of an integral-based mathematical model.

A five-level decomposition process was performed using the high-pass and low-pass filters with daubachies4 mother wavelet, and then the energy values of the detail

TABLE 8.2

Mathematical Model of PQDs [39]

Class	PQD Signals	Equations	Threshold Parameters
C1	Nominal	$x(t) = \sin(2\pi ft - \varphi)$	$49.8 \le f \le 50.2, -\pi \le \varphi \le \pi$
C2	Sag	$x(t) = \left[1 - \alpha\left(u(t-t_1) - u(t-t_2)\right)\right]\sin(2\pi ft - \varphi)$	$T \le t_2 - t_1 \le 9T$ $0.1 \le \alpha < 0.9$
C3	Swell	$x(t) = \left[1 + \beta\left(u(t-t_1) - u(t-t_2)\right)\right]\sin(2\pi ft - \varphi)$	$T \le t_2 - t_1 \le 9T$ $0.1 \le \beta < 0.8$
C5	Transient	$x(t) = \sin(2\pi ft - \varphi) + \beta\exp\left(-(t-t_1)/\tau \sin\left(2\pi f_n(t-t_1) - \vartheta\right)\right)\left(u(t-t_2) - u(t-t_1)\right)$	$300\,\text{Hz} \le f_n \le 900\,\text{Hz};$ $8\,\text{ms} \le \tau \le 40\,\text{ms}; \ -\pi \le \vartheta \le \pi x$ $0.5T \le t_2 - t_1 \le 3T$
C7	Harmonics	$x(t) = \sin(2\pi ft - \varphi) + \displaystyle\sum_{n=3}^{7} \alpha_n \sin(n\pi ft - \vartheta_n)$	$n = \{3,5,7\};\ 0.05 \le \alpha_n \le 0.15$ $-\pi \le \vartheta_n, \vartheta_n', \vartheta_n'' \le \pi$ $n' = \{3,5,7\};\ 0.05 \le \alpha_{n'} \le 0.15$ $n'' = \{1,3,5\}$ $\alpha_{n'} = 1 \mid n'' = 1; 0.05 \le \alpha_{n'} \le 0.15 \mid n'' = \{3,5\}$

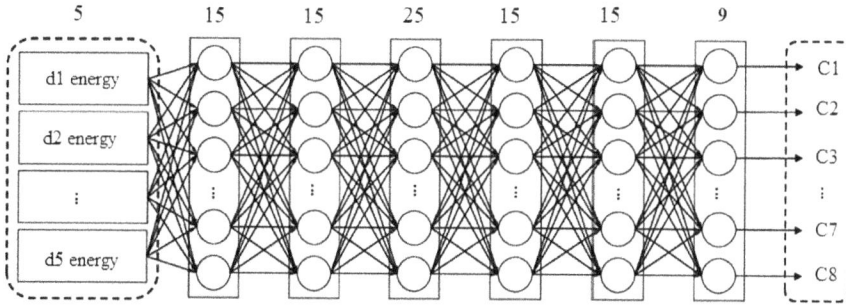

FIGURE 8.9 FDNN structure.

coefficients were calculated. FDNN structure is shown in Figure 8.9. Energy values of five detail coefficients are given as input to the network. At the output of the system, there are labelled 8 class outputs determined.

The energy feature extraction technique is applied to DWT coefficients of the single-phase PQD signal. The energy feature extraction technique is used to DWT coefficients of the single-phase PQD signal. The energy of the DWT coefficients is found in Equation 8.10 to represent the level of j: decomposition, d: detail coefficients, N: a number of detail coefficients.

$$E_j = \sum_{n=1}^{N} |D_{jn}| \qquad j = 1,\ldots,l \tag{8.10}$$

8.4   RESULTS

The performance scores of FDNN are given in Table 8.3. FDNN classifier has the best score with a training accuracy (Train Acc) of 100% and with a test accuracy

TABLE 8.3
Detailed Performance Report for FDNN Classifier Test Set

Classes	Precision	Recall	F1 Score	Support
C1	95.17%	99.11%	97.10%	338
C2	100%	97.20%	98.76%	321
C3	99.64%	97.89%	98.55%	285
C4	99.00%	100%	99.50%	298
C5	100%	100%	100%	301
C6	100%	100%	100%	298
C7	99.66%	99.66%	99.66%	292
C8	99.06%	98.13%	98.59%	321
C9	99.64%	99.64%	99.64%	290
Accuracy			99.09%	2744
Macro avg	99.17%	99.11%	99.13%	2744

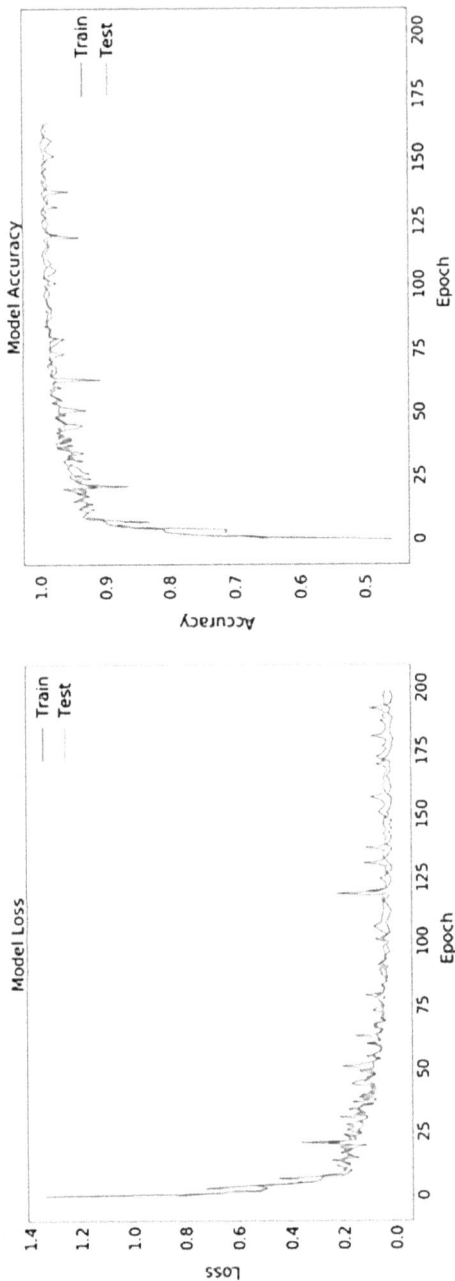

FIGURE 8.10 Learning curves.

TABLE 8.4
Performance Comparison of Existing Methods

Method	No. of Futures	Number of PQDs	Accuracy
Reference [23]	5	11	98.2%
Reference [41]	5	14	97.28%
Reference [42]	19	9	99.66%
Reference [43]	10	5	98.51%
Reference [44]	45	22	99.09%
Reference [45]	2	10	94.2%
Reference [21]	6	9	97.22%
Proposed method	5	9	99.13%

(Test Acc) of 99.13%. Training and test set were randomly distributed, and the test set was determined to be approximately 30%.

The training lasted 200 Epoch – about 608 sec. Learning curves are shown in Figure 8.10.

As shown in Table 8.4, the PQDs classification accuracy corresponding to the WT and FDNN-based algorithm is higher than all the other investigated methods.

8.5 CONCLUSION

This chapter presents SP-based and AI-based PQD detection and classification methods used in smart grids. In Section 8.1, SP-based PQD detection methods used in smart grids are classified, and then wavelet transform (WT)-based detection methods are presented. Current WT-based methods were introduced for the detection of PQDs. Some PQD events are investigated in MATLAB Simulink by using the developed discrete WT method.

The classification of PQDs with AI-based methods are also discussed in the last section. Deep neural networks are presented in the chapter to show the effectiveness of AI-based methods. A hybrid method based on DWT and feed-forward deep neural network (FDNN) is proposed for the classification of PQDs occurring in grid-connected PV systems. In the developed process, the nominal and PQD signals were subjected to five-level DWT decomposition, and then the energy values of the detail coefficients were calculated. The calculated energy values constitute the inputs of the multilayer FDNN. At the last stage, PQDs were classified using the softmax activation function. The results show that the proposed hybrid WT and FDNN methods have 99.13% accuracy in classifying PQDs.

The proposed detection and classification methods for detecting PQDs are suitable for smart grids, and the obtained results in this chapter will contribute to selecting the proper detection or classification methods of PQDs using AI-based methods in smart grids.

REFERENCES

1. Biswas, S., Goswami, S. K., & Chatterjee, A. (2012). Optimum distributed generation placement with voltage sag effect minimization. *Energy Conversion and Management*, 53(1), 163–174.
2. Hung, D. Q., Mithulananthan, N., & Bansal, R. C. (2010). Analytical expressions for DG allocation in primary distribution networks. *IEEE Transactions on Energy Conversion*, 25(3), 814–820.
3. Bollen, Math H. J. (2000). Understanding power quality problems. In *Voltage Sags and Interruptions*. IEEE Press.
4. Chapman, David. (2001). The cost of poor power quality. *Power Quality Application Guide* 1–4.
5. IEEE. Std. 929 (2000). *IEEE Recommended Practice for Utility Interface of Photovoltaic (PV) Systems*. Institute of Electrical and Electronics Engineers, Inc., New York.
6. F II I (1993). *IEEE Recommended Practices and Requirements for Harmonic Control in Electrical Power Systems*. New York.
7. IEEE (2014). *Std. IEEE Recommended Practices and Requirements for Harmonic Control in Electrical Power Systems*, 519.
8. Chen, C. I., & Chen, Y. C. (2014). Intelligent identification of voltage variation events based on IEEE std 1159–2009 for SCADA of distributed energy system. *IEEE Transactions on Industrial Electronics*, 62(4), 2604–2611.
9. Mahela, O. P., Shaik, A. G., & Gupta, N. (2015). A critical review of detection and classification of power quality events. *Renewable and Sustainable Energy Reviews*, 41, 495–505.
10. Khokhar, S., Zin, A. A. B. M., Mokhtar, A. S. B., & Pesaran, M. (2015). A comprehensive overview on signal processing and artificial intelligence techniques applications in classification of power quality disturbances. *Renewable and Sustainable Energy Reviews*, 51, 1650–1663.
11. Latran, M. B., & Teke, A. (2015). A novel wavelet transform based voltage sag/swell detection algorithm. *International Journal of Electrical Power and Energy Systems*, 71, 131–139.
12. Karegar, H. K., & Sobhani, B. (2012). Wavelet transform method for islanding detection of wind turbines. *Renewable Energy*, 38(1), 94–106.
13. Mohammadnian, Y., Amraee, T., & Soroudi, A. (2019). Fault detection in distribution networks in presence of distributed generations using a data mining–driven wavelet transform. *IET Smart Grid*, 2(2), 163–171.
14. Bayrak, G. (2018). Wavelet transform-based fault detection method for hydrogen energy-based distributed generators. *International Journal of Hydrogen Energy*, 43(44), 20293–20308.
15. Deokar, S. A., & Waghmare, L. M. (2014). Integrated DWT–FFT approach for detection and classification of power quality disturbances. *International Journal of Electrical Power and Energy Systems*, 61, 594–605.
16. Lopez-Ramirez, M., Cabal-Yepez, E., Ledesma-Carrillo, L. M., Miranda-Vidales, H., Rodriguez-Donate, C., & Lizarraga-Morales, R. A. (2018). FPGA-based online PQD detection and classification through DWT, mathematical morphology, and SVD. *Energies*, 11(4), 769.
17. Yılmaz, A., & Bayrak, G. (2019). A real-time UWT-based intelligent fault detection method for PV-based microgrids. *Electric Power Systems Research*, 177. http://www.ncbi.nlm.nih.gov/pubmed/105984.
18. Dhimish, M., Holmes, V., Mehrdadi, B., & Dales, M. (2018). Comparing Mamdani Sugeno fuzzy logic and RBF ANN network for PV fault detection. *Renewable Energy*, 117, 257–274.

19. Bangalore, P., & Patriksson, M. (2018). Analysis of SCADA data for early fault detection, with application to the maintenance management of wind turbines. *Renewable Energy*, 115, 521–532.
20. Zhang, L., Zhou, W., & Jiao, L. (2004). Wavelet support vector machine. *IEEE Transactions on Systems, Man, and Cybernetics, Part B (Cybernetics)*, 34(1), 34–39.
21. Thirumala, K., Pal, S., Jain, T., & Umarikar, A. C. (2019). A classification method for multiple power quality disturbances using EWT based adaptive filtering and multiclass SVM. *Neurocomputing*, 334, 265–274.
22. Puliyadi Kubendran, A. K., & Loganathan, A. K. (2017). Detection and classification of complex power quality disturbances using S-transform amplitude matrix–based decision tree for different noise levels. *International Transactions on Electrical Energy Systems*, 27(4), e2286.
23. Lee, C. Y., & Shen, Y. X. (2011). Optimal feature selection for power-quality disturbances classification. *IEEE Transactions on Power Delivery*, 26(4), 2342–2351.
24. Khokhar, S., Zin, A. A. M., Memon, A. P., & Mokhtar, A. S. (2017). A new optimal feature selection algorithm for classification of power quality disturbances using discrete wavelet transform and probabilistic neural network. *Measurement*, 95, 246–259.
25. Kong, X., Xu, X., Yan, Z., Chen, S., Yang, H., & Han, D. (2018). Deep learning hybrid method for islanding detection in distributed generation. *Applied Energy*, 210, 776–785.
26. Liu, H., Hussain, F., Shen, Y., Arif, S., Nazir, A., & Abubakar, M. (2018). Complex power quality disturbances classification via curvelet transform and deep learning. *Electric Power Systems Research*, 163, 1–9.
27. Wang, S., & Chen, H. (2019). A novel deep learning method for the classification of power quality disturbances using deep convolutional neural network. *Applied Energy*, 235, 1126–1140.
28. Balouji, E., & Salor, O. (2017). Classification of power quality events using deep learning on event images. In *2017 3rd International Conference on Pattern Recognition and Image Analysis (IPRIA)*, 216–221. IEEE.
29. Mohan, N., Soman, K. P., & Vinayakumar, R. (2017). Deep power: Deep learning architectures for power quality disturbances classification. In *2017 International Conference on Technological Advancements in Power and Energy (TAP Energy)*, 1–6. IEEE.
30. Bayrak, G., & Yilmaz, A. (2019). Assessment of power quality disturbances for grid integration of PV power plants. *Sakarya University Journal of Science*, 23(1), 1–1.
31. Heydt, G. T., Fjeld, P. S., Liu, C. C., Pierce, D., Tu, L., & Hensley, G. (1999). Applications of the windowed FFT to electric power quality assessment. *IEEE Transactions on Power Delivery*, 14(4), 1411–1416.
32. Cohen, L. (1995). *Time-Frequency Analysis*. Prentice-Hall.
33. Samantaray, S. R., Samui, A., & Babu, B. C. (2011). Time-frequency transform-based islanding detection in distributed generation. *IET Renewable Power Generation*, 5(6), 431–438.
34. Strang, G., & Nguyen, T. (1996). *Wavelets and Filter Banks*. Wellesley-Cambridge Press.
35. Shensa, M. J. (1992). The discrete wavelet transform: Wedding the a trous and Mallat algorithms. *IEEE Transactions on Signal Processing*, 40(10), 2464–2482.
36. Zafar, T., & Morsi, W. G. (2013). Power quality and the un-decimated wavelet transform: An analytic approach for time-varying disturbances. *Electric Power Systems Research*, 96, 201–210.
37. Hinton, G. E., & Salakhutdinov, R. R. (2006). Reducing the dimensionality of data with neural networks. *Science*, 313(5786), 504–507.

38. Lei, Y., Yang, B., Jiang, X., Jia, F., Li, N., & Nandi, A. K. (2020). Applications of machine learning to machine fault diagnosis: A review and roadmap. *Mechanical Systems and Signal Processing*, 138. http://www.ncbi.nlm.nih.gov/pubmed/106587.

39. Igual, R., Medrano, C., Arcega, F. J., & Mantescu, G. (2018). Integral mathematical model of power quality disturbances. In *2018 18th International Conference on Harmonics and Quality of Power (ICHQP)*, 1–6. IEEE.

40. Igual, R., Medrano, C., Arcega, F. J., & Mantescu, G. (2017). *Mathematical Model of Power Quality Disturbances*. Mendeley Data, 1.

41. Thirumala, K., Prasad, M. S., Jain, T., & Umarikar, A. C. (2016). Tunable-Q wavelet transform and dual multiclass SVM for online automatic detection of power quality disturbances. *IEEE Transactions on Smart Grid*, 9(4), 3018–3028.

42. Abdoos, A. A., Mianaei, P. K., & Ghadikolaei, M. R. (2016). Combined VMD-SVM based feature selection method for classification of power quality events. *Applied Soft Computing*, 38, 637–646.

43. Erişti, H., & Demir, Y. (2012). Automatic classification of power quality events and disturbances using wavelet transform and support vector machines. *IET Generation, Transmission and Distribution*, 6(10), 968–976.

44. Dalai, S., Dey, D., Chatterjee, B., Chakravorti, S., & Bhattacharya, K. (2014). Cross-spectrum analysis-based scheme for multiple power quality disturbance sensing device. *IEEE Sensors Journal*, 15(7), 3989–3997.

45. Borrás, M. D., Bravo, J. C., & Montaño, J. C. (2016). Disturbance ratio for optimal multi-event classification in power distribution networks. *IEEE Transactions on Industrial Electronics*, 63(5), 3117–3124.

9 Robust Design of Artificial Neural Network Methodology to Solve the Inverse Kinematics of a Manipulator of 6 DOF

Ma. del Rosario Martínez-Blanco,
Teodoro Ibarra-Pérez, Fernando Olivera-Domingo,
and José Manuel Ortiz-Rodríguez

CONTENTS

9.1   INTRODUCTION

With recent advances in electronics, mechanics, computer science, and robotics, robotic manipulators have become the main focus of interest for the development of industrial applications [1]. Among the multiple articular configurations available, generally, the manipulators that have more degrees of freedom (DOF) offer greater control flexibility for complicated tasks; so the scientific community has taken an interest in this type of robotic manipulators during the last decades [2–4].

During the last years, several investigations have been carried out in the field of robotics due to its high impact in various areas such as space exploration [5], industrial [6], military [7], and medical applications [8], among others. The use of these manipulators has contributed considerably to the development and application of new methods and technologies applied to the field of robotics, mainly in the control of movement in real time [9].

In the field of robotics manipulators, the main challenge of motion control is to find a precise and reliable solution for inverse kinematics. The calculation necessary for the resolution of inverse kinematics requires expensive processing and is computationally complex [10].

In general, the solutions of the inverse kinematics of a robotic manipulator are based on three methods: the geometric, iterative, and analytical or algebraic methods. The singularities and uncertainties, or configurations where manipulator mobility is reduced, are the main complications in the aforementioned methods for the control of movement in robotic manipulators. For example, for the algebraic method, closed-form solutions are not guaranteed and such solutions must exist for the first three joints if the geometric method is used. In the case of an iterative method, it converges to a single solution that depends on the initial point; so it requires high-performance hardware without guaranteeing the accuracy of the calculations [3, 11].

Therefore, the traditional methods used in inverse kinematics problems with geometric, iterative, and algebraic approaches are sometimes complex and generally unsuitable for multiple joint configurations that can present robotic manipulators nowadays. As a consequence, various approaches based on artificial neural networks (ANNs) have been proposed due to the great advantages they present in their parallel distribution, non-linear mapping, the ability to learn through examples, and the high performance in the ability to generalization, among others [4, 12].

Generally speaking, the application of ANNs has reduced the error in controlling movement in real time, allowing increasing accuracy [8]. A well-trained neural network can work with very fast response times, so it is ideal for real-time applications, compared to other conventional methods where the response time is longer [13].

Most of the research in ANNs has focused on specific applications of model development and training algorithms to improve the convergence and accuracy of the results obtained; however, finding the parameters to achieve adequate learning in the training of neural networks remains a difficult and complex task [9, 10, 14–17].

In general, the quality in the development and training of ANNs is highly reliable; however, structural parameters, just like the number of hidden layers and neurons per hidden layer, represent an important role in the accuracy of the expected results and these are usually proposed due to the previous experience of the researcher in trial and error procedures, consuming time and resources without guaranteeing that the optimal configuration of the parameters to achieve a better performance in the neural network is obtained [3, 10, 18–21].

In this work, the use of a robust design methodology based on the fractional factorial design of experiments is proposed to get the optimal parameters of an ANN architecture for the calculation of the robotic manipulator of six DOF.

The chapter is organized as follows: First, the introduction is addressed, where the bases of the kinematics of robotic manipulators are presented, as well as those of the artificial neural networks, to approach the solution of the inverse kinematics solution with artificial neural networks. The robust design of artificial neural networks applied in this chapter is also addressed in the introduction. In the second section, the proposed methodology and its different phases are presented. The third section describes Ketzal – the robotic manipulator used, performing a kinematic analysis of it, which allows obtaining a data set to be used in training. The data set description and the reduction data filter algorithm used are also presented, followed by data set analysis of training and test. Finally, the different phases of the proposed methodology are addressed in Planning and Experimentation Stage and Analysis and Confirmation Stage subsections. In the fourth section, the conclusions are analyzed and finally, in the fifth section, the future work is presented.

9.1.1 KINEMATICS OF ROBOTIC MANIPULATORS

The morphology of robotic manipulators refers to the description of the components, parts, and mechanical structure [22]. A robotic manipulator is usually made using consecutive rigid mechanisms called links and connected by joints to form an open kinematic chain that allows the robot to move [23].

A robot can be modelled as an articulated open chain of several rigid links connected in series by prismatic or rotating joints actioned by actuators. The interconnection between two links consecutively represents an articulation [4, 21].

From the mechanical point of view, it is an open kinematic chain as long as each of the links is connected through joints to the next and previous links, except for the first and last, where the first is usually always fixed to support and the latter is usually free, that is to say, that a closed loop is not formed, otherwise it would be a closed kinematic chain [22].

The last link is usually a terminal element as a clamping tool or final actuator. Each of the joints forms a DOF. The joints can produce rotational and linear movement, where the latter ones are called linear prismatic joints [22].

The kinematics of a manipulator can be represented analytically as the description of the movement of the robot arm as a function of time according to a fixed reference system without considering the forces/torques that cause them, that is, the dynamics of the system [24, 25].

The kinematics of the manipulator describes the relationship between the joint angle, the position, and the orientation of the robot's end. The robot kinematics analysis includes the solution of two aspects: direct kinematics and inverse kinematics [3], as shown in Figure 9.1.

Direct kinematics consists of obtaining the position and orientation of the robot's end given the value of the articular coordinates. The inverse kinematics consists in determining the configuration that the robot must adapt for a known position

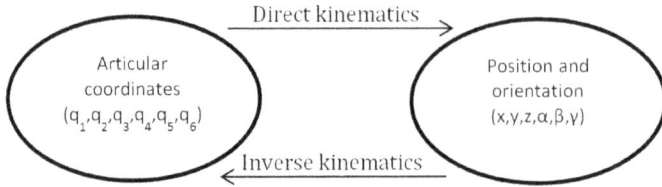

FIGURE 9.1   Direct and inverse kinematics.

and orientation of its end, so the complexity in the solution to solve this problem is mainly due to the manipulator geometry, being able to obtain multiple solutions and singularities that do not always correspond to physical solutions [12, 24].

In this work, the direct kinematics model applied to a six DOF manipulator was developed, obtaining a set of data on position, orientation, and joint values corresponding to the robotic manipulator's workspace. A filter was used to decrease the volume of the data and the methodology of Robust Design of Artificial Neural Networks (RDANN) was applied to obtain the optimal parameters of an ANN architecture to solve the problem of inverse kinematics for the proposed manipulator.

9.1.2 ARTIFICIAL NEURAL NETWORKS

An artificial neuron represents the mathematical model of the functioning of a biological neuron. The ANNs are inspired in the functioning of the human brain that processes the information very efficiently, particularly these represent a computational structure inspired by the biological neurons in our brains that make complex systems and can learn by experience [26].

A standard neural network consists of the interconnection of several simple processors called neurons, each producing an activation. Input neurons receive information from the environment and also get activated by weighted connections from previously active neurons [27].

Achieving learning in a neural network consists of finding the synaptic weights between the interconnections to minimize the cost function and ensure that the neural network shows the desired behaviour. The cost or loss function is regularly optimized with the gradient descent method. Although the method is very effective, a globally optimized solution is not always guaranteed because it can fall to a local minimum [20].

An artificial neuron can be conceived as a mathematical model of the functioning of a biological neuron, as shown in Figure 9.2, where the inputs x_i represent the

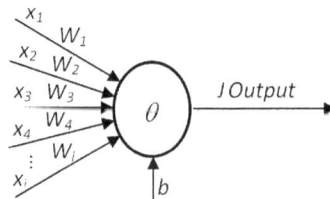

FIGURE 9.2   Artificial neuron model.

signals captured by the dendrite that come from another neuron, where the weights W_i represent the intensity of the synapse that connects two neurons that may be positive (excitatory) or negative (inhibitory). θ is the transfer function or threshold that the neuron must overcome to enter the activation state. Finally, the accumulated sum in the node from all the input signals and multiplied by the synaptic weights go to the output signal through the transfer function or activation function [28].

In essence, an artificial neuron mimics the behaviour of a biological neuron, that is, it receives varying signals from other neighbouring neurons and can easily process the information from the other neurons to let it pass to another neuron with a certain level of energy [29].

The signals that enter the neuron are weighted by the multiplication of weights, which represents a process similar to the strength of the synaptic connection in a biological neuron; the weighting is added by a neural node and the output is calculated as the sum of the weighted inputs plus a value b called bias. The output of the neuronal node is used as input to a transfer function that responds to the artificial neuron [30]. The output signal of a neuron *net* can be represented as described below.

Figure 9.3 shows the simplification of an artificial neuron model, where it can be observed that the inputs to a neuron, x_j $(j = 1, 2, ..., n)$, are represented by the input array component, the synaptic weights are represented by the matrix $W_{i,j}$ $(j = 1, 2, ..., n)$ and b represents the bias. The response of the artificial neuron to the input signals can be described mathematically as follows:

$$y_i = f(net) = f\left(\sum_{i=1}^{n} x_j w_{ij} + b \right) \qquad (9.1)$$

where y_i is the value obtained at the end of the artificial neuron and f is the transfer function used. Usually, a neuron can have one or more inputs.

A model of neurons with multiple inputs is shown in Figure 9.4. In this diagram, the inputs $x_1, x_2, ..., x_i$ are shown, which are multiplied by the corresponding weights $w_{1,1}, w_{1,2}, ..., w_{i,j}$ belonging to the synaptic weight matrix. The subscripts of the matrix represent the neurons involved in the link, where the first subscript represents the target neuron and the second represents the source of the input signal to the target neuron [14]. For example, $w_{2,5}$ indicates that this weight is the connection from the fifth input and the second neuron.

$$net = \sum_{j=1}^{n} x_j w_{ij} + b$$

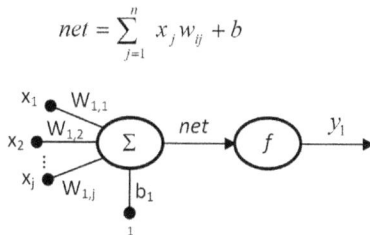

FIGURE 9.3 Artificial neuron model with transfer function.

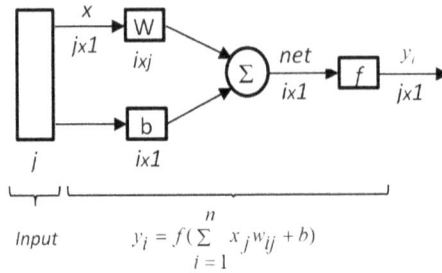

$$y_i = f(\sum_{i=1}^{n} x_j w_{ij} + b)$$

FIGURE 9.4 Neuron model with multiple inputs.

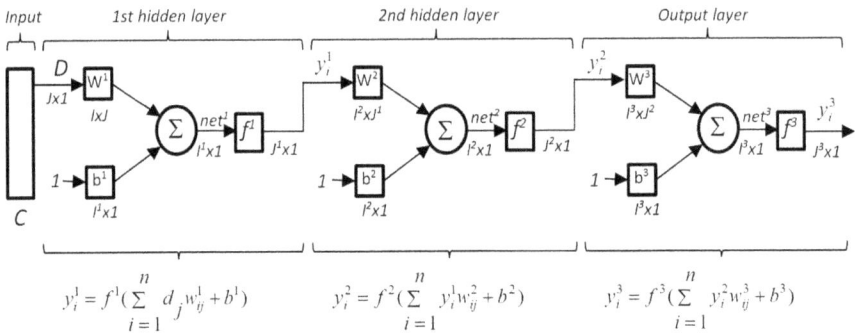

$$y_i^1 = f^1(\sum_{i=1}^{n} d_j w_{ij}^1 + b^1) \qquad y_i^2 = f^2(\sum_{i=1}^{n} y_i^1 w_{ij}^2 + b^2) \qquad y_i^3 = f^3(\sum_{i=1}^{n} y_i^2 w_{ij}^3 + b^3)$$

FIGURE 9.5 Architecture with two hidden layers.

An ANN is a set of interconnected neurons distributed in parallel that acts as an information processing system. Networks achieve learning due to an external source or set of training data by making changes in their structure to predict linear or non-linear trends [18].

Generally, an ANN is formed by the interconnection of several neurons. The form of connection between neurons varies according to the type of network, but they are usually grouped by layers that can be classified according to the location of the layer in the network as the input layer, hidden layer, and output layer. The architecture of the neural network refers to the arrangement of neurons within layers and the patterns of connection between them [28].

A simplified descriptive model of an ANN with three layers is shown in Figure 9.5, where it is noticeable that the network has R^1 inputs, i^1 neurons in the first layer, i^2 neurons in the second layer, and i^3 neurons in the third layer. An input constant is known as bias and a value of 1 is added to each neuron. Each intermediate output connects to the input of the next layer. In this way, a layered approach can be taken to analyze the entire network [29].

According to the existence of feedback connections or not in the neural network, two architectures can be described: forward propagation (known as perceptron) and backpropagation. The first architecture has no feedback and does not maintain a record of the previous values of its output and the activation state of the neurons. The second architecture has a closed-loop connection and therefore it maintains a record of its preceding states, so the next state depends on the preceding states in addition to the input signals [28].

The Artificial Neural Propagation Network (ANNBP) was proposed by Rumelhart and McClelland in 1986 and it is a widely used algorithm. The fundamental architecture of the network consists of three layers: the input layer, the hidden layer, and the output layer. For such networks, there are no restrictions on the number of hidden layers and has a solid mathematical foundation based on the algorithm of the gradient descent. The backpropagation refers to the method in which the synaptic weights between the connections change after being processed by the neuron layer through an activation function, producing outputs for the next layer [20, 28, 31].

The problem with this type of network is mainly related to the parameters that must be initially established before performing any training. The user must choose, according to their experience, the network architecture to be used and determine several of the parameters of the chosen network. These parameters determine success or failure in network training. There is currently no procedure that fully guarantees the optimal configuration of these parameters [14].

Currently, most of the selection of these parameters is based on the practice of trial and error, so a large amount of ANN models are developed and compared to choose the best model, as described in Figure 9.6. If the value of a parameter is modified and it does not have a significant effect on the network performance, another parameter is changed to improve network performance until the parameters that guarantee the desired performance are found for a specific problem.

The main disadvantage of this procedure is that while one parameter is evaluated, the others remain at a single level. Consequently, the best parameter evaluated in a particular design variable may not necessarily be the best parameter for the final model. Regularly, this practice does not guarantee the interaction and effects of the variables between their different levels, because it only combines one parameter at the same time [32].

This problem has motivated researchers in the search for design solutions and strategies to optimize design parameters in ANNs. One of the alternatives for this

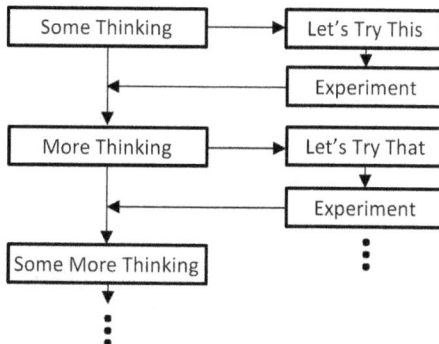

FIGURE 9.6 The traditional approach based on trial and error.

problem could be to evaluate all possible combinations of levels in the variables involved, that is, a complete factorial design. This practice could be complex and computationally expensive due to the multiple combinations that could be generated even for a small number of parameters and levels [33].

However, the number of experiments could be substantially reduced with the fractional factorial method, which is a statistical method based on the robust design of the Taguchi philosophy. This method makes it possible to find the optimal adjustment parameters, involving design and noise variables, allowing the noise factors involved in the design to be insensitive during the process. This technique, based on the philosophy of robust design, is a very powerful optimization method that differs from traditional practices [14, 34].

9.1.3 Inverse Kinematics Solution with Artificial Neural Networks

In recent years, the ANNs have contributed significantly to the development of various fields of engineering and science [10, 35–38]. The main interest in the use of ANNs is due to its characteristics related to non-linearity, high robustness, parallelism, fault tolerance, and its great capacity for learning and generalization, based on the learning process through complex, non-linear, and multiple examples of input and output relations [13].

The multilayer perceptron (MLP) algorithm trained with backpropagation (BP) is one of the most used techniques in modelling, classification, and optimization applications. The cost or loss function is regularly optimized with the gradient descent method; although the algorithm is very effective, it tends to converge very slowly and its main disadvantage is that an optimal and global solution is not always guaranteed, because the gradient descent method runs the risk of getting caught in a local minimum [14, 16, 39].

The quality in the data sets is another of the most important factors to consider in the training of ANNs, because a low quality generally leads to low quality in the knowledge extracted, so the efficiency in the algorithms of extraction of the knowledge largely depends on the quality of the data sets [40].

In general, the quality in the development and training of artificial neural networks is highly reliable; however, structural parameters, just like the number of hidden layers and neurons per hidden layer, represent an important role in the accuracy of the expected results and generally these are usually proposed based on the previous experience of the investigator in trial and error procedures, consuming time and resources without guaranteeing the optimal configuration of the parameters to achieve a better performance in the neural network [2, 3, 18, 19, 21].

The process optimization, Taguchi methods, and experiment design constitute some of the most appropriate methods to achieve this objective since they can be applied to artificial neural networks design and training. Robust design is a statistical technique widely used to analyze the relationship between the variability factors that influence the results of a process and that can be used to systematically find the optimal configuration to obtain the desired result [14, 32, 41].

This chapter proposes a systematic and experimental strategy according to the requirements of the problem. The RDANN methodology was used to find the optimal parameters in an artificial neural network architecture of inverse propagation applied to the solution of the inverse kinematics of a robotic manipulator of six DOF.

9.1.4 ROBUST DESIGN OF ARTIFICIAL NEURAL NETWORKS

In the 1920s, the design of experiments by multifactor intervention by Sir Ronald A. Fisher was used for the first time to determine the results in agricultural practices [42].

This technique is known as factorial design of experiments, where a full factorial design allows identifying all possible combinations, implying a significant amount of experiments concerning the number of factors or variables involved and where the variables involved determining the performance or functionality of a system or product that can be controlled [43].

The robust parameter design method was proposed by Genichi Taguchi. This engineering method applied to the design of products or processes is focused on the decrease or sensitivity to noise. The method has proven to be a powerful and efficient procedure in the design of products or processes allowing optimal and consistent operation under certain conditions [44].

The main objective of the method is to find the selection of the factors involved that minimize the variability of the response to different inputs of the system by properly selecting the levels in the controllable design variables. The robust design approach allows studying all the parameter space with a small number of experiments using one orthogonal array (OA) and fractionated factorial design [45].

An OA represents a lesser part of the conventional complete factorial design and allows a well-adjusted comparison of the levels in the variables involved in the design of experiments. The design parameters involved represent the columns of the OA and the rows are the individual experiments by combining levels in the design parameters [46].

To determine the combination of levels, Taguchi proposed a two-phase procedure. First, the most significant control factors are determined to reduce variability. The factors that are significant to affect the sensitivity of the system are chosen below. The second phase aims to adjust the response to the desired values [47].

The Taguchi method can be described in four stages:

1. **Quality characteristics and parameter design**. Generate a brainstorm about the most important design parameters and quality characteristics in the process. At this stage, there are variables that the user can control and others that cannot. These are called design and noise factors, respectively.
2. **Design of experiments**. Choose the most appropriate OA by selecting the levels of the design factors that reduce the effects produced by noise, that is, that the product response maintains the minimum variation and its average approximates the desired objective.

An OA can evaluate several factors involved in a small number of experiments, allowing to reduce the time in conducting experiments and obtaining more information with fewer experiments. The mean and variance of the response in each configuration proposed by the OA are combined into a single performance measure known as the signal-to-noise ratio (*S/N*).

3. **Analysis of results for the determination of the optimal conditions**. An excellent quality indicator is the S/N ratio, in which the effect produced on the performance of the process or product can be evaluated by changing a parameter. Taguchi used the *S/N* ratio to assess the variation in system performance. For the static characteristic, Taguchi classified them into three types of relationship:

(a)   Smaller-the-better (STB)
(b)   Larger-the-better (LTB)
(c)   Nominal-the-best (NTB)

To obtain the optimal factors, for the NTB case, Taguchi developed a two-phase optimization procedure. For both other cases, Taguchi recommends the direct minimization of the expected loss.

In dynamic characteristics the relationship

$$SN_i = 10 \cdot \log_{10} \left(\beta_i \, / \, MSE_i \right) \tag{9.2}$$

is used to evaluate the *S/N* ratio, and *MSE* represents the average of the squared errors, that is, the difference between the measured response and what is estimated.

4. **Confirmation test in optimal conditions**. The main objectives in the design of parameters are to reduce the design variation of the processes or products to different environments or conditions. At this stage, a confirmation experiment is performed using the optimal design conditions.

Currently, the ANNs can be trained to solve complex and difficult problems for conventional computers and humans, including various fields such as pattern recognition and speech, vision, or control systems. In the field of robotics, ANNs have been successfully applied in recent decades [13, 48], mainly in the solution to the problem of the inverse kinematics in robotic manipulators with a high number of joints [1–3, 49].

9.2   ROBUST DESIGN OF ARTIFICIAL NEURAL NETWORKS METHODOLOGY

Determining the solution of inverse kinematics in manipulators with a high DOF number is a complex task [2]. To solve this problem, three methods are generally used: the geometric method, the algebraic method, and the iterative method [3].

Generally, most solutions using traditional methods with geometric, iterative, and algebraic approaches sometimes turn out to be complex and generically inadequate for the multiple articular configurations that robotic manipulators may present due to the high degree of precision required in the manipulator positioning [50].

In geometric and algebraic methods, a closed solution can only be guaranteed as long as the manipulator meets some special conditions, such as the spherical wrist [51]. For their part, iterative methods tend to be too slow and do not meet the requirement for implementation in real-time control systems [49].

This problem has been addressed by researchers during the last decades, who have proposed various methods of solution through the use of ANNs, so they constitute one of the most appropriate methods in solving this problem. An important problem in the calculation of the inverse kinematics using ANNs is the determination of the parameters of the network. The choice of basic ANN parameters determines success in ANN training [9, 15, 52].

However, there are no rules to identify the optimal selection of these parameters. Generally, the traditional trial and error technique produces a low capacity for generalization and low performance, consuming a great deal of time and computational resources [3].

To overcome this disadvantage easily and efficiently, the RDANN methodology is used. This methodology, proposed in Ref. [14] and shown in Figure 9.7, describes a systematic process to find the optimal parameters of an ANNBP architecture applied to the inverse kinematics solution of a six DOF robotic manipulator.

According to Figure 9.8, the steps to obtain the optimal ANN parameters are described below:

1. **Planning stage**

    In this stage, the objective function and the design and noise variables are identified.

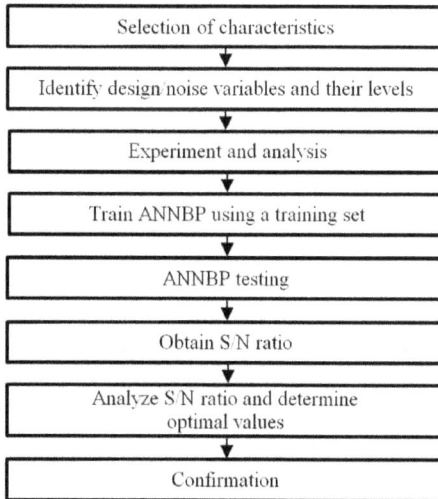

FIGURE 9.7 RDANN methodology.

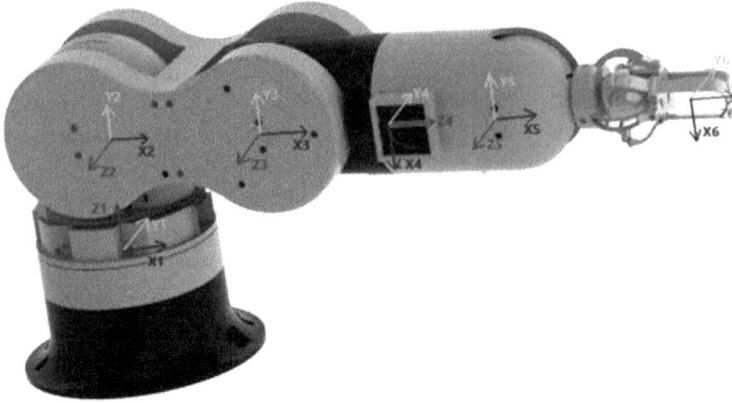

FIGURE 9.8 Ketzal robot manipulator.

(a) *The objective function.* The requirements of the problem allow defining the objective function. The objective function in this research is the prediction or classification of errors between the expected values and the output values of the ANNBP in the testing stage. The performance or value of the mean square quadratic error (MSE) of the ANNBP is used to achieve this goal, which is expressed mathematically as follows:

$$MSE = \sqrt{\frac{1}{N} \sum_{i=1}^{N} \left(\theta_i^{predicted} - \theta_i^{Deduced} \right)^2}$$ (9.3)

where N is the number of attempts, $\theta_i^{Predicted}$ is the array of values predicted by the neuronal network, and $\theta_i^{Deduced}$ is the array of the values originally inferred or calculated by the forward kinematics [14, 18].

(b) *Design and noise variables.* Users can select some of the design variables that may vary during the optimization process or iteration process according to the requirements of the problem. Among the different parameters that affect the performance of an ANNBP, four design variables were chosen as they are the variables that can be controlled by the user, as shown in Table 9.1.

TABLE 9.1
Design Variables and Their Levels

Design Variables	Level 1	Level 2	Level 3
A	$L1$	$L2$	$L3$
B	$L1$	$L2$	$L3$
C	$L1$	$L2$	$L3$
D	$L1$	$L2$	$L3$

TABLE 9.2

Noise Variables and Their Levels

Noise Variables	Level 1	Level 2
U	Set 1	Set 2
V	9:1	8:2
W	Tr-1/Ts-1	Tr-2/Tst-2

Here A is the number of neurons in the first hidden layer, B is the number of neurons in the second hidden layer, C is the momentum, and D is the learning rate.

The noise variables are shown in Table 9.2. Under various conditions of noise or distortion in the process, three variables were determined. These variables in most cases are not controlled by the user as is the case with the initialization of the set of weights U, which is normally selected randomly. In training and test data sets V, the designer must decide the size of the training set versus the test size. Once V is determined, the designer makes the random selection to decide the data to be included in the training and test set W.

U represents the initial set of random weights, V is the size of the training set versus the size of the test set, i.e. V=90%/10%, 80%/20%, and W is the random selection of the set of training and set of tests, that is, W=Training1/Test1, Training2/Test2. These types of variables cannot be controlled by the designer because they are randomly determined.

2. **Experimentation Stage**

Choosing the right elements in an OA is critical to success at this stage. An OA allows evaluating the main effects of interaction between factors involved through a small number of experiments. Taguchi suggests the use of two cross-configurations of OAs with an $L_9(3^4)$ and $L_4(3^2)$, as shown in Table 9.3.

TABLE 9.3

OA with $L_9(3^4)$ and $L_4(3^2)$ Configuration

Trial No.	A	B	C	D	S1	S2	S3	S4	Mean	S/N
1	1	1	1	1						
2	1	2	2	2						
3	1	3	3	3						
4	2	1	2	3						
5	2	2	3	1						
6	2	3	1	2						
7	3	1	3	2						
8	3	2	1	3						
9	3	3	2	1						

3. **Analysis Stage**

The S/N ratio considers the mean and the variation in the measured responses allowing a quantitative evaluation of the design parameters. The unit of measure is a decibel and the formula used is

$$S/N = 10 \cdot \log_{10}(MSD) \tag{9.4}$$

where MSD is the measure of the mean square deviation in the performance of the neural network.

Since it is desirable to have a higher signal and less noise, the highest S/N ratio indicates the best design parameters. A statistical program (JMP) was used at this stage to determine the optimal parameters of the network.

4. **Confirmation Stage**

The robust measurement value is predicted by the optimal design condition. A confirmation experiment is performed by calculating the performance robustness measurement and verifying if the value of the robust measurement is close to the value deduced.

Once the aspects involved in each of the stages of the RDANN methodology were identified, the training data set regarding the position, orientation, and articular values of the robotic manipulator was generated, based on the analysis of the direct kinematics of the six DOF robotic manipulator, to subsequently perform the 36 workouts described in the proposed OA.

9.3 KINEMATICS ANALYSIS OF ROBOTIC MANIPULATOR CALLED KETZAL

The solution to the direct kinematic problem refers to finding a homogeneous transformation matrix T that relates the position and orientation of the end of the robot concerning a fixed reference system located at the base of the robot since the movement of the robotic manipulator describes the relationship between the angles of the joints, the position, and orientation of the final effector [53].

Figure 9.8 shows the structure and reference coordinates of the Ketzal robot used in this work, which is a six DOF robotic manipulator taken from an open-source, 3D printable, and low-cost project [54].

The direct kinematics problem is to find the position and orientation of the final effector with respect to a reference system, given the array of joint angles, $\Theta = (\Theta_1, \Theta_2, \Theta_3, \Theta_4, \Theta_5, \Theta_6)$ of the robotic manipulator.

The objective of inverse kinematics is to calculate the array of joint angles Θ given the position and orientation of the final effector with respect to a reference coordinate system. Direct kinematics is represented as a 4×4 homogeneous transformation matrix that represents the position and orientation of the final effector of the robotic manipulator with respect to a reference coordinate system and which is given by the following equation [24].

$$T_6^0 = \begin{bmatrix} R_6^0 & P_6^0 \\ 0 & 1 \end{bmatrix} = \begin{bmatrix} n_x & o_x & a_x & p_x \\ n_y & o_y & a_y & p_y \\ n_z & o_z & a_z & p_z \\ 0 & 0 & 0 & 1 \end{bmatrix} \tag{9.5}$$

where R_6^0 is the 3×3 rotation matrix which contains the orientation array $[n\ o\ a]$ of the final effector, and P_6^0 is the position array $[p]$ of the final effector in the reference coordinate system.

The Denavit–Hartenberg (DH) method was used to analyze the direct kinematics of the Ketzal robot by implementing four basic transformations that depend exclusively on the geometric characteristics of the links [55]. The DH parameters of the Ketzal robot are shown in Table 9.4.

These transformations consist of a succession of rotations and translations that allow the reference system of element i to be related to the system of element $i - 1$, where the homogeneous transformation matrix for each joint is given by the following equation [10].

$$^{i-1}A_i = Rot_{z,\theta i} Trans_{x,di} Trans_{x,ai} Rot_{z,\alpha i} \tag{9.6}$$

Thus,

$$^{i-1}A_i = \begin{bmatrix} c_i & -s_i c\alpha_i & s_i s\alpha_i & a_i c_i \\ s_i & c_i c\alpha_i & -c_i s\alpha_i & a_i s_i \\ 0 & s\alpha_i & c\alpha_i & d_i \\ 0 & 0 & 0 & 1 \end{bmatrix} \tag{9.7}$$

TABLE 9.4
DH Parameters of the Ketzal Robot

Link Offset (cm)	Joint Angle (rad)	Link Length (cm)	Twist Angle (rad)
$d_1 = 20.2$	$\theta_1 = q_1$	$a_1 = 0$	$\alpha_1 = \dfrac{\pi}{2}$
$d_2 = 0$	$\theta_2 = q_2$	$a_2 = 16$	$\alpha_2 = 0$
$d_3 = 0$	$\theta_3 = q_3 + \dfrac{\pi}{2}$	$a_3 = 0$	$\alpha_3 = \dfrac{\pi}{2}$
$d_4 = 19.5$	$\theta_4 = q_4$	$a_4 = 0$	$\alpha_4 = -\dfrac{\pi}{2}$
$d_5 = 0$	$\theta_5 = q_5$	$a_5 = 0$	$\alpha_5 = \dfrac{\pi}{2}$
$d_6 = 6.715$	$\theta_6 = q_6$	$a_6 = 0$	$\alpha_6 = 0$

where i is the number of the link, α_i is the joint rotation, a_i is the length of the link, d_i is the displacement of the link, $c_i = \cos(\Theta_i)$, and $s_i = \sin(\Theta_i)$; therefore, Θ_i is the rotation angle of the joint.

The matrix that describes the location of the final system with respect to a reference system located at the base of the robot, also known as the homogeneous transformation matrix, is obtained by making the product of the six matrices obtained from (9.5) [10].

$$T_6^0 = {}^0A_1 \cdot {}^1A_2 \cdot {}^2A_3 \cdot {}^3A_4 \cdot {}^4A_5 \cdot {}^5A_6 \tag{9.8}$$

Therefore, each of the transformation matrices of the links can be calculated as shown below:

$$
{}^0A_1 = \begin{bmatrix} c_1 & 0 & s_1 & 0 \\ s_1 & 0 & -c_1 & 0 \\ 0 & 1 & 0 & a_0 + a_1 \\ 0 & 0 & 0 & 1 \end{bmatrix} \tag{9.9}
$$

$$
{}^1A_2 = \begin{bmatrix} c_2 & -s_2 & 0 & a_2 * c_2 \\ s_2 & c_2 & 0 & a_2 * s_2 \\ 0 & 0 & 1 & 0 \\ 0 & 0 & 0 & 1 \end{bmatrix} \tag{9.10}
$$

$$
{}^2A_3 = \begin{bmatrix} -s_3 & 0 & c_3 & 0 \\ c_3 & 0 & s_3 & 0 \\ 0 & 1 & 0 & 0 \\ 0 & 0 & 0 & 1 \end{bmatrix} \tag{9.11}
$$

$$
{}^3A_4 = \begin{bmatrix} c_4 & 0 & -s_4 & 0 \\ s_4 & 0 & c_4 & 0 \\ 0 & -1 & 0 & a_3 + a_4 \\ 0 & 0 & 0 & 1 \end{bmatrix} \tag{9.12}
$$

$$
{}^4A_5 = \begin{bmatrix} c_5 & 0 & s_5 & 0 \\ s_5 & 0 & -c_5 & 0 \\ 0 & 1 & 0 & 0 \\ 0 & 0 & 0 & 1 \end{bmatrix} \tag{9.13}
$$

$$
{}^5A_6 = \begin{bmatrix} c_6 & -s_1 & 0 & 0 \\ s_6 & c_6 & 0 & 0 \\ 0 & 0 & 1 & a_5 + a_6 \\ 0 & 0 & 0 & 1 \end{bmatrix} \tag{9.14}
$$

$$T_0^6 = \begin{bmatrix} n_x & o_x & a_x & p_x \\ n_y & o_y & a_y & p_y \\ n_z & o_z & a_z & p_z \\ 0 & 0 & 0 & 1 \end{bmatrix} \tag{9.15}$$

where

$$n_x = -(c_1 \cdot (c_2 \cdot (c_3 \cdot c_6 \cdot s_5 + (c_4 \cdot c_5 \cdot c_6 - s_4 \cdot s_6) \cdot s_3) + (c_3 \cdot (c_4 \cdot c_5 \cdot c_6 - s_4 \cdot s_6) \tag{9.16}$$
$$-c_6 \cdot s_3 \cdot s_5) \cdot s_2) - (c_4 \cdot s_6 + c_5 \cdot s_4) \cdot s_1)$$

$$n_y = -(c_1 \cdot (c_4 \cdot s_6 + c_5 \cdot c_6 \cdot s_4) + (c_2 \cdot (c_3 \cdot c_6 \cdot s_5 + (c_4 \cdot c_5 \cdot c_6 - s_4 \cdot s_6) \cdot s_3) \tag{9.17}$$
$$+ (c_3 \cdot (c_4 \cdot c_5 \cdot c_6 - s_4 \cdot s_6) - c_6 \cdot s_3 \cdot s_5) \cdot s_2) \cdot s_1)$$

$$n_z = c_2 \cdot (c_3 \cdot (c_4 \cdot c_5 \cdot c_6 - s_4 \cdot s_6) - c_6 \cdot s_3 \cdot s_5) \tag{9.18}$$
$$- (c_3 \cdot c_6 \cdot s_5 + (c_4 \cdot c_5 \cdot c_6 - s_4 \cdot s_6) \cdot s_3) \cdot s_2$$

$$o_x = c_1 \cdot (c_2 \cdot (c_3 \cdot s_5 \cdot s_6 + (c_4 \cdot c_5 \cdot s_6 + c_6 \cdot s_4) \cdot s_3) \tag{9.19}$$
$$+ (c_3 \cdot (c_4 \cdot c_5 \cdot s_6 + c_6 \cdot s_4) - s_3 \cdot s_5 \cdot s_6) \cdot s_2) + (c_4 \cdot c_6 - c_5 \cdot s_4 \cdot s_6) \cdot s_1$$

$$o_y = -(c_1 \cdot (c_4 \cdot c_6 - c_5 \cdot s_6 \cdot s_4) - (c_2 \cdot (c_3 \cdot s_6 \cdot s_5 + (c_4 \cdot c_5 \cdot s_6 + s_4 \cdot c_6) \cdot s_3) \tag{9.20}$$
$$+ (c_3 \cdot (c_4 \cdot c_5 \cdot s_6 + s_4 \cdot c_6) - s_3 \cdot s_5 \cdot s_6) \cdot s_2) \cdot s_1)$$

$$o_z = -(c_2 \cdot (c_3 \cdot (c_4 \cdot c_5 \cdot s_6 + c_6 \cdot s_4) - s_3 \cdot s_5 \cdot s_6) \tag{9.21}$$
$$- (c_3 \cdot s_5 \cdot s_6 + (c_4 \cdot c_5 \cdot s_6 + c_6 \cdot s_4) \cdot s_3) \cdot s_2)$$

$$a_x = c_1 \cdot (c_2 \cdot (c_3 \cdot c_5 - c_4 \cdot s_3 \cdot s_5) - (c_3 \cdot c_4 \cdot s_5 + c_5 \cdot s_3) \cdot s_2) + s_1 \cdot s_4 \cdot s_5 \tag{9.22}$$

$$a_y = -(c_1 \cdot s_4 \cdot s_5 - (c_2 \cdot (c_3 \cdot c_5 - c_4 \cdot s_3 \cdot s_5) - (c_3 \cdot c_4 \cdot s_5 + c_5 \cdot s_3) \cdot s_2) \cdot s_1 \tag{9.23}$$

$$a_z = c_2 \cdot (c_3 \cdot c_4 \cdot s_5 + c_5 \cdot s_3) + (c_3 \cdot c_5 - c_4 \cdot s_3 \cdot s_5) \cdot s_2 \tag{9.24}$$

$$p_x = a_2 \cdot c_1 \cdot c_2 + a_3 \cdot c_1 \cdot (c_2 \cdot c_3 - s_2 \cdot s_3) + a_4 \cdot c_1 \cdot (c_2 \cdot c_3 - s_2 \cdot s_3) + (a_5 + a_6) \tag{9.25}$$
$$\cdot (c_1 \cdot (c_2 \cdot (c_3 \cdot c_5 - c_4 \cdot s_3 \cdot s_5) - (c_3 \cdot c_4 \cdot s_5 + c_5 \cdot s_3) \cdot s_2) + s_1 \cdot s_4 \cdot s_5)$$

$$p_y = a_2 \cdot c_2 \cdot s_1 + a_3 \cdot (c_2 \cdot c_3 - s_2 \cdot s_3) \cdot s_1 + a_4 \cdot (c_2 \cdot c_3 - s_2 \cdot s_3) \cdot s_1 - (a_5 + a_6) \tag{9.26}$$
$$\cdot (c_1 \cdot s_4 \cdot s_5 - (c_2 \cdot (c_3 \cdot c_5 - c_4 \cdot s_3 \cdot s_5) - (c_5 \cdot c_4 \cdot s_5 + c_5 \cdot s_3) \cdot s_2) \cdot s_1)$$

$$p_z = a_0 + a_1 + a_2 \cdot s_2 + a_3 \cdot (c_2 \cdot s_3 + c_3 \cdot s_2) + a_4 \cdot (c_2 \cdot s_3 + c_3 \cdot s_2) + (a_5 + a_6)$$

$$\cdot (c_2 \cdot (c_3 \cdot c_4 \cdot s_5 + c_5 \cdot s_3) + (c_3 \cdot c_5 - c_4 \cdot s_3 \cdot s_5) \cdot s_2$$

$$(9.27)$$

It can be seen that Equations 9.16–9.24 represent the orientation array $[n \ o \ a]$ of the end of the robot based on joint coordinates $(q_1, q_2, q_3, q_4, q_5, q_6)$ and Equations 9.25–9.27 represent the position array $[p]$ according to the joint coordinates and the lengths of the links $(a_0, a_1, a_2, a_3, a_4, a_5, a_6)$.

When solving (9.8), the position and initial orientation of the final effector is obtained with respect to the reference system located at the base of the robot as shown below:

$$T_0^6 = \begin{bmatrix} 0 & 0 & 1 & 42.215 \\ 0 & -1 & 0 & 0 \\ 1 & 0 & 0 & 20.2 \\ 0 & 0 & 0 & 1 \end{bmatrix} \tag{9.28}$$

Where the initial position of the final effector of the Ketzal robot was $p_x = 42.215$ cm, $p_y = 0$ cm, and $p_z = 20.2$ cm and the orientation obtained was $n_x = 0$, $o_x = 0$, $a_x = 1$, $n_y = 0$, $o_y = -1$, $a_y = 0$, $n_z = 1$, $o_z = 0$, and $a_z = 0$. This confirms that the data obtained coincide with the initial position and orientation of the robot arm with the calculation of direct kinematics.

The simulation of the graphic representation of the Ketzal robot was carried out with the Robotics Toolbox for MATLAB software [56]. Figure 9.9 shows the initial position and orientation of the final effector of the robotic manipulator.

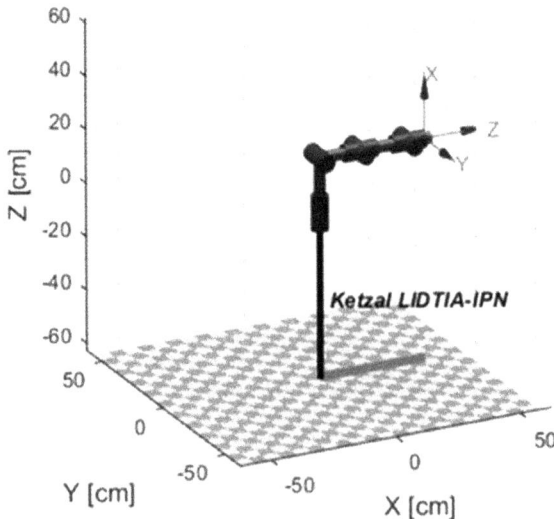

FIGURE 9.9 Graphic representation of the robotic manipulator.

9.3.1 Data Set Description

Based on the geometry and dimensions of the manipulator, the workspace of the robotic manipulator can be represented by the position array, orientation array, and joint values. This representation can be defined as an infinite set of coordinates of position, orientation, and articular values.

Therefore, it is very important to define the size of the training data set for the neural network, since the volume of the data is an important factor that has a decisive influence on the available processing capabilities [57].

In this study, the possibility of obtaining representative samples in the entire workspace was analyzed, so that the selected data guaranteed a representative sample of the entire original set, minimizing the risk that some or a large part of the population was not represented [58, 59].

The data set was generated from Equations 9.16–9.27 and according to the geometry of the robot, the articular values of $(\Theta_1 \cdots \Theta_6)$ were established in the range shown in Table 9.5.

Generally, the amount of data depends on the spatial resolution used in the range of joint values. For example, if a jump of $\Delta\Theta_1 = \pi/2$ is used to generate the range of values in Θ_1, only five values $[0, \pi/2, \pi, 3\pi/2, 2\pi]$ will be considered; on the contrary, if a jump of $\Delta\Theta_1 = \pi/5$ is used, 11 values are considered, allowing to have a better spatial resolution in the data set.

In this study, four data sets were generated by applying the direct kinematics model described above. The data sets generated can be described by a matrix of *rows* × *columns* represented by the following equation:

$$Data\_set = \begin{bmatrix} a_{11} & a_{12} & \cdots & a_{1c} \\ a_{21} & a_{22} & \cdots & a_{2c} \\ \cdots & \cdots & \cdots & \cdots \\ a_{r1} & a_{r2} & \cdots & a_{rc} \end{bmatrix} \tag{9.29}$$

where the subscript r represents the number of data generated and the subscript c represents the number of variables used in this work, where the elements a_{r1}, a_{r2}, and a_{r3} correspond to the data of the position array $[p] = \{p_x, p_y, p_z\}$, the elements $a_{r4} \cdots a_{r12}$ correspond to the data of the three-orientation array $[n\ o\ a] = \{n_x, n_y, n_z, o_x, o_y, o_z, a_x,$

TABLE 9.5

Angular Ranges in the Joints of the Ketzal Robot

rad	θ_1	θ_2	θ_3	θ_4	θ_5	θ_6
Minimum	0	0	2π	0	2π	0
Maximum	2π	π	$\dfrac{\pi}{2}$	2π	$\dfrac{\pi}{2}$	2π

a_y, a_z}, and finally the elements $a_{r13}\cdots a_{r18}$ correspond to the array of articular values $[\Theta] = \{\Theta_1, \Theta_2, \Theta_3, \Theta_4, \Theta_5, \Theta_6\}$.

The spatial resolution used in this work was $17 \times 17 \times 17 \times 17 \times 17 \times 17 \times 17$ with a total of 24 137 569 data with 18 variables. The jump values where set to $\Delta\Theta_1 = \pi/8$, $\Delta\Theta_2 = \pi/16$, $\Delta\Theta_3 = \pi/16$, $\Delta\Theta_4 = \pi/8$, $\Delta\Theta_5 = \pi/16$, and $\Delta\Theta_6 = \pi/8$. The 18 variables considered in the data set generated correspond to the position array $[p]$, orientation array $[n\ o\ a]$, and joint value array $[\Theta]$.

To guarantee better performance in the knowledge extraction algorithms, the data pre-processing stage is fundamental, since the efficiency in the results of performance and knowledge extraction depends largely on the quality of the data, and low quality regularly leads to low quality in the knowledge extracted [57].

The techniques generally used in data pre-processing are divided into two areas: data preparation and data reduction. Data preparation is generally an obligatory technique and consists of a series of algorithms whose purpose is to prepare the data so that the knowledge extraction algorithms can be executed without difficulties during processing. This area includes data transformation, normalization, cleaning, and probably the recovery of lost values [40].

On the other hand, data reduction techniques are not always mandatory, but sometimes they help to obtain better results, where the main objective in this type of techniques is to obtain a reduced amount of the original data, keeping as much as possible the integrity of the information in the data, highlighting the methods of feature selection, instance selection, grouping, compaction, and sampling [60, 61].

Theoretically, obtaining a complete data set will allow for a broader understanding of the problem; however, if the population is too large, the processing time required will be much more expensive, so it would be unfeasible.

In practice, sampling is one of the most appropriate processes to solve this problem due to the advantages obtained in the performance of time and cost of processing, even better results can be obtained from the knowledge extracted compared to the processing of data sets without sampling [57].

Therefore, data sampling can be widely applied in various engineering fields such as statistics, machine learning, and data mining [59].

Systematic sampling, either alone or in combination with any other technique, is one of the most used methods because of its simplicity and ease of use; in its simplest form, it is also known as linear systematic sampling (LSS) and was introduced in 1944 in Ref. [62].

The LSS method can be seen as a sampling technique by cluster in its simple form, where the procedure is to divide the population N into k groups of n elements each. Taking only one group of the sample accordingly, each of the k groups has a probability of $1/k$ to be selected; in this case, the application of systematic sampling is feasible because $N = k \cdot n$ [63].

LSS allows all units of the data set to have the same opportunities to be selected, for example, if you have an initial set of size N from which you want to select a new set of size n, a random number is selected for sampling between 1 and k to select the instance k of the data set where k represents the sampling interval. From the selected instance, the kth item is selected forward up to the n instances proposals.

This procedure ensures that each unit has the same possibility of being included in the sample [64].

The sampling interval constant is usually taken as the integer closest to N/n, known as the inverse of the sampling fraction. This method has two advantages mainly. Firstly, the selection of the first unit is determined by a sampling interval or sampling period chosen, guaranteeing a random sample among a set of samples of a given interval. Secondly, systematic samples are distributed in a good way among the population, guaranteeing a representative sample of the entire original data set, that is to say, there is less risk that some or a large part of the population will not be represented, maintaining a constant and uniform distribution between the data [57].

Although it is one of the most common and simple methods, it has two main drawbacks. Firstly, the sampling variance cannot be taken impartially based on the only systematic sample taken. Secondly, when population size N cannot be divided equally by the desired sample size n, systematic sampling cannot be performed, that is, when N is not an integer multiple of the desired sample size n where $N \neq n \cdot k$.

In this case, k is not an integer value and an inefficient sampling effect can probably be produced, that is, if at any time the characteristics of the population were periodic and they coincide with a selected sampling interval, the representativeness of the sample could be biased [63].

In this work, the data set proposed N_1 is not an integer multiple of the desired size n due to the spatial resolution used in generating the data set. It is clear that in this case the criteria for applying the method LSS is not met, where $N = n \cdot k$.

To solve this problem, the reduction data filter (RDF) algorithm was designed based on the LSS method in such a way that it allows obtaining a representative distribution of the original data set, minimizing the risk that some or part of the population is not represented and that allows obtaining a constant and uniform distribution for training and test data sets.

9.3.2 DESCRIPTION OF REDUCTION DATA FILTER ALGORITHM

Working with a volume of data greater than 24 million × 18 variables could be a difficult task for a common computer. An RDF algorithm was designed based on the LSS method for solving the problem above, where the task processing is much easier for a conventional processor.

Considering the data set A with a population of $N_1 = 24\ 137\ 569$ where the sample size desired is $n = 1000$, it is observed that N_1 is not an integer multiple of n, due to $k = N_1/n = 24\ 137\ 569$, where k is not an integer that meets the criteria for applying the LSS method.

To solve the problem, the RDF algorithm, shown in Figure 9.10, helps to fulfil this purpose.

In general, the steps to follow for the implementation of the filter are described below:

1. The values of the original population size N and the desired sample size n are set.

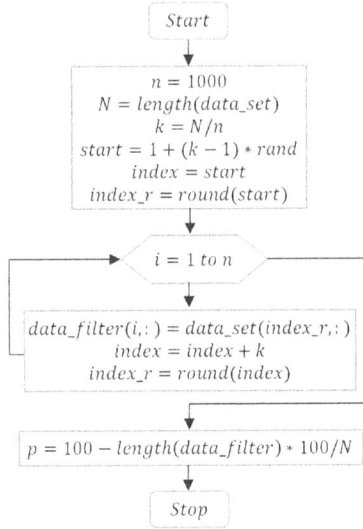

FIGURE 9.10 RDF algorithm based on linear systematic sampling.

2. The sampling interval constant $k = N/n$ is calculated, where k must not be an integer value.
3. A random number between 1 and k is generated.
4. The random value is rounded to an integer $r1$.
5. Element $r1$ of the original data set is selected as the first filtered data.
6. The cumulative sum between the random number and the constant k is made. In this step, the cumulative sum is a positive non-integer value.
7. The cumulative value of the sum is rounded to an integer $r2$.
8. Element $r2$ of the original data set is selected as the second filtered data and steps 6, 7, and 8 are repeated until reaching the desired n samples.

According to the flowchart shown in Figure 9.11, in the first section the constant $n = 1000$, which represents the amount of desired samples, is initialized. The variable N represents the number of samples for the data set previously generated by Equations 9.16–9.27.

The variable k represents the sampling interval, also known as the inverse of the sampling fraction with a value of $k = 24\ 137\ 569$. The *start* variable represents the random value between 1 and k. The *index* variable is the decimal value of the index in the instance, while the *index_r* variable is the rounded-up integer value of the index in the instance.

In the loop section of the iterative control structure *for*, the instance selection is made from the kth element forward until reaching the 1000 instances.

In this process, the *index* variable accumulates the values of the indexes with decimals and the value of the sampling interval k. Finally, the p variable calculates the percentage of reduction applied by the proposed algorithm.

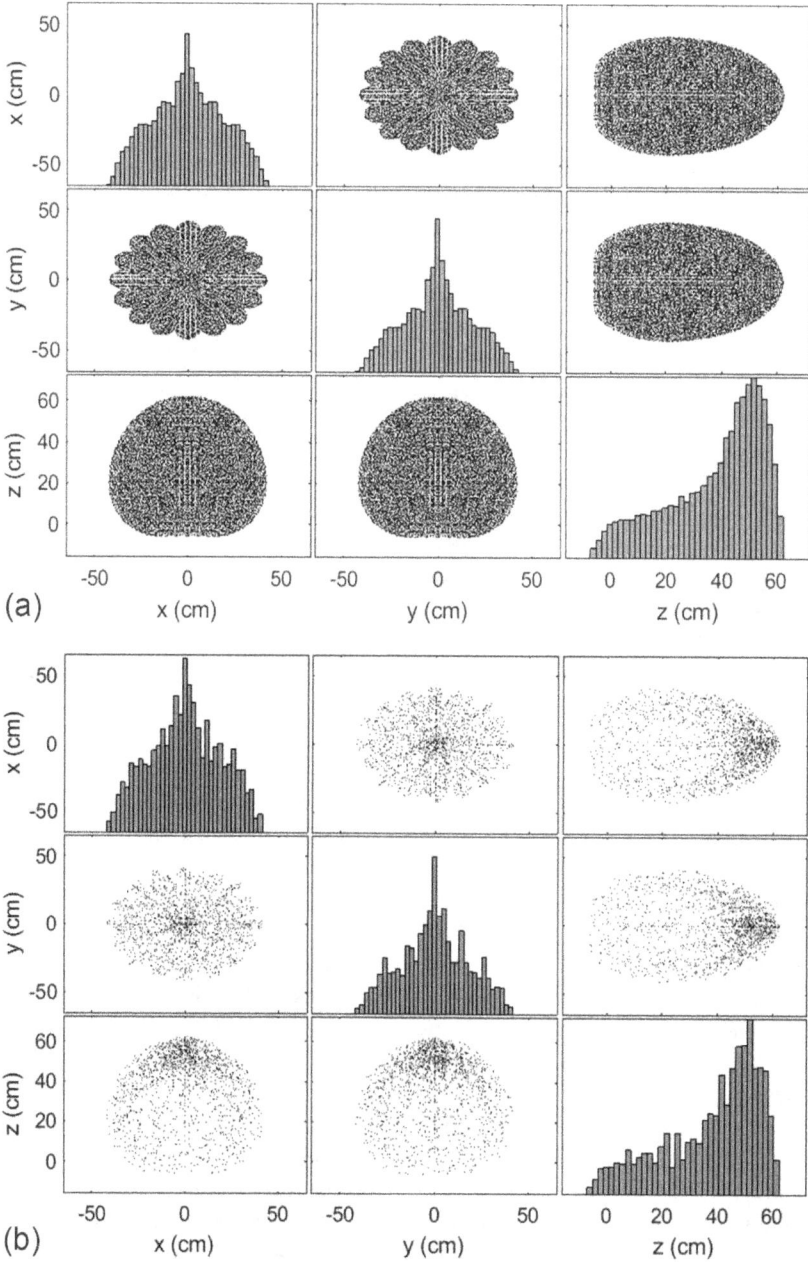

FIGURE 9.11 Dispersion matrix of the position data set D: (a) Before filtering (b) After filtering.

9.3.3 DATA SET ANALYSIS OF TRAINING AND TEST

Once the RDF was applied to data set A, a reduction to 1000 data with 18 variables was obtained. According to the pre-established range of motion for each of the joints, the distribution of the position data was analysed before and after applying the filter by means of scatter matrix diagrams shown in Figure 9.11, where it was observed that the position data maintained a constant and uniform distribution in their workspace with respect to the originally generated data set, as shown in Figure 9.11b, allowing to obtain a representative distribution of the original population.

Figure 9.12 shows the analysis of the distribution of the position and orientation data set, corresponding to the input data for the training of the ANN. As shown in Figure 9.12b, after applying the RDF algorithm, it was observed that the position and orientation data maintained a constant and uniform distribution in their workspace with respect to the data set A, allowing to obtain a representative distribution of the original data set.

The distribution of the set of input data used for training and testing in the ANN is shown in Figure 9.13. As described previously in Section 9.2, one of the noise variables that are not controlled by the user is the random selection of training and test sets, in this case, two sets with proportions of 80:20 and 90:10 were selected, respectively.

Figure 9.13a shows the distribution of the input data of the position array [p]; at the top, three graphs of the distribution of the position data without a filter are shown, of which the first corresponds to the position data, the second to the training data, and the third to the test data with an 80:20 proportion, respectively.

In the lower part of the same figure, three graphs of data distribution with filters are shown, of which the first corresponds to the position data, the second to the training data, and the third to the test data with a proportion of 90:10, respectively.

Figure 9.13b shows the distribution of the input data of the orientation array [$n\ o\ a$]; at the top, three graphs of the distribution of the orientation data without a filter are shown, of which the first corresponds to the orientation data, the second to the training data, and the third to the test data with an 80:20 proportion, respectively.

In the lower part of the same figure, three graphs of data distribution with filters are shown, of which the first corresponds to the orientation data, the second to the training data, and the third to the test data with a proportion of 90:10, respectively.

Figure 9.14 shows the distribution of the output data set corresponding to the joint array [Θ]. The first three graphs in the upper part of the figure correspond to the distribution of joint data without a filter in training and test data with proportions of 80:20, respectively, and the next three graphs in the lower part correspond to the distribution of the data of the joints with the filter applied in training and tests data with proportions of 90:10, respectively.

In this graph it can be seen that the filtered data maintains a constant and uniform distribution in the data regardless of the proportion used, maintaining a representative distribution with respect to the original population of the data.

The filtered data set was used to perform the experimentation stage and to determine the optimal parameters of the neural network as described in the next section.

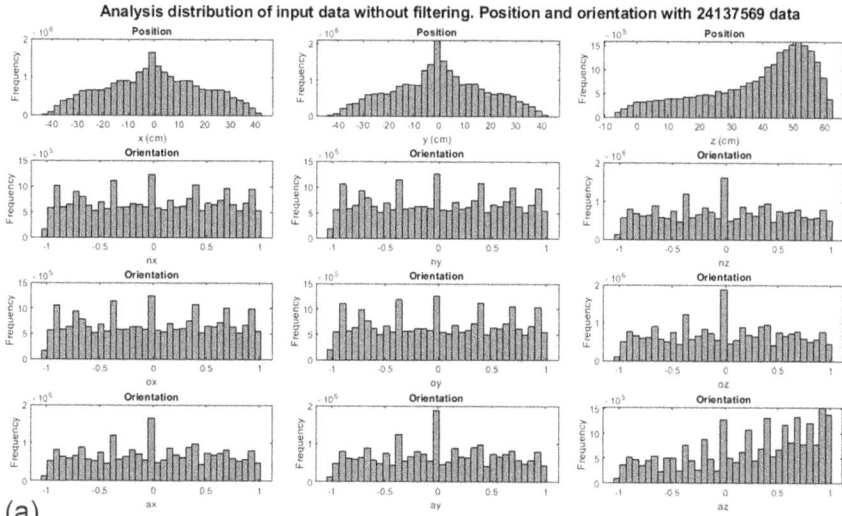

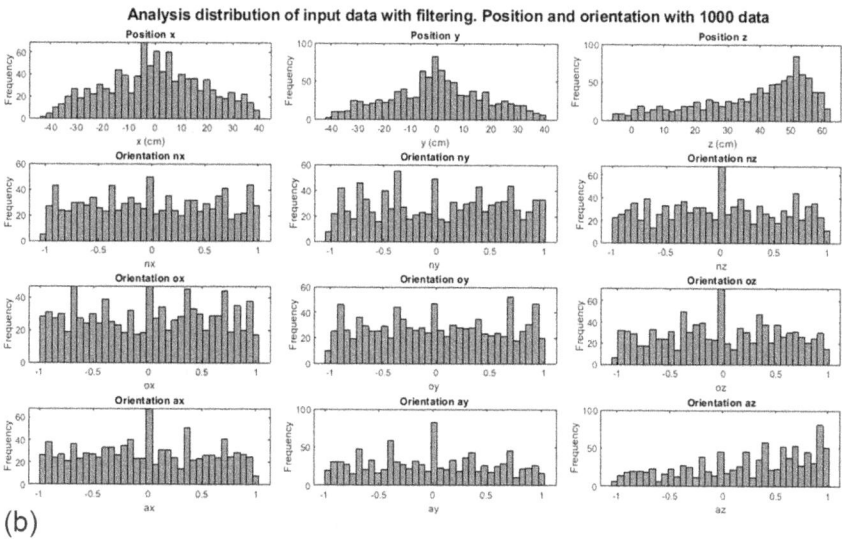

FIGURE 9.12 Distribution analysis of position and orientation data set *A*: (a) Before filtering (b) After filtering.

9.3.4 PLANNING AND EXPERIMENTATION STAGE

In the planning stage, the design and noise variables with their respective levels were chosen, as shown in Table 9.6.

Table 9.7 shows the noise variables and their respective levels, where U is the initial set of random weights, V is the size of the training set versus the size of the test set, that is, $V = 90\%/10\%$, $80\%/20\%$, and W is the random selection of the training set and test set, that is, $W = $ Training1/Test1, Training2/Test2.

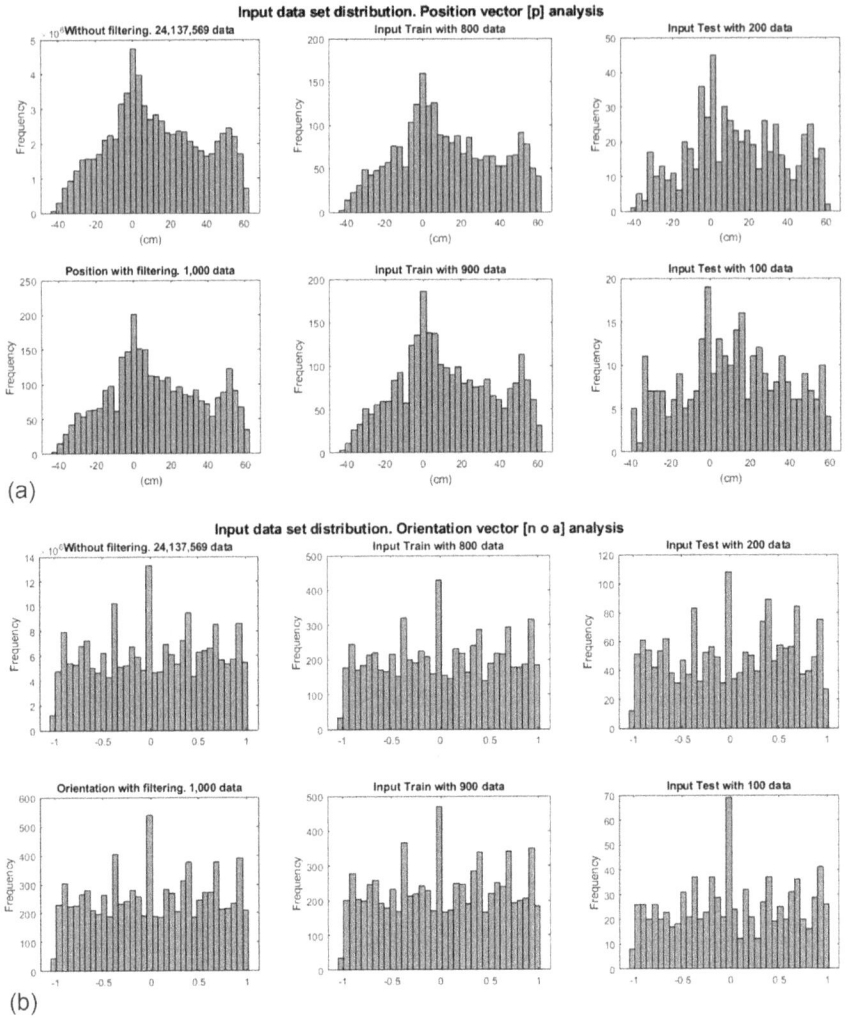

FIGURE 9.13 Distribution analysis of input data set with training and test: (a) Position (b) Orientation.

OA cross-configuration was used in the experimental stage with $L_9(3^4)$ and $L_4(3^2)$, where 36 different ANN architectures were trained and tested, as shown in Table 9.8.

The analysis of variance (ANOVA) was used with the JMP statistical program to obtain the signal-to-noise ratio.

9.3.5 ANALYSIS AND CONFIRMATION STAGE

Once the results of the MSE were obtained in the 36 trainings of the proposed OA, a statistical analysis in JMP was performed to identify the possible optimal values corresponding to the possible best network topologies, as shown in Figure 9.15.

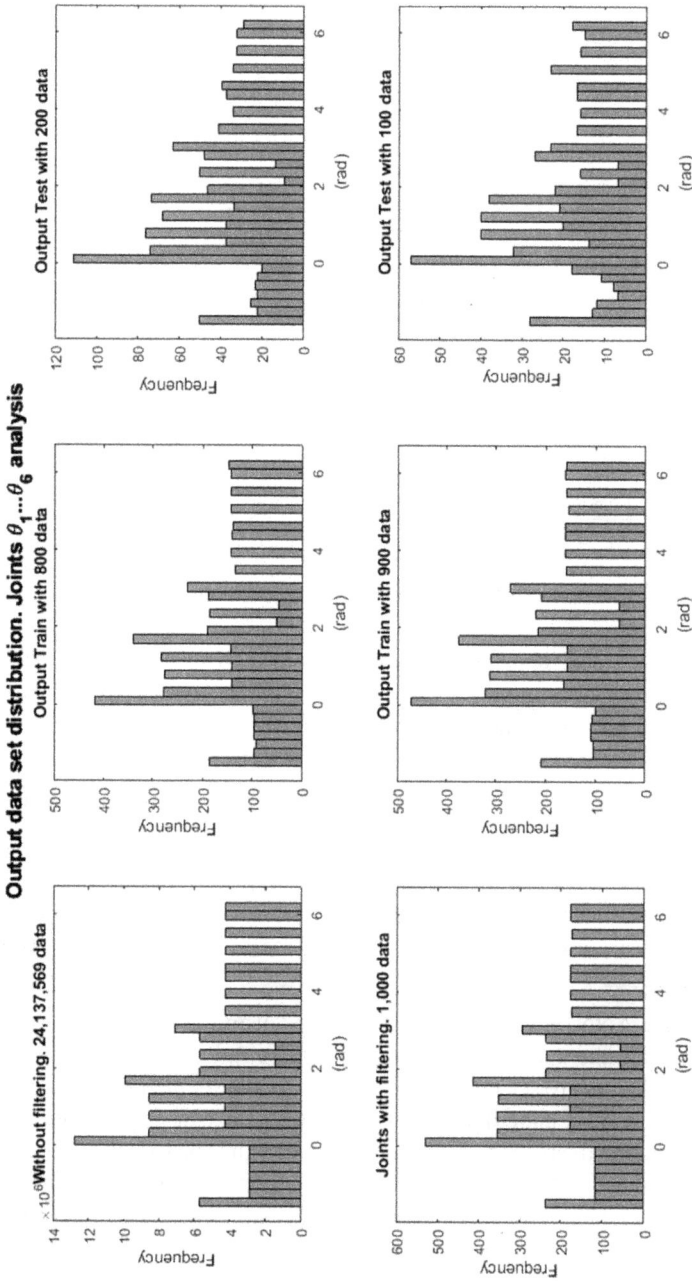

FIGURE 9.14 Distribution analysis of output data set with training and test.

TABLE 9.6
Design Variables and Their Levels

Design Variables	Level 1	Level 2	Level 3
A	6	9	12
B	0	3	0
C	0.1	0.2	0.3
D	0.01	0.1	0.2

TABLE 9.7
Noise Variables and Their Levels

Noise Variables	Level 1	Level 2
U	Set 1	Set 2
V	9:1	8:2
W	Tr-1/Ts-1	Tr-2/Tst-2

TABLE 9.8
OA with $L_9(3^4)$ and $L_4(3^2)$ Configuration

Trial No.	S1	S2	S3	S4	Mean	S/N
1	0.34028409	0.33660998	0.33523831	0.33120539	0.3358344	−39.042614
2	0.34294367	0.34775807	0.34162374	0.34110013	0.3433564	−41.071318
3	0.33780647	0.33966576	0.34055964	0.33274657	0.3376946	−39.708489
4	0.31766289	0.32279780	0.32111454	0.31413515	0.3189276	−38.378724
5	0.33082154	0.32892629	0.32892339	0.31509698	0.3259421	−33.015696
6	0.32950029	0.31278398	0.31908650	0.31464315	0.3190035	−32.598799
7	0.30203761	0.29094929	0.30443676	0.29524616	0.2981675	−33.657940
8	0.32171345	0.31260463	0.32582416	0.30687483	0.3167543	−31.330813
9	0.30484374	0.30907752	0.31075605	0.29823255	0.3057275	−34.659736

The optimal ANN architecture is determined using the signal-to-noise ratio. After carrying out the analysis of the signal-to-noise ratio, the values closest to the red line on the X-axis are selected, whose values are highlighted in Table 9.9.

Figure 9.16 shows a cube diagram showing the effect on the interaction of the factors involved and the response in the obtained design. The two cubes show all the combinations of the configuration variables of the proposed factors and the adjusted mean of each combination. The cube on the left shows the measurements of the responses when the learning rate is equal to 0.1 and the cube on the right shows the responses when the learning rate is 0.03.

The average value obtained from the MSE and the S/N ratio are two of the factors that are considered to determine the appropriate levels in the variables involved in

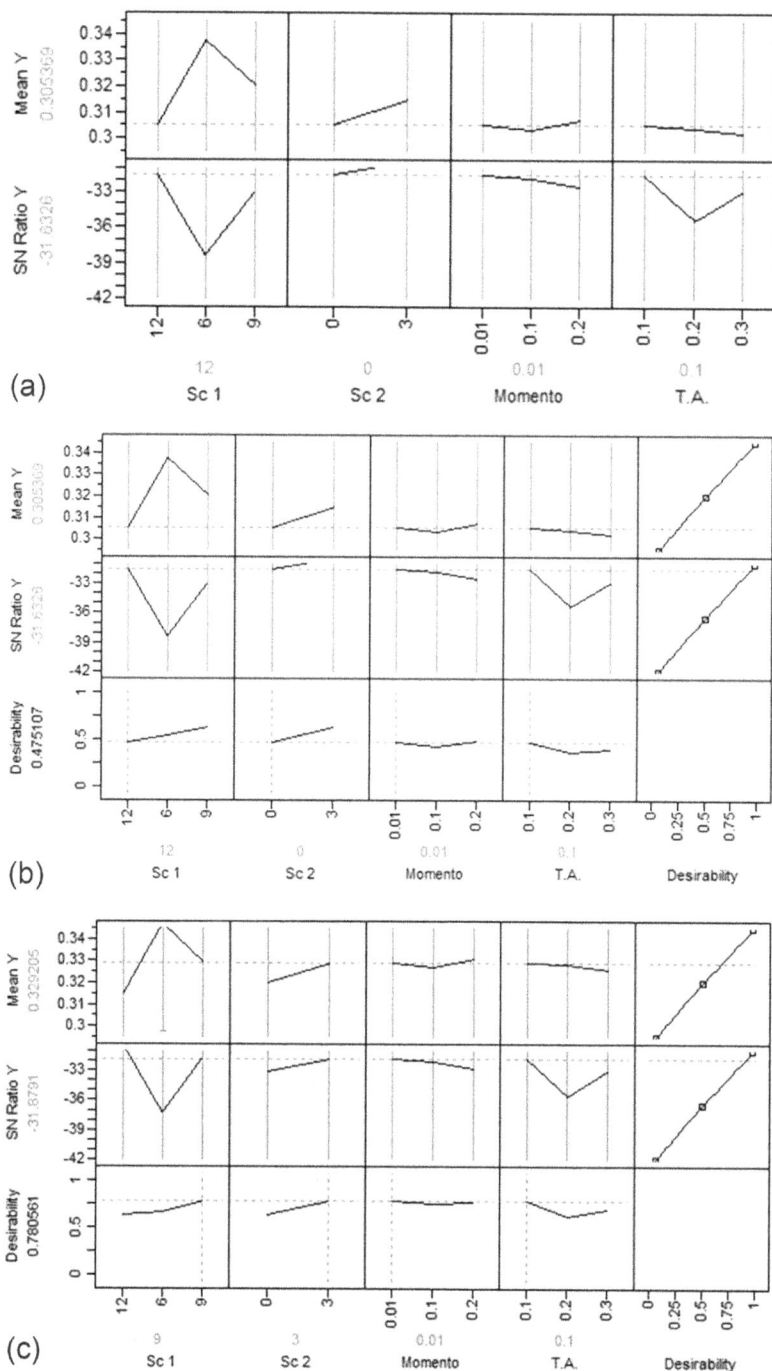

FIGURE 9.15 *S/N* analysis for the determination of optimal network parameters: (a) Best topology with normal profile (b) Best topology with desirability profile (c) Best topology with maximizing desirability profile.

TABLE 9.9

Best Values to Design the ANN with Normal, Desirability and Maximized Profiles

Trial No.	A	B	C	D
1	12	0	0.01	0.1
2	12	0	0.01	0.1
3	9	3	0.01	0.1

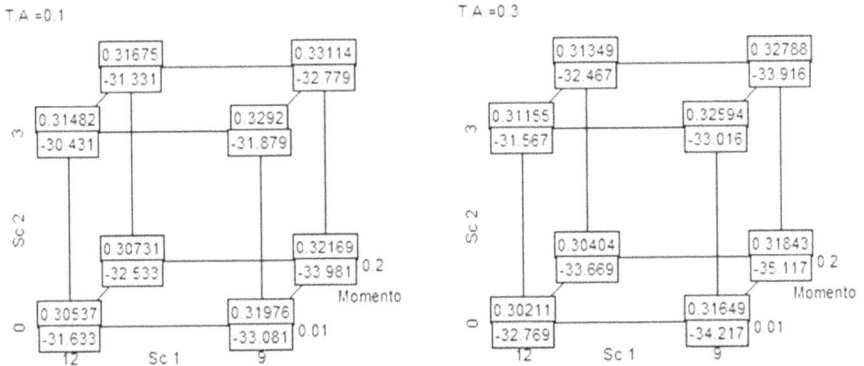

FIGURE 9.16   Variable response and interaction analysis.

the design of the network. The cube on the left describes that the best average value of the MSE obtained is given by 12 neurons in the first layer, a moment equal to 0.01, and a learning rate equal to 0.1. However, an increase in neurons in the second layer can improve the S/N ratio but tends to decrease the yield by approximately 3% above the MSE average.

On the other hand, in the cube on the right can be seen that 12 neurons in the first layer, a moment equal to 0.01, and a learning rate equal to 0.3 are sufficient to obtain the best performance. However, there is a tendency that coincides with the cube on the left in including a second hidden layer to improve the S/N ratio, but the network performance tends to decrease by 3% above the MSE average.

From the results obtained in the JMP statistical analysis, three trainings were carried out with the parameters obtained in each of the profiles, and tests were performed to validate the data obtained with the designed ANN. In the final validation, the statistical correlation and chi-square tests were performed for the best and worst prediction, as shown in Figures 9.17 and 9.18, respectively.

The cross-validation method was used with three data sets *A*, *B*, and *C* with data sizes of 600, 1000, and 1500 elements with 18 variables each, respectively. Each data set was divided into four subsets of the same size, as shown in Figure 9.19. The validation set in each training was used to estimate the generalization error, that is,

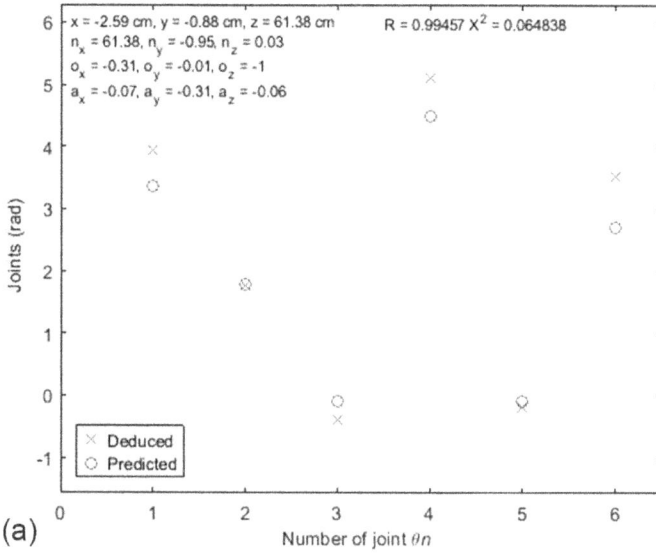

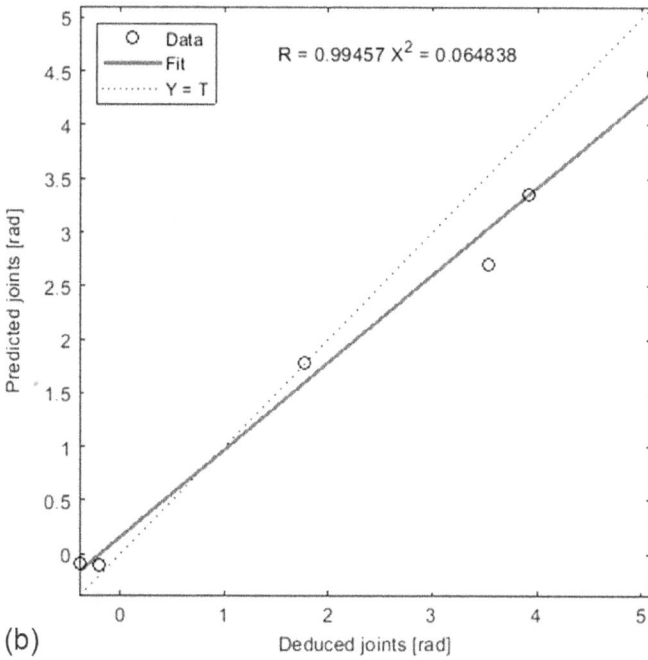

FIGURE 9.17 Best joints predicted and correlation test: (a) Best joints predicted (b) Correlation test.

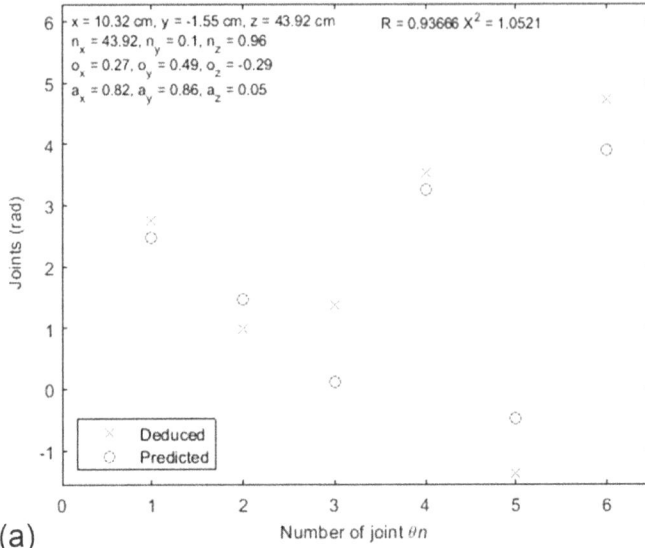

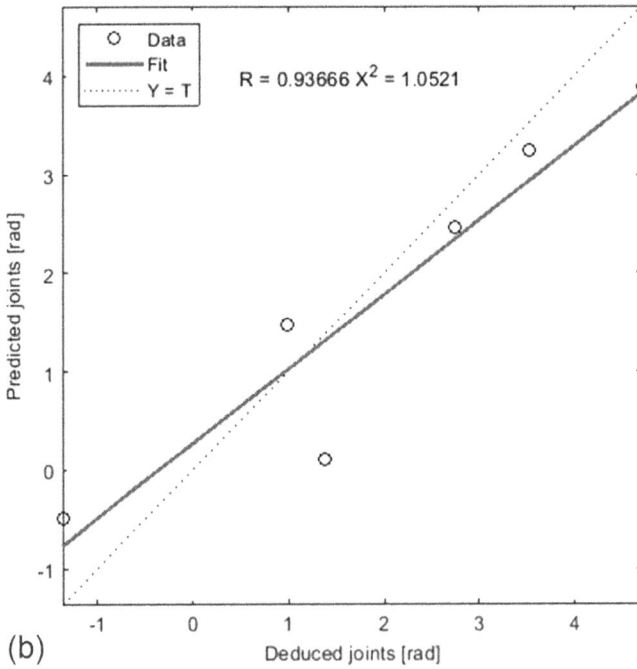

FIGURE 9.18 Worst joints predicted and correlation test: (a) Worst joints predicted (b) Correlation test.

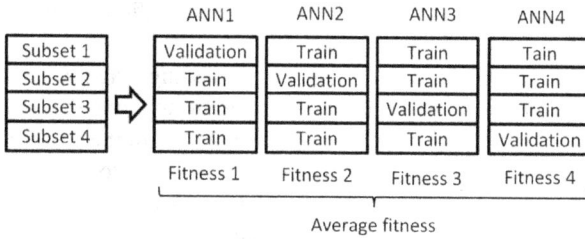

FIGURE 9.19   Cross-validation for data set *n*.

TABLE 9.10
Results of the Fourfold Cross-Validation

Fold	MSE		
Number	A	B	C
1	0.30089341	0.31660459	0.33148271
2	0.32114314	0.32018527	0.33704913
3	0.30725039	0.32869244	0.34501222
4	0.31404578	0.32941529	0.34597482
Average	0.31083318	**0.32372439**	0.33987972
Std Deviation	0.00872260	0.00633206	0.00688028

---

the incorrect classification rate of the model with data different from those used in the training process.

Table 9.10 shows the average MSE value and the standard deviation obtained during the workouts performed for each of the sets proposed during the validation of the neural network using the cross-validation method.

The average value of MSE obtained, and highlighted in Table 9.10 for data set *B*, is approximately equal to the value obtained in the confirmation step. However, as it is shown in Table 9.10, the average MSE value obtained during training with data set *A* is approximately 4% less than that obtained with set *B* and the average MSE value is approximately 5% higher than that obtained with the set *B*.

The use of the RDANN methodology allowed solving problems in the design of the ANN critical parameters. The density of the data used was 80% for training data and 20% for the test. The best architecture obtained in this work was 12:12:6, a momentum$=0.01$, a learning rate$=0.1$, *trainrp* training algorithm and a MSE$=1$E-4.

9.4   CONCLUSIONS AND DISCUSSIONS

In this research work, an RDF filter based on the LSS method was designed with the aim of reducing the data set, and the RDANN methodology was applied to solve the inverse kinematics problem in a 6 DOF robotic manipulator. During the optimization process, an OA was selected in L_9 (3^4) and L_4 (3^2) configuration considering four design variables and three noise variables.

Proper preparation of the data in the pre-processing stage was essential during ANNs training. Consequently, the effect produced by applying the data filter allowed obtaining a wider range and better predictive resolution in the workspace.

The RDANN methodology was successfully applied to solve the problem of inverse kinematics in a six DOF robotic manipulator, where the most significant factors in this study were the number of hidden neurons in the first layer, the learning rate, and the momentum constant.

The best topology obtained in this study was 12:12:6 with a pulse constant = 0.01, a learning rate = 0.1, MSE = 1E-4, and the *trainrp* learning algorithm. The hyperbolic sigmoid tangent transfer function was used in the input and output layers. The parameters obtained for the optimized network architecture showed high performance and generalization capacity.

The integration of ANNs and optimization techniques is a powerful tool in the design of structural parameters and improvement in the performance of ANNs. In this work, the RDANN methodology allowed determining the optimal parameters in the ANN design.

To validate the quality in the measurement of the results in the prediction of the model, the cross-validation method confirmed that the proposed data set, the structural and learning parameters in the ANN architecture obtained by the RDANN methodology, is statistically reliable, because the average MSE value obtained and highlighted in Table 9.10 with data set B is approximately equal to the value obtained in the confirmation stage.

According to the results obtained by the method of cross-validation, the performance of the neural network could be improved to the use a smaller data set; however, the corresponding statistical tests must be performed to confirm the generalization capacity, since, as observed in the cross-validation method for this case, the network performance is independent of the size and partition between the training and test data.

The systematic and experimental strategy used in this work is an alternative that considers the concept of robustness in the ANNs design process, due to the simultaneous incorporation of noise and design variables.

The RDANN methodology significantly decreases training time and effort in the modelling phase compared to the traditional trial and error method that is usually proposed by the previous experience of the researcher, where he can spend several days to months testing different architectures and consuming a lot of time and resources without guaranteeing the optimal configuration and the network performance. Consequently, the RDANN methodology may take a few hours to determine the optimal parameters of the network, allowing more time to solve the problem in question.

It is advisable to use the RDANN methodology to obtain an optimal architecture and minimize the training time instead of using a more complex architecture that uses a long training time and possibly the effect of over-adjustment in the data and reduced capacity in the generalization of the network; so it is not advisable to use a great architecture and stop the training of the network in a certain time to avoid overfitting the data.

The initialization of the weights significantly affects the performance of the ANN because a large amount of noise is introduced into the training data and has an impact on the effects produced at the start, specifically on the joints Θ_1, Θ_2, Θ_4, and Θ_6. In this case, the negative value does not make physical sense for the joints, because they cannot make movements outside the mechanically realizable range.

FUTURE SCOPE

The robust design methodology can be used to find a better artificial neural network configuration with high performance and generalization capability, reducing the time spent determining the optimal architecture of the artificial neural network compared to the trial and error approach. In the field of robotics, it is possible to direct future work towards the use of the robust design of artificial neural networks methodology presented here for the solution of the inverse kinematics of robots that do not have an algebraic solution or even to find the direct kinematics of parallel robots, which is very complex, unlike in serial manipulators.

ACKNOWLEDGEMENTS

This work was partially supported by CONACYT – Becas Nacionales de Posgrado con la Industria under contract 431101/640582 and paid salary license program for research IPN-COTEBAL under assignments CPE/COTEBAL/14/2018, CPE/COTEBAL/33/2019, and CPE/COTEBAL/23/2020. This work was supported by OMADS S.A. of C.V., an enterprise dedicated to innovation and technological development.

REFERENCES

1. S. Li, Y. Zhang, and L. Jin, "Kinematic Control of Redundant Manipulators Using Neural Networks," *IEEE Transactions on Neural Networks and Learning Systems*, vol. 28, no. 10, pp. 2243–2254, Oct. 2017, doi:10.1109/TNNLS.2016.2574363.
2. Z. Zhou, H. Guo, Y. Wang, Z. Zhu, J. Wu, and X. Liu, "Inverse Kinematics Solution for Robotic Manipulator Based on Extreme Learning Machine and Sequential Mutation Genetic Algorithm," *International Journal of Advanced Robotic Systems*, vol. 15, no. 4, p. 1729881418792992, Jul. 2018, doi:10.1177/1729881418792992.
3. G. Jiang, M. Luo, K. Bai, and S. Chen, "A Precise Positioning Method for a Puncture Robot Based on a PSO-Optimized BP Neural Network Algorithm," *Applied Sciences*, vol. 7, no. 10, p. 969, 2017, doi:10.3390/app7100969.
4. R. Köker, "Reliability-Based Approach to the Inverse Kinematics Solution of Robots Using Elman's Networks," *Engineering Applications of Artificial Intelligence*, vol. 18, no. 6, pp. 685–693, Sep. 2005, doi:10.1016/j.engappai.2005.01.004.
5. M. Ono, M. Pavone, Y. Kuwata, and J. Balaram, "Chance-Constrained Dynamic Programming with Application to Risk-Aware Robotic Space Exploration," *Autonomous Robots*, vol. 39, no. 4, pp. 555–571, Dec. 2015, doi:10.1007/s10514-015-9467-7.

6. W. Ji, and L. Wang, "Industrial Robotic Machining: A Review," *The International Journal of Advanced Manufacturing Technology*, vol. 103, no. 1, pp. 1239–1255, Jul. 2019, doi:10.1007/s00170-019-03403-z.
7. G. R. Lucas, "Industrial Challenges of Military Robotics," *Journal of Military Ethics*, vol. 10, no. 4, pp. 274–295, Dec. 2011, doi:10.1080/15027570.2011.639164.
8. B. Long, J. Yang, X. Chen, Y. Sun, and X. Li, "Medical Robotics in Bone Fracture Reduction Surgery: A Review," *Sensors*, vol. 19, no. 16, Feb. 2019, doi:10.3390/s19163593.
9. A.-M. Zou, Z.-G. Hou, S.-Y. Fu, and M. Tan, "Neural Networks for Mobile Robot Navigation: A Survey," in *Advances in Neural Networks - ISNN 2006*, 2006, pp. 1218–1226.
10. A. R. J. Almusawi, L. C. Dülger, and S. Kapucu, "A New Artificial Neural Network Approach in Solving Inverse Kinematics of Robotic Arm (Denso VP6242)," *Computational Intelligence and Neuroscience: CIN*, vol. 2016, 2016, doi:10.1155/2016/5720163.
11. D. Martins, and R. Guenther, "Hierarchical Kinematic Analysis of Robots," *Mechanism and Machine Theory*, vol. 38, no. 6, pp. 497–518, Jun. 2003, doi:10.1016/S0094-114X(03)00005-3.
12. B. Karlik, and S. Aydin, "An Improved Approach to the Solution of Inverse Kinematics Problems for Robot Manipulators," *Engineering Applications of Artificial Intelligence*, vol. 13, no. 2, pp. 159–164, Apr. 2000, doi:10.1016/S0952-1976(99)00050-0.
13. L. Jin, S. Li, J. Yu, and J. He, "Robot Manipulator Control Using Neural Networks: A Survey," *Neurocomputing*, vol. 285, pp. 23–34, Apr. 2018, doi:10.1016/j.neucom.2018.01.002.
14. J. M. Ortiz, M. del R. Martínez, J. M. C. Viramontes, and H. R. Vega, "Robust Design of Artificial Neural Networks Methodology in Neutron Spectrometry," *Artificial Neural Networks - Architectures and Applications*, Jan. 2013, doi:10.5772/51274.
15. X. Wu, and Z. Xie, "Forward Kinematics Analysis of a Novel 3-DOF Parallel Manipulator," *Scientia Iranica. Transaction B, Mechanical Engineering*, vol. 26, no. 1, pp. 346–357, 2019, doi:10.24200/sci.2018.20740.
16. A. Hasan, Ali T. Hasan, and H. M. A. A. Al-Assadi, "Performance Prediction Network for Serial Manipulators Inverse Kinematics Solution Passing Through Singular Configurations," *International Journal of Advanced Robotic Systems*, vol. 7, no. 4, pp. 11–24, Jan. 2011.
17. M. I. Petra, and L. C. da Silva, "Inverse Kinematic Solutions Using Artificial Neural Networks," *Applied Mechanics and Materials*, vol. 534, pp. 137–143, Feb. 2014, doi:10.4028/www.scientific.net/AMM.534.137.
18. L. Aggarwal, K. Aggarwal, and R. J. Urbanic, "Use of Artificial Neural Networks for the Development of an Inverse Kinematic Solution and Visual Identification of Singularity Zone(s)," *Procedia CIRP*, 2014, vol. 17, pp. 812–817, doi:10.1016/j.procir.2014.01.107.
19. R. R. Kumar, and P. Chand, "Inverse Kinematics Solution for Trajectory Tracking Using Artificial Neural Networks for SCORBOT ER-4u," in *2015 6th International Conference on Automation, Robotics and Applications (ICARA)*, Queenstown, New Zealand, Feb. 2015, pp. 364–369, doi:10.1109/ICARA.2015.7081175.
20. Y. Zhang, D. Guo, and Z. Li, "Common Nature of Learning Between Back-Propagation and Hopfield-Type Neural Networks for Generalized Matrix Inversion with Simplified Models," *IEEE Transactions on Neural Networks and Learning Systems*, vol. 24, no. 4, pp. 579–592, Apr. 2013, doi:10.1109/TNNLS.2013.2238555.
21. R. Köker, T. Çakar, and Y. Sari, "A Neural-Network Committee Machine Approach to the Inverse Kinematics Problem Solution of Robotic Manipulators," *Engineering with Computers*, vol. 30, no. 4, pp. 641–649, Oct. 2014, doi:10.1007/s00366-013-0313-2.

22. F. Reyes, *Robótica - Control de Robots Manipuladores*. Alfaomega Grupo Editor, 2011.
23. A. O. Baturone, *Robótica: Manipuladores y Robots Móviles*. Marcombo, 2005.
24. Lee, "Robot Arm Kinematics, Dynamics, and Control," *Computer*, vol. 15, no. 12, pp. 62–80, Dec. 1982, doi:10.1109/MC.1982.1653917.
25. J. J. Craig, *Robótica*. Pearson Educación, 2006.
26. X. He, S. Xu, and SpringerLink (Online Service), *Process Neural Networks: Theory and Applications*. Springer Berlin Heidelberg, 2010.
27. S. A. Kalogirou, "Artificial Neural Networks and Genetic Algorithms in Energy Applications in Buildings," *Advances in Building Energy Research*, vol. 3, no. 1, pp. 83–119, Jan. 2009, doi:10.3763/aber.2009.0304.
28. A. K. Jain, Jianchang Mao, and K. M. Mohiuddin, "Artificial Neural Networks: A Tutorial," *Computer*, vol. 29, no. 3, pp. 31–44, Mar. 1996, doi:10.1109/2.485891.
29. J. Zupan, "Introduction to Artificial Neural Network (ANN) Methods: What They Are and How to Use Them," *Acta Chimica Slovenica*, vol. 41. 1994.
30. S. A. Ziaee, E. Sadrossadat, A. H. Alavi, and D. Mohammadzadeh Shadmehri, "Explicit Formulation of Bearing Capacity of Shallow Foundations on Rock Masses Using Artificial Neural Networks: Application and Supplementary Studies," *Environmental Earth Sciences*, vol. 73, no. 7, pp. 3417–3431, Apr. 2015, doi:10.1007/s12665-014-3630-x.
31. T. Ozaki, T. Suzuki, T. Furuhashi, S. Okuma, and Y. Uchikawa, "Trajectory Control of Robotic Manipulators Using Neural Networks," *IEEE Transactions on Industrial Electronics*, vol. 38, no. 3, pp. 195–202, Jun. 1991, doi:10.1109/41.87587.
32. J. M. Ortiz, Ma. del R. Martinez, and H. R. Vega, "Robust Design of Artificial Neural Networks Applying the Taguchi Methodology and DoE," in *Electronics, Robotics and Automotive Mechanics Conference (CERMA'06)*, Sep. 2006, vol. 2, pp. 131–136, doi:10.1109/CERMA.2006.83.
33. Ş. Karabulut, "Optimization of Surface Roughness and Cutting Force During AA7039/Al2O3 Metal Matrix Composites Milling Using Neural Networks and Taguchi Method," *Measurement*, vol. 66, pp. 139–149, Apr. 2015, doi:10.1016/j.measurement.2015.01.027.
34. S. Sholahudin, and H. Han, "Simplified Dynamic Neural Network Model to Predict Heating Load of a Building Using Taguchi Method," *Energy*, vol. 115, pp. 1672–1678, Nov. 2016, doi:10.1016/j.energy.2016.03.057.
35. T. P. Teixeira, C. M. Salgado, R. S. de F. Dam, and W. L. Salgado, "Inorganic Scale Thickness Prediction in Oil Pipelines by Gamma-Ray Attenuation and Artificial Neural Network," *Applied Radiation and Isotopes*, vol. 141, pp. 44–50, Nov. 2018, doi:10.1016/j.apradiso.2018.08.008.
36. V. Uraikul, C. W. Chan, and P. Tontiwachwuthikul, "Artificial Intelligence for Monitoring and Supervisory Control of Process Systems," *Engineering Applications of Artificial Intelligence*, vol. 20, no. 2, pp. 115–131, Mar. 2007, doi:10.1016/j.engappai.2006.07.002.
37. K. López-Linares et al., "Fully Automatic Detection and Segmentation of Abdominal Aortic Thrombus in Post-Operative CTA Images Using Deep Convolutional Neural Networks," *Medical Image Analysis*, vol. 46, pp. 202–214, May 2018, doi:10.1016/j.media.2018.03.010.
38. R. A. Teixeira, A. de P. Braga, and B. R. de Menezes, "Control of a Robotic Manipulator Using Artificial Neural Networks with On-line Adaptation," *Neural Processing Letters*, vol. 12, no. 1, pp. 19–31, Aug. 2000, doi:10.1023/A:1009694129740.
39. B. K. Bose, "Neural Network Applications in Power Electronics and Motor Drives—An Introduction and Perspective," *IEEE Transactions on Industrial Electronics*, vol. 54, no. 1, pp. 14–33, Feb. 2007, doi:10.1109/TIE.2006.888683.

40. S. García, J. Luengo, and F. Herrera, *Data Preprocessing in Data Mining*, vol. 72. Springer International Publishing, 2015.

41. S. S. Mahapatra, and A. Patnaik, "Parametric Optimization of Wire Electrical Discharge Machining (WEDM) Process Using Taguchi Method," *Parametric Optimization of Wire Electrical Discharge Machining (WEDM) Process Using Taguchi Method*, no. Journal, Electronic, 2006.

42. R. K. Roy, *A Primer on the Taguchi Method, Second Edition*. Society of Manufacturing Engineers, 2010.

43. M. Ibrahim, N. Zulikha, Z. Abidin, N. R. Roshidi, N. A. Rejab, and Mohd Faizal Johari, "Design of an Artificial Neural Network Pattern Recognition Scheme Using Full Factorial Experiment," *Applied Mechanics and Materials*, 465–466, pp. 1149–1154, Dec. 2013, doi:10.4028/www.scientific.net/AMM.465-466.1149.

44. J. Limon-Romero, D. Tlapa, Y. Baez-Lopez, A. Maldonado-Macias, and L. Rivera-Cadavid, "Application of the Taguchi Method to Improve a Medical Device Cutting Process," *The International Journal of Advanced Manufacturing Technology*, vol. 87, no. 9–12, pp. 3569–3577, Dec. 2016, doi:10.1007/s00170-016-8623-3.

45. T. Y. Lin, and C. H. Tseng, "Optimum Design for Artificial Neural Networks: An Example in a Bicycle Derailleur System," *Engineering Applications of Artificial Intelligence*, vol. 13, no. 1, pp. 3–14, Feb. 2000, doi:10.1016/S0952-1976(99)00045-7.

46. M. S. Packianather, and P. R. Drake, "Modelling Neural Network Performance through Response Surface Methodology for Classifying Wood Veneer Defects," *Proceedings of the Institution of Mechanical Engineers*, vol. 218, no. 4, pp. 459–466, Apr. 2004.

47. T. Y. Lin, and C. H. Tseng, "Optimum Design for Artificial Neural Networks: An Example in a Bicycle Derailleur System," *Engineering Applications of Artificial Intelligence*, vol. 13, no. 1, pp. 3–14, 2000, https://doi.org/10.1016/S0952-1976(99)00 045-7.

48. K. Vahab, "An Optimized Artificial Neural Network for Human-Force Estimation: Consequences for Rehabilitation Robotics," *Industrial Robot: An International Journal*, vol. 45, no. 3, pp. 416–423, Jan. 2018, doi:10.1108/IR-10-2017-0190.

49. S. Kucuk, and Z. Bingul, "Inverse Kinematics Solutions for Industrial Robot Manipulators with Offset Wrists," *Applied Mathematical Modelling*, vol. 38, no. 7, pp. 1983–1999, 2014, doi:https://doi.org/10.1016/j.apm.2013.10.014.

50. L. Zhang, and N. Xiao, "A Novel Artificial Bee Colony Algorithm for Inverse Kinematics Calculation of 7-DOF Serial Manipulators," *Soft Computing*, vol. 23, no. 10, pp. 3269–3277, May 2019, doi:10.1007/s00500-017-2975-y.

51. D. L. Pieper, "The Kinematics of Manipulators under Computer Control," Physiol.D., Stanford University, Ann Arbor, 1969.

52. M. Tarokh, and M. Kim, "Inverse Kinematics of 7-DOF Robots and Limbs by Decomposition and Approximation," *IEEE Transactions on Robotics*, vol. 23, no. 3, pp. 595–600, Jun. 2007, doi:10.1109/TRO.2007.898983.

53. S. Tejomurtula, and S. Kak, "Inverse Kinematics in Robotics Using Neural Networks," *Information Sciences*, vol. 116, no. 2, pp. 147–164, Jan. 1999, doi:10.1016/S0020-0255(98)10098-1.

54. A. Larrañaga, "3D Printable Robotic Arm," *GitHub*, 2018. https://github.com/AngelLM (accessed Sep. 18, 2019).

55. J. Denavit, and R. S. Hartenberg, "A Kinematic Notation for Lower-Pair Mechanisms Based on Matrices.," *Transactions of the ASME Journal of Applied Mechanics*, vol. 22, pp. 215–221, 1955.

56. P. Corke, *Robotics, Vision and Control: Fundamental Algorithms in MATLAB*, vol. 73. Springer, 2011.

57. H. Liu, and H. Motoda, *Instance Selection and Construction for Data Mining.* Springer US, 2001.
58. P. R. Krishnaiah and C. R. Rao, Eds., *Sampling.* Elsevier Science Publisher, 1988.
59. B. Gu, F. Hu, and H. Liu, "Sampling and Its Application in Data Mining," Technical Report TRA6/00, Department of Computer Science, National University of Singapur, 2000.
60. H. Liu, and H. Motoda, "On Issues of Instance Selection," *Data Mining and Knowledge Discovery*, vol. 6, no. 2, pp. 115–130, Apr. 2002, doi:10.1023/A:1014056429969.
61. H. Brighton, and C. Mellish, "Advances in Instance Selection for Instance-Based Learning Algorithms," *Data Mining and Knowledge Discovery*, vol. 6, no. 2, pp. 153–172, Apr. 2002, doi:10.1023/A:1014043630878.
62. W. G. Madow, and L. H. Madow, "On the Theory of Systematic Sampling, I," *The Annals of Mathematical Statistics*, vol. 15, no. 1, pp. 1–24, 1944.
63. S. A. Mostafa and I. A. Ahmad, "Remainder Linear Systematic Sampling with Multiple Random Starts," *Journal of Statistical Theory and Practice*, vol. 10, no. 4, pp. 824–851, Oct. 2016, doi:10.1080/15598608.2016.1231094.
64. L. H. Madow, "Systematic Sampling and its Relation to Other Sampling Designs," *Journal of the American Statistical Association*, vol. 41, no. 234, pp. 204–217, Jun. 1946, doi:10.1080/01621459.1946.10501864.

10 Generative Adversarial Network and Its Applications

A. Vijayalakshmi and Deepa V. Jose

CONTENTS

10.1 INTRODUCTION

Generative adversarial networks (GAN) is a deep learning unsupervised machine learning technique that was introduced in 2014 by Ian J. Goodfellow and co-authors [1]. The objective of GAN is generating data from scratch wherein it consists of two neural networks that generate new data points from random uniform distribution to get an accurate prediction. GANs are modelled as a methodology to generative modelling that uses deep learning techniques like convolutional neural networks. In order to generate new samples, the model automatically learns the patterns in the input image data. GANs consist of two models as represented in Figure 10.1: generator and discriminator.

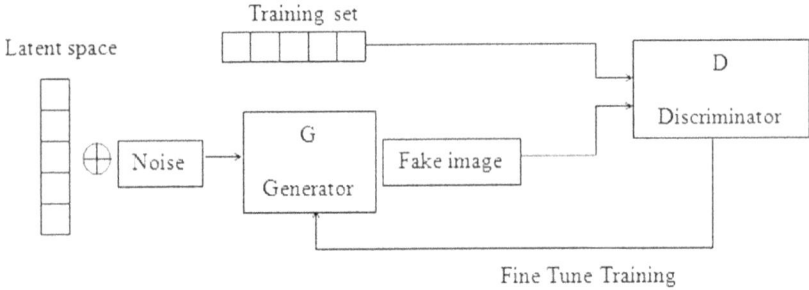

FIGURE 10.1   Generator and discriminator model.

The generator is a neural network that generates new data instances whose objective is to fool the discriminator. The discriminator evaluates the instances generated by the generator for authenticity of the instances so as not to be fooled by the generator. The generator feeds as the input a random vector at each training step, thereby preventing the model from producing the same sample every single time. The discriminator receives the synthetic sample produced by the generator and outputs the probability that the input image is a real training sample or a fake sample that is generated by the generator predicting the authenticity of the input image.

Table 10.1 describes the steps taken by a generative adversarial network based on the diagram depicted in Figure 10.1.

GAN is a deep learning method where two neural networks compete with each other. Deep learning is a subset of machine learning based on artificial neural networks wherein there is a learning involved by improving its own computer algorithms [2]. Neural network consists of layers made up of different numbers of nodes which are connected to nodes of adjacent layers. Each node within the layer is assigned a weight and the node with heavier weight will exert more effect on the next layer of nodes. Deep learning is used in discovering hierarchical models that epitomize probability distribution on data that is involved in artificial intelligence applications.

10.2   DISCRIMINATIVE LEARNING VS. GENERATIVE LEARNING

The two different approaches to classification techniques in machine learning are generative and discriminative models. Discriminative model builds the output based

---

TABLE 10.1

Steps taken by GAN

- A generator takes random numbers as input and returns an image.
- The image so generated is fed to the discriminator together with the actual images from the data set.
- The real and fake images are taken by the discriminator and returns a probability
- The discriminator is in a feedback loop with the training set of images.
- The generator is in a feedback loop with the discriminator.

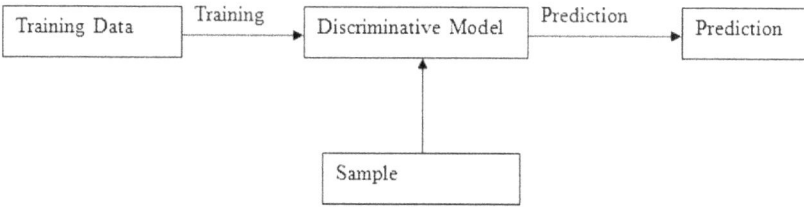

FIGURE 10.2   Process of discriminative model.

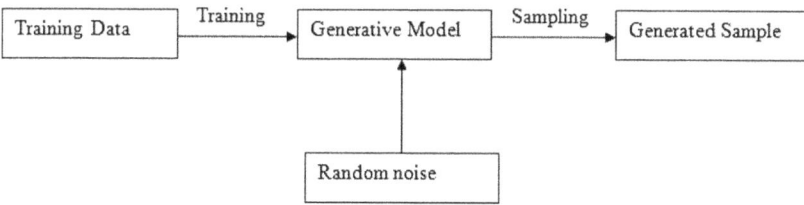

FIGURE 10.3   Process of generative modelling.

on the observed data and quality of data rather than depending on the probabilistic distribution. When provided with a good amount of data, discriminative models tend to give better results. Figure 10.2 describes the process of a discriminative model.

Generative model understands the distribution of data and performs the task of classification. Generative model creates similar data as the training data after learning the distribution of data. Figure 10.3 shows the process of generative modelling.

The training data consists of observations with many features for a problem of image generation. Generative models can generate new sets of features that look similar to the one that is created using the original data. In order to accomplish this, the model should include a random element that builds a probabilistic distribution model which affects each of the individual samples generated by the model.

10.3   DEEP GENERATIVE MODEL

In a deep learning model, the major challenge is the scarcity of training data. Generative models are profoundly used in generating new data. For a healthcare industry, where there is scarcity of data, generative models can be helpful in generating data for training the model and hence will play a major role in increasing the size of data set. Translating satellite images to Google maps, converting a black and white image to colour etc. are some of the applications of a generative model.

The generative model generates new data points by learning the data distribution from the training data set. To make it learn the exact distribution of data, a neural network is used [3]. Deep generative model has difficulty in approximating many inflexible probabilistic calculations in the generative context [1]. Deep generative models build generative models and also algorithms that are used for learning

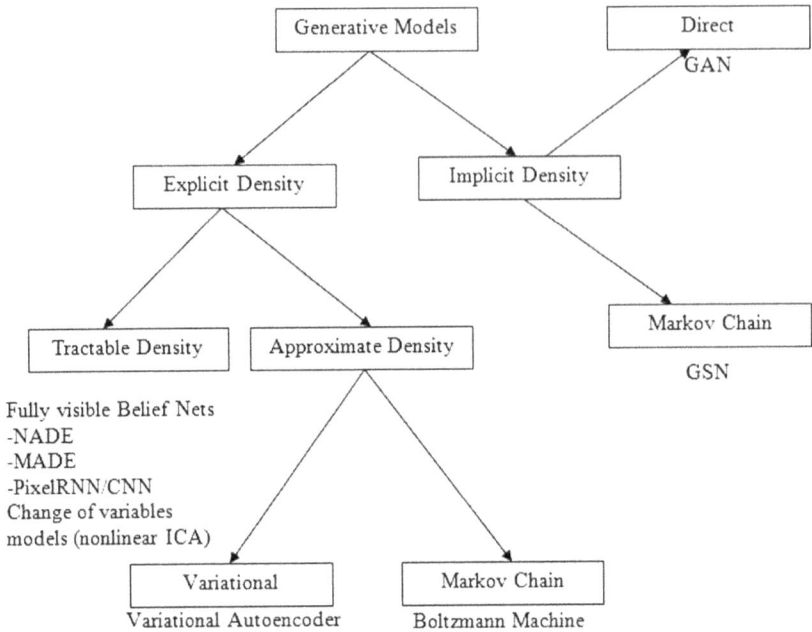

FIGURE 10.4 Types of generative models.

the model using ideas from deep learning. Different types of generative models are shown in Figure 10.4.

Explicit probabilistic model provides explicit parametric specification of the distribution of data that have controllable functions. Figure 10.5 shows a simple explicit probabilistic model.

The explicit probabilistic model for Figure 10.4 is

$$p\left(x,z|\alpha\right) = p\left(x|z\right)p\left(z|\alpha\right)$$

Implicit probabilistic models define a stochastic process wherein after training of the model, the samples are generated based on the underlying distribution of data. Variational auto encoders (VAEs) and generative adversarial networks are examples of generative model approaches.

10.4   VARIATIONAL AUTO ENCODERS

An artificial neural network that is engaged in training models that produce a reconstructed input as the output of a network is termed as auto encoders. Auto encoders consist of two connected networks, encoder and decoder, wherein an encoder takes an input and produces a feature, as demonstrated in Figure 10.6.

Incorporating a decoder network and loss function in order to learn the features and train the model, the auto encoder network is presented in Figure 10.7.

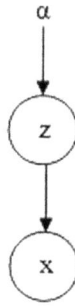

FIGURE 10.5   Explicit probabilistic model.

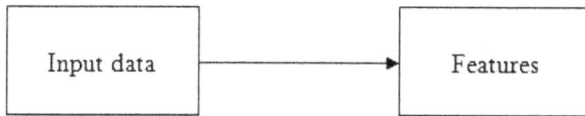

FIGURE 10.6   Block diagram of auto encoder

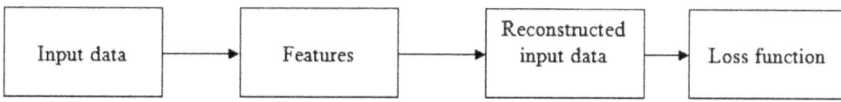

FIGURE 10.7   Auto encoder network.

The neural network uses dimensionality reduction techniques in order to restructure the inputs so that input values will be better for a training model. A variational auto encoder (VAE) is an architecture that is developed consisting of an encoder, a decoder, and a loss function which in turn is trained in order to minimize the error between the encoded–decoded data and the initial data that is generated while reconstruction. The purpose of the decoder is to reconstruct the data and encoder represents a neural network [4].

The objective of both encoders and variational auto encoders is to generate an input data of lower dimension. Considering the input, $x \in Rn$, the objective of an auto encoder is to produce $h \in Rd$ where $d < n$. The output is designed such that it contains the very vital features of the input, x, in order to reconstruct it. The difference between an auto encoder and a VAE lies in the fact that VAE is probabilistic and a generative model. Hence, by sampling from the distribution, samples $x \in R^n$ can be obtained [5].

10.5   GENERATIVE ADVERSARIAL NETWORK

The objective of GAN is to generate images (mostly) from scratch [1, 6–8]. GAN belongs to the category of generative model and it has the capability to handle sharp density functions and generate preferred samples. GAN achieves this with the help

---

Algorithm 1 Minibatch stochastic gradient descent training of generative adversarial nets. The number of steps to apply to the discriminator, k is a hyperparameter. We used k=1, the least expensive option, in our experiments.

---

for number of training iteration do
for k steps do
- Sample minibatch of m noise sample $\{z^{(1)},.....,z^{(m)}\}$ from noise prior $p_z(z)$.
- Sample minibatch of m examples $\{x^{(1)},...,x^{(m)}\}$ from data generating distribution $p_{data}(x)$.
- Update the discriminator by ascending its stochastic gradient:

$$\nabla_{\theta_d} \frac{1}{m} \sum_{i=1}^{m} \left[\log D\left(x^{(i)}\right) + \log\left(1 - D\left(G(z^{(i)})\right)\right) \right]$$

end for
- Sample minibatch of m noise samples $\{z^{(1)},...,z^{(m)}\}$ from noise prior $p_z(z)$
- Update the generator by descending its stochastic gradient:

$$\nabla_{\theta_g} \frac{1}{m} \sum_{i=1}^{m} \log\left(1 - D\left(G\left(z^{(i)}\right)\right)\right)$$

end for
The gradient based approach can use any standard gradient-based learning rule. We used momentum in our experiments.

FIGURE 10.8   Summary of GAN training algorithm.

of a pair of networks with backpropagation signals. The samples generated by GAN can be used in applications like image synthesis, image editing, image super-resolution and classification, etc. [8]. The algorithm for generative adversarial networks taken from original paper by Goodfellow, 2014, is depicted in Figure 10.8. The algorithm consists of the process of training both the generator model and the discriminator model in parallel.

10.6   ARCHITECTURE OF GENERATIVE ADVERSARIAL NETWORK

Generative adversarial network belongs to the class of generative models that can generate new content from the available data [1]. GAN is well accepted by the deep learning community and its application spreads across in the domains of computer vision, natural language processing, time series synthesis, semantic segmentation, etc. [8].

Figure 10.9 represents the architecture of GAN that involves two major modules: a discriminator (D) that helps in differentiating between a real image and generated image and a generator (G) that produces images to deceive the discriminator in the model. For a given distribution $z \sim pz$, G defines a probability distribution pg as the distribution of the samples $G(z)$. The objective of GAN is to learn the generator's distribution pg that estimates the real data distribution pr. The joint loss function for the discriminator D and generator G leads to the optimization of GAN.

$$\min_G \max_D V\left(D,G\right) = E_{x \sim pdata}\left[\log D(x)\right] + E_{z \sim pz(z)}\left[\log 1 - D\left(G\left(Z\right)\right)\right]$$

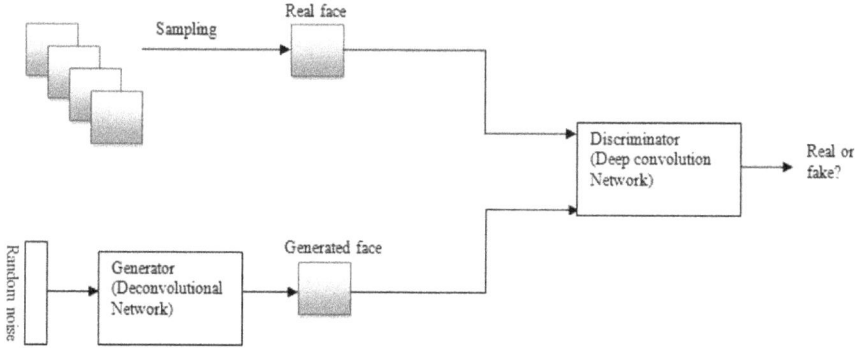

FIGURE 10.9 Architecture of GAN.

10.7 VARIATIONS OF GAN ARCHITECTURES

10.7.1 FULLY CONNECTED GAN (FCGAN)

The first architecture for GAN depicted a fully connected neural network for both the components, namely generator and discriminator [7], and this architecture was implemented on simple image data sets like MNSIT handwritten digit data set.

10.7.2 LAPLACIAN PYRAMID OF ADVERSARIAL NETWORKS (LAPGAN)

CNN is considered to be best for image data [7]. The Laplacian pyramid of adversarial networks (LAPGAN) offers a solution to the problem of difficulty in training the generator and discriminator using convolutional neural networks. LAPGAN decomposes the process of generation using multiple scales. The image is decomposed into a Laplacian pyramid. Further, a conditional, convolutional GAN is trained to produce each layer [7, 8].

10.7.3 DEEP CONVOLUTIONAL GAN (DCGAN)

Deep convolutional GAN is one of the effective network models of GAN. This network works with continuous training of a pair of deep convolutional discriminator and generator. DCGAN applied deconvolutional neural network architecture for the generator and this model has demonstrated a good performance for CNNs visualization [8]. This model consists of convolutional layers without max pooling and the convolution in this model allows spatial down-sampling up-sampling operators that would be learned during training the sample data [7]. Figure 10.10 illustrates the architecture of the deep convolutional network.

10.7.4 CONDITIONAL GAN

GAN framework is extended to conditional setting [9] in which both generator and discriminator class is conditioned with extra auxiliary information like class labels

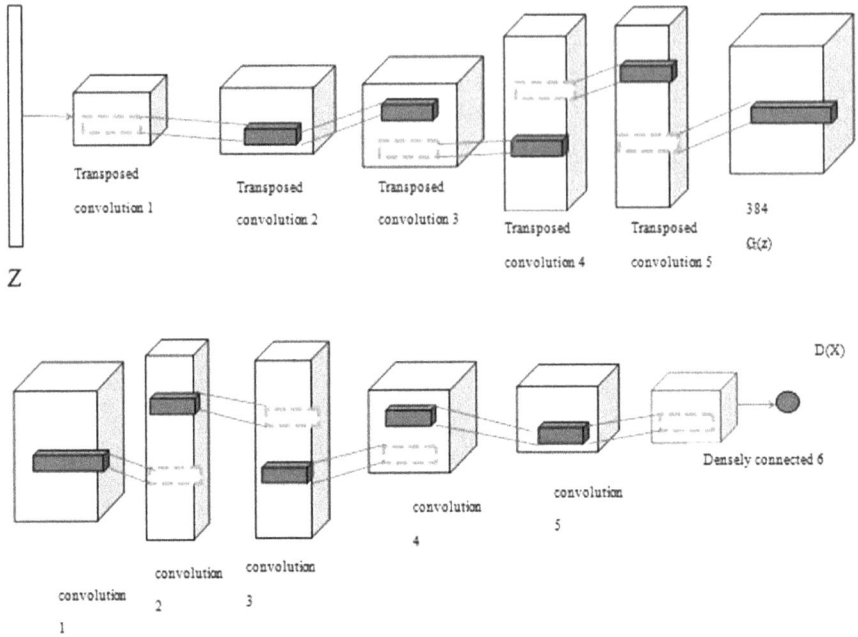

FIGURE 10.10   Components of GAN architecture.

by feeding this information as an extra layer. This model provides better representation for multimodal data generation. Figure 10.11 illustrates the structure of a conditional adversarial network.

10.7.5   Least-Square GAN

GAN works with the basic concept of simultaneously training the discriminator and generator where the function of the generator lies in generating fake image samples that look real. There are innumerable works showing the application of GAN in generation of new image samples but the quality of such generated images is limited. GAN uses sigmoid cross-entropy loss function in the discriminator. LSGAN is proposed based on the assumption that the sigmoid loss function will lead to vanishing gradient when updating the generator [10]. Least-square GAN (LSGAN) transforms the GAN through L2 loss function as a replacement for log loss function and hence we could say that LSGAN uses the least square loss function for the discriminator. In this method labels for fake and real data, a and b, are used as a–b coding scheme is used in the discriminator. The objective function can be written as

$$\min_D V_{\text{LSGAN}}(D) = \frac{1}{2} E_{x \sim pdata(x)} \left[\left(D(x) - b \right)^2 \right] + \frac{1}{2} E_{z \sim pz(z)} \left[\left(D\left(G(Z) \right) - a \right)^2 \right]$$

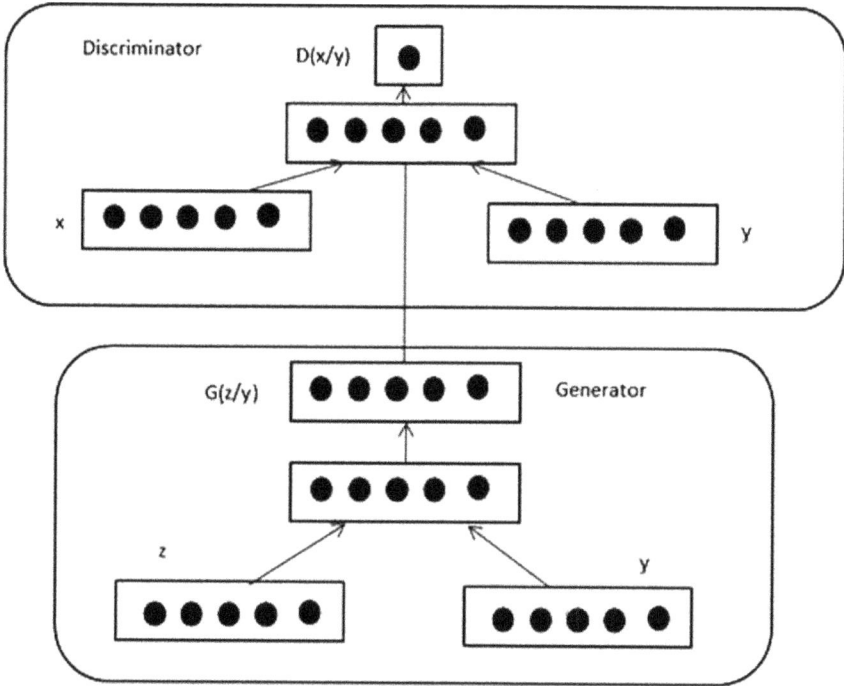

FIGURE 10.11 Architecture of GAN.

$$\min_G V_{\text{LSGAN}}(G) = \frac{1}{2} E_{z \sim pz(z)} \left[\left(D\big(G(z)\big) - c \right)^2 \right]$$

In the above example, c represents the value of fake data that the generator and discriminator want to believe. The advantage of LSGAN is that unlike regular GAN, LSGAN produce superior quality images and they are more stable in the learning procedure.

10.7.6 AUXILIARY CLASSIFIER GAN

The auxiliary GAN (AC-GAN) aimed at improving the training of generative adversarial networks for synthesizing images [11]. In this proposed method, a class label in addition to the noise is available for each of the generated samples. The objective functions defined by L_S and L_C are described as follows:

$$L_S = E\left[\log P\big(S = real | X_{real}\big) \right] + E\left[\log P\big(S = fake | X_{fake}\big) \right]$$

$$L_C = E\left[\log P\big(C = c | X_{real}\big) \right] + E\left[\log P\big(C = c | X_{fake}\big) \right]$$

The discriminator is trained so that $LS+LC$ are maximized, and the generator is trained so that the value of $LC\text{-}LS$ is maximized. This method provided a new metric for image discriminability. The result of the work proves to be more discriminable than that of a model that generates images with low resolution.

10.7.7 INFOGAN

InfoGAN incorporates the information theory for the transformation of noise in an image to latent codes that have predictable consequences in the output [12]. In this method, the generator is split into two parts: traditional noise vector and latent code vector. Further, the mutual information between the codes is maximized. This algorithm is implemented by including a regularization term $-\lambda I(c;G(z,c))$ – to the objective function of GAN.

$$\min_G \max_D V_I(D,G) = V(D,G) - \lambda I(c;G(z,c))$$

Lambda is the regularization term, $I(c;G(z,c))$ is the mutual information between the latent code c and $G(z,c)$, which is the generator output. InfoGAN is unsupervised and it learns interpretable and untangled information on images. The advantage of this method is that it adds negligible computation cost on top of GAN and hence makes the whole system very easy to train the images.

10.8 APPLICATIONS OF GAN

The ability of GAN to generate immense new samples on demand led to its use in a variety of real-time applications. The currently existing GAN applications can be segregated into three broad categories: image generation, image translation, and anomaly detection. This section gives an overview of the various applications under these three main categories.

10.8.1 IMAGE GENERATION

The major applications related to this category belong to the generation of image data sets of handwritten digits, photographs of small objects, and human faces based on the MNIST, CIFAR-10, and Toronto Face Database [1]. Generation of quite realistic human faces is described in Ref. [13], which was later adopted to generate images of other objects. Realistic photographs [14] is another commonly used and widely accepted application. It seemed to be so perfect and was unable to differentiate between the synthetically generated and the real-time ones. Creation of impressive cartoon characters [15], especially characters from the Japanese comic books, was extensively used.

Similarly, another fancy application is the photos to emojis [16] where the human faces or signboards, etc. will be converted to various emojis. Face frontal view generation [17] found immense use in human face identification and verification applications. Generating new human poses [18] seems to be useful in textile and modelling industries.

10.8.2 Image Translation

Another quite useful application of GAN is the image translation and editing. In many scenarios, we need to convert images into different image formats or perform various operations on it to extract information. Image-to-Image translation [19] is used in many areas like creation of Google maps from satellite images, converting sketches or black and white images to colour, etc.

Image-to-image translation includes producing another engineered rendition of a given picture with a particular adjustment, for example, making an interpretation of a late spring scene to winter. Training a model for picture to-picture interpretation commonly requires an enormous data set of combined models. These data sets can be troublesome and costly to make, and at times infeasible. The CycleGAN [20] is a technique that does the automatic training of the image to-image interpretation models without combined models. The models are prepared in an unsupervised way, utilizing an assortment of pictures from the source and target domain that should not be connected in any way. This procedure is an effective method for accomplishing outwardly amazing outcomes on a scope of a variety of use areas.

Similarly, text-to-image translation [21] and semantic image-to-photo translation [22] are quite useful in many scientific and learning applications. Photograph editing is another vibrant application for reconstructing photographs [23]. This also helps in enhancing specific features and adding or removing features. To visualize the faces based on age, face ageing [17] application is useful. Blending the features from different images is possible with photo blending [24]. Besides, all the best quality view of images can be assured through the high resolution using super resolution [25] methods. If we want to fill some areas of images which were removed for some reason, it is possible with photo inpainting [26]. This will retain the originality of the image. Clothing translation [27] is now a popular application installed in cloth marts where the customer can view how the attire looks on them. In fact, video prediction [28], especially in sports and entertainment, is another application related to this area. We can also see a variety of 3D object generation [29] currently used in design applications to augment the user experience.

10.8.3 Anomaly Detection

Enhancing the security for comfortable living is always the need of the hour, especially in this era of smart manufacturing. To ensure security and prevent intrusion from any sort through predictions is currently in practice in many scenarios wherein we are completely connected to the internet. Security analysis [30] can be achieved through GANs. Similarly, intrusion detection [31–33] is another area where GANs are of much help where models can be devised to predict the chances of intrusions by monitoring the continuous data from the actuators and the sensors.

10.9 CONCLUSION

The enormous attraction received by the generative adversarial network is not only because of the ability of the model to learn deep mappings but also because of its capability to make use of unlabelled image data. This chapter gives a brief introduction on GAN and its variations with possible applications. It is very much clear that the influence of deep neural networks offers enormous openings for novel applications.

REFERENCES

1. I. J. Goodfellow et al., "Generative Adversarial Networks," *ArXiv14062661 Cs Stat*, Jun. 2014, Accessed: Dec. 16, 2020. [Online]. Available: http://arxiv.org/abs/1406.2661.

2. X. Dong, J. Wu, and L. Zhou, "How deep learning works –The geometry of deep learning," *ArXiv171010784 Cs Stat*, Oct. 2017, Accessed: Dec. 16, 2020. [Online]. Available: http://arxiv.org/abs/1710.10784.

3. R. Salakhutdinov, "Learning Deep Generative Models," *Annu. Rev. Stat. Its Appl.*, vol. 2, no. 1, pp. 361–385, Apr. 2015, doi: 10.1146/annurev-statistics-010814-020120.

4. C. Doersch, "Tutorial on Variational Autoencoders," *ArXiv160605908 Cs Stat*, Aug. 2016, Accessed: Dec. 16, 2020. [Online]. Available: http://arxiv.org/abs/1606.05908.

5. R. Wei, C. Garcia, A. El-Sayed, V. Peterson, and A. Mahmood, "Variations in Variational Autoencoders - A Comparative Evaluation," *IEEE Access*, vol. 8, pp. 153651–153670, 2020, doi: 10.1109/ACCESS.2020.3018151.

6. K. Wang, C. Gou, Y. Duan, Y. Lin, X. Zheng, and F.-Y. Wang, "Generative Adversarial Networks: Introduction and Outlook," *IEEECAA J. Autom. Sin.*, vol. 4, no. 4, pp. 588–598, 2017, doi: 10.1109/JAS.2017.7510583.

7. A. Creswell, T. White, V. Dumoulin, K. Arulkumaran, B. Sengupta, and A. A. Bharath, "Generative Adversarial Networks: An Overview," *IEEE Signal Process. Mag.*, vol. 35, no. 1, pp. 53–65, Jan. 2018, doi: 10.1109/MSP.2017.2765202.

8. Z. Wang, Q. She, and T. E. Ward, "Generative Adversarial Networks in Computer Vision: A Survey and Taxonomy," *ArXiv190601529 Cs*, Jun. 2020, Accessed: Dec. 16, 2020. [Online]. Available: http://arxiv.org/abs/1906.01529.

9. M. Mirza and S. Osindero, "Conditional Generative Adversarial Nets," *ArXiv14111784 Cs Stat*, Nov. 2014, Accessed: Dec. 16, 2020. [Online]. Available: http://arxiv.org/abs/1411.1784.

10. X. Mao, Q. Li, H. Xie, R. Y. K. Lau, Z. Wang, and S. P. Smolley, "Least Squares Generative Adversarial Networks," p. 9.

11. A. Odena, C. Olah, and J. Shlens, "Conditional Image Synthesis With Auxiliary Classifier GANs," *ArXiv161009585 Cs Stat*, Jul. 2017, Accessed: Dec. 16, 2020. [Online]. Available: http://arxiv.org/abs/1610.09585.

12. X. Chen, Y. Duan, R. Houthooft, J. Schulman, I. Sutskever, and P. Abbeel, "InfoGAN: Interpretable Representation Learning by Information Maximizing Generative Adversarial Nets," *ArXiv160603657 Cs Stat*, Jun. 2016, Accessed: Dec. 16, 2020. [Online]. Available: http://arxiv.org/abs/1606.03657.

13. T. Karras, T. Aila, S. Laine, and J. Lehtinen, "Progressive Growing of GANs for Improved Quality, Stability, and Variation," *ArXiv171010196 Cs Stat*, Feb. 2018, Accessed: Dec. 16, 2020. [Online]. Available: http://arxiv.org/abs/1710.10196.

14. A. Brock, J. Donahue, and K. Simonyan, "Large Scale GAN Training for High Fidelity Natural Image Synthesis," *ArXiv180911096 Cs Stat*, Feb. 2019, Accessed: Dec. 16, 2020. [Online]. Available: http://arxiv.org/abs/1809.11096.

15. Y. Jin, J. Zhang, M. Li, Y. Tian, H. Zhu, and Z. Fang, "Towards the Automatic Anime Characters Creation with Generative Adversarial Networks," *ArXiv170805509 Cs*, Aug. 2017, Accessed: Dec. 16, 2020. [Online]. Available: http://arxiv.org/abs/1708.05509.

16. Y. Taigman, A. Polyak, and L. Wolf, "Unsupervised Cross-Domain Image Generation," *ArXiv161102200 Cs*, Nov. 2016, Accessed: Dec. 16, 2020. [Online]. Available: http://arxiv.org/abs/1611.02200.

17. G. Antipov, M. Baccouche, and J.-L. Dugelay, "Face Aging With Conditional Generative Adversarial Networks," *ArXiv170201983 Cs*, May 2017, Accessed: Dec. 16, 2020. [Online]. Available: http://arxiv.org/abs/1702.01983.

18. L. Ma, X. Jia, Q. Sun, B. Schiele, T. Tuytelaars, and L. V. Gool, "Pose Guided Person Image Generation," p. 11.

19. P. Isola, J.-Y. Zhu, T. Zhou, and A. A. Efros, "Image-to-Image Translation with Conditional Adversarial Networks," in *2017 IEEE Conference on Computer Vision and Pattern Recognition (CVPR)*, Honolulu, HI, Jul. 2017, pp. 5967–5976, doi: 10.1109/CVPR.2017.632.

20. J.-Y. Zhu, T. Park, P. Isola, and A. A. Efros, "Unpaired Image-to-Image Translation using Cycle-Consistent Adversarial Networks," *ArXiv170310593 Cs*, Aug. 2020, Accessed: Dec. 16, 2020. [Online]. Available: http://arxiv.org/abs/1703.10593.

21. H. Zhang et al., "StackGAN: Text to Photo-realistic Image Synthesis with Stacked Generative Adversarial Networks," *ArXiv161203242 Cs Stat*, Aug. 2017, Accessed: Dec. 16, 2020. [Online]. Available: http://arxiv.org/abs/1612.03242.

22. T.-C. Wang, M.-Y. Liu, J.-Y. Zhu, A. Tao, J. Kautz, and B. Catanzaro, "High-Resolution Image Synthesis and Semantic Manipulation with Conditional GANs," *ArXiv171111585 Cs*, Aug. 2018, Accessed: Dec. 16, 2020. [Online]. Available: http://arxiv.org/abs/1711.11585.

23. G. Perarnau, J. van de Weijer, B. Raducanu, and J. M. Álvarez, "Invertible Conditional GANs for image editing," *ArXiv161106355 Cs*, Nov. 2016, Accessed: Dec. 16, 2020. [Online]. Available: http://arxiv.org/abs/1611.06355.

24. H. Wu, S. Zheng, J. Zhang, and K. Huang, "GP-GAN: Towards Realistic High-Resolution Image Blending," *ArXiv170307195 Cs*, Aug. 2019, Accessed: Dec. 16, 2020. [Online]. Available: http://arxiv.org/abs/1703.07195.

25. C. Ledig et al., "Photo-Realistic Single Image Super-Resolution Using a Generative Adversarial Network," in *2017 IEEE Conference on Computer Vision and Pattern Recognition (CVPR)*, Honolulu, HI, Jul. 2017, pp. 105–114, doi: 10.1109/CVPR.2017.19.

26. D. Pathak, P. Krahenbuhl, J. Donahue, T. Darrell, and A. A. Efros, "Context Encoders: Feature Learning by Inpainting," in *2016 IEEE Conference on Computer Vision and Pattern Recognition (CVPR)*, Las Vegas, NV, USA, Jun. 2016, pp. 2536–2544, doi: 10.1109/CVPR.2016.278.

27. D. Yoo, N. Kim, S. Park, A. S. Paek, and I. S. Kweon, "Pixel-Level Domain Transfer," *ArXiv160307442 Cs*, Nov. 2016, Accessed: Dec. 16, 2020. [Online]. Available: http://arxiv.org/abs/1603.07442.

28. C. Vondrick, H. Pirsiavash, and A. Torralba, "Generating Videos with Scene Dynamics," *ArXiv160902612 Cs*, Oct. 2016, Accessed: Dec. 16, 2020. [Online]. Available: http://arxiv.org/abs/1609.02612.

29. J. Wu, C. Zhang, T. Xue, W. T. Freeman, and J. B. Tenenbaum, "Learning a Probabilistic Latent Space of Object Shapes via 3D Generative-Adversarial Modeling," *ArXiv161007584 Cs*, Jan. 2017, Accessed: Dec. 16, 2020. [Online]. Available: http://arxiv.org/abs/1610.07584.

30. S. R. Chhetri, A. B. Lopez, J. Wan, and M. A. Al Faruque, "GAN-Sec: Generative Adversarial Network Modeling for the Security Analysis of Cyber-Physical Production Systems," in *2019 Design, Automation & Test in Europe Conference & Exhibition (DATE)*, Florence, Italy, Mar. 2019, pp. 770–775, doi: 10.23919/DATE.2019.8715283.

31. H. Zenati, C. S. Foo, B. Lecouat, G. Manek, and V. R. Chandrasekhar, "Efficient GAN-Based Anomaly Detection," *ArXiv180206222 Cs Stat*, May 2019, Accessed: Dec. 16, 2020. [Online]. Available: http://arxiv.org/abs/1802.06222.

32. H. Zenati, M. Romain, C. S. Foo, B. Lecouat, and V. R. Chandrasekhar, "Adversarially Learned Anomaly Detection," *ArXiv181202288 Cs Stat*, Dec. 2018, Accessed: Dec. 16, 2020. [Online]. Available: http://arxiv.org/abs/1812.02288.

33. D. Li, D. Chen, J. Goh, and S. Ng, "Anomaly Detection with Generative Adversarial Networks for Multivariate Time Series," *ArXiv180904758 Cs Stat*, Jan. 2019, Accessed: Dec. 16, 2020. [Online]. Available: http://arxiv.org/abs/1809.04758.

11 Applications of Artificial Intelligence in Environmental Science

Praveen Kumar Gupta, Apoorva Saxena,
Brahmanand Dattaprakash, Ryna Shireen Sheriff,
Surabhi Hitendra Chaudhari, Varun Ullanat,
V. Chayapathy

CONTENTS

11.1 INTRODUCTION

11.1.1 ARTIFICIAL INTELLIGENCE

In 1956, artificial intelligence (AI) was established as an educational field of study [1]. In the field of computer applications and science, artificial intelligence can be described as the intelligence that is indicated by machines, as compared to the natural intelligence manifested in humans [2]. The idea was based on the assumption that human intelligence "can be described so accurately that a machine can be trained to simulate it" [3]. The human mind is capable of conjuring logical reasons and explanations for events that occur beyond the measure of their control. Through logic and emotions, some problems may be straightforward for humans to solve, but rather difficult to computationally reciprocate. For this reason, there are two classes of models: functionalist and structuralist. Where the functionalist data correlates data to the equivalent obtained computationally, a structural model casually impersonates the cardinal functioning of the mind [4]. Artificial intelligence is broken into three domains: mundane, expert, and formal tasks. Each of these three domains, in turn, contains the fields artificial intelligence plays a role in. Formal tasks are known for the role they play in the mathematical and gaming fields. Mathematics is not exclusive to arithmetic but includes geometry, calculus, and logical math as well. Artificial intelligence has been used to verify existing theorems as well as theorems being worked on. Expert tasks include artificial intelligence applications in diagnosis in the medical field, analysis of scientific data, analysis of financial data as well as in engineering. In engineering, artificial intelligence is applied to error probing and analysis, manufacturing processes, efficiency testing, and monitoring all manufactured items and processes. Mundane tasks describe the role of artificial intelligence in language processing, perception, logic, and reasoning. Perception includes speech and voice applications as well as computer vision [5].

11.1.2 ARTIFICIAL INTELLIGENCE AND ITS ROLES IN THE ENVIRONMENT

Artificial intelligence is now being used to improve environmental conditions, which is in a dire state and requires improvement as soon as possible. From water quality testing to air quality testing, the applications of artificial intelligence are limitless and we just need to create suitable algorithms for it to be achieved. Currently, IBM is in the lead in attempting to reach new heights in the field of research related to improving environmental conditions. From their new radical recycling process, to obtaining real-time data from oceans, to testing soil and water conditions, IBM methods of bettering the environment using artificial intelligence show the capability of machine intelligence. The further subtopics in this chapter will describe the various methods IBM and the rest of the world have undertaken to ameliorate the environmental state [6].

11.2 TECHNOLOGICAL SOLUTIONS

11.2.1 AUTONOMOUS AND CONNECTED ELECTRIC VEHICLES

The automotive industry is on the verge of the age of automated vehicles. Connected and Autonomous Electric Vehicles (CAEV) works on three underlying technologies: Connected Vehicles (CV), Autonomous Vehicles (AV), and Electric Vehicles (EV). The amalgamation of facilities offered by CAEVs plays a significant function. CAEVs can provide services for individuals and products shipped, such as self-driving, advanced communications, and greatly enhanced mobility. Vehicle knowledge and networking are part of an evolving response to the current problems of travel in the world. These advance swiftly to provide fully controlled and partly autonomous vehicles. By 2030, roads are going to be dominated by these vehicles, generating significant advantages such as improvised equity, safer roads, and transport access. Thus, a considerable reduction of greenhouse gases and congestion of traffic can be expected.

The Society of Automotive Engineers states that there are six levels of driving autonomy from level 0 (no automation) to level 5 (self-driving). Intermediate levels (levels 1, 2, 3, and 4) are supposed to gain positive advantages, such as safer roads. The automation level 5 is necessary to achieve the maximum economic benefit of AV. That's when cars don't need drivers. An autonomous vehicle must be adaptive, technical, and networking to its environment. Another significant consideration is that energy is required to reduce GHG emissions as desired. An electricity source should be incorporated in a vehicle to grid [V2G], a grid to the vehicle [G2V] to coordinate its charging and unload operations and autonomously reduce its energy usage. Many advantages can be seen as on-demand technologies such as mobility as a service, dissipation of traffic congestion as a service, electricity storage as a service, cellular charging as a service, sensing as a service, computation as a service, and content distribution as a service. Such programs may be used by citizens, municipal governments, and non-governmental organizations, and even by corporations. Such services include a portal for accepting demands for facilities, prices, bill processing, and vehicle assignment. Yet, in comparison to current systems for the software administration of automotive clouds and V2G, this framework handles heterogeneous vehicle service distribution. Therefore, the platform has to use appropriate strategic planning and customized selection plans by integrating service characterizations and accurate pricing specifications to achieve effective management [7].

11.2.2 CONSERVATION BIOLOGY

The revolutionary advances in artificial intelligence have unlocked the ability to quickly process a variety of signals, accurately identify risks, and provide real-time alerts to the conservationists. Several AI models have been developed to take in these signals, such as images, video, and audio, and use it to monitor the threat faced by different wildlife populations. Machine learning (ML), a subset of AI, is consistently being employed to develop algorithms for predicting the risk of extinction of several species [8], assess the overall footprint of fisheries [9], and identify animals using sensor data obtained from biodiversity hotspots [10]. Initiatives by tech giants such as Microsoft's "AI for Earth" and Google's "AI for Social Good" are also contributing greatly to wildlife and biodiversity conservation. Deep learning (DL), yet another subset of AI, is highly effective in identifying patterns. Unlike machine learning, deep learning can work on unlabelled data sets, such as video, audio, and images, and hence is becoming a powerful tool in conservation biology. For example, a fully trained deep learning model can be fed hours of signals collected from the field to recognize unique species of organisms with remarkable accuracy. The data can also be collected from open-source repositories such as Google

Earth for satellite data [11] and Move Bank for animal tracking data [12] and fed into ML or DL algorithms. In 2019, Google along with several organizations such as Conservation International, developed "Wildlife Insights," an AI-powered cloud-based platform to store and process the millions of photos taken by biologists and conservationists to streamline the conservation monitoring process. With the help of AI, Wildlife Insights can label up to 3.6 million photos every hour, several times faster than human experts [13].

Many projects are being conducted around the world using a combination of AI and signals from the field to address specific problems. For example, Microsoft AI for Earth and Gramener has developed a deep learning algorithm to assist the Nisqually Foundation in the United States to identify salmon species from underwater camera trap videos. The algorithm was able to process several hours of video to first identify the frames with salmon in it, and then subsequently identify the type of salmon in those frames [14].

Automated browser-based tools for detecting and counting animals in images and identifying individual animal species are also developed. One such tool is Wildbook, a software framework produced by Wild Me in Portland, Oregon. Wildbook uses a combination of neural networks and computer vision for animal detection and identification. The team also uses AI to gather pictures from YouTube and Twitter of a particular animal. Researchers and scientists can start projects for a particular species by submitting many manually annotated images to train the algorithm. Some examples of these projects include that for whale sharks (*Rhincodon typus*), manta rays (*Manta birostris* and *M. alfredi*), Iberian lynxes (*Lynx pardinus*), and giraffes (*Giraffa* sp.) (Figure 11.1) [15].

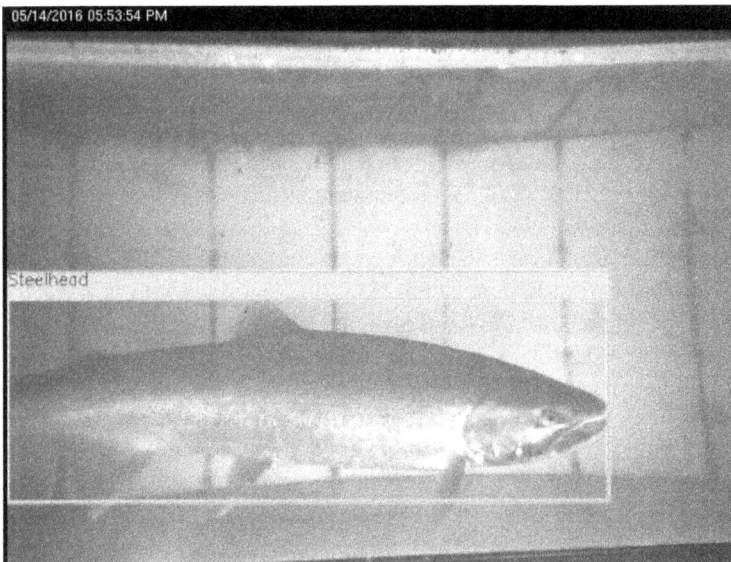

FIGURE 11.1 Salmon species identification solution developed by Gramener and Microsoft [17].

Another example is "iNaturalist" one of the world's most powerful citizen science platforms. It was recently improved by the collaboration between the California Academy of Sciences (the Academy) and the National Geographic Society. Crowdsourced images, which can be clicked even with the simplest of cameras, can be submitted on to this platform to help scientists all over the world track the response of various species to climate change and devise solutions for the same. The iNaturalist AI has learned to recognize more than 24,000 types of plants, animals, and fungi by analyzing more than 65 million images submitted by ordinary citizens. The network also shares its findings with the Global Biodiversity Information Facility (GBIF), enabling scientists' access to verified data [16].

AI is especially helpful in providing quick solutions for high-risk situations such as poaching and climate change. For example, in the African subcontinent, monitoring elephant population can be cumbersome due to its very dispersed population across a large surface area. Here, AI can be used along with aerial images generated by organizations such as "Save the Elephants" to identify concentrations of elephants most susceptible to poaching [17].

The new tools presented by AI along with the rapid influx of data generated have presented new opportunities for conservation biologists all over the world. AI has enabled us to process large amounts of data in very less amount of time with high accuracy, sometimes even better than human experts. It saves time and effort for these experts, enabling them to focus on devising solutions based on the information generated by the AI algorithms.

11.2.3 Next-Generation Weather and Climate Prediction

Severe thunderstorms, tornadoes, and hurricanes are high-impact weather phenomena that cause severe damages, loss of property, and life losses. High-impact activities may have a positive effect on society, such as the effect of renewable energy on savings. Predicting these incidents has substantially improved with expanded computational capacities, increasing computational power, and enhanced model mechanics; however, there is a scope for improvement. By enhancing precision, artificial intelligence and data science technology, especially machine learning and data mining, fill the difference between the predictive system forecasts and feedback in real-time. Artificial intelligence strategies often derive potentially inaccessible knowledge from prediction models by merging model performance with studies to get the forecasters and consumers with additional decision help. Applying AI strategies perhaps with a physical knowledge of the atmosphere will significantly improve the predictive ability for multiple high effect weather forms. The AI method is also a key contributor to computational sustainability in the growing field. AI techniques will analyze big data," offer clarity into weather patterns of high effect, and strengthen our awareness of high-impact weather [18]. It has helped to achieve 90% accuracy in predicting cyclones, transition zones between masses of air at Earth's surface called weather fronts, and narrow corridors of moisture in the atmosphere called atmospheric rivers which cause high precipitation and is not possibly predicted by humans (Figure 11.2).

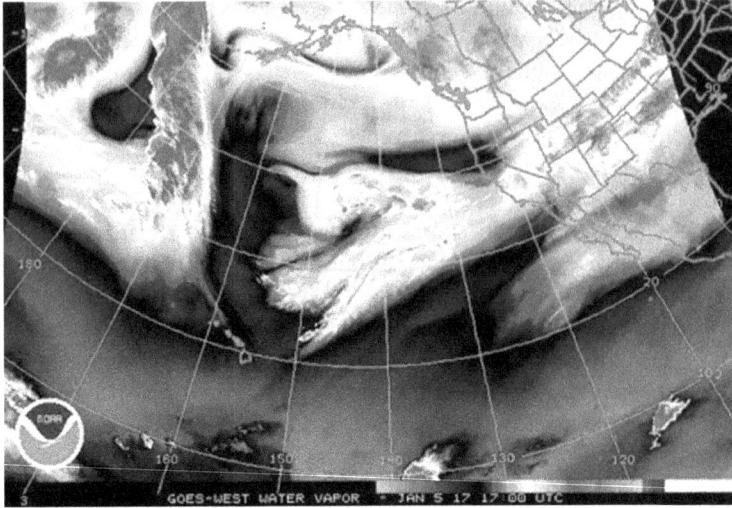

FIGURE 11.2   An atmospheric river over California [19].

The huge data sets needed and the Earth atmosphere's intrinsic unpredictability make forecasting upcoming events very difficult indeed. Existing computer systems are needed to make decisions on several phenomena on a massive scale.

These include issues such as how the Sun heats the Earth's atmosphere, how pressure changes affect wind patterns, and how temperature-changing processes (ice to water to vapor) impact the flow of energy through the atmosphere.

We also need to remember the rotation of the Earth in space which helps to churn the atmosphere all day long. Some tiny shifts in one aspect will affect the potential events profoundly.

AI may be used to find correlations and create a valid hypothesis, generalizing the data, utilizing computer-generated mathematical programs, and analytical problem-solving techniques on broad results sets.

Given the inherent difficulty of weather forecasting, scientists are now using AI for climate prediction to achieve improved and precise performance. By using deep mathematical machine learning, AI may learn to predict the future from past weather data.

The Numerical Weather Prediction (NWP) is an example. This model analyses and examines large satellite data sets as well as other sensor data sets to produce short-term weather predictions and long-term temperature projections.

Many companies currently also invest heavily in forecasting AI environment. IBM, for instance, recently acquired the Weather Service, integrating its data with its in-house Watson AI growth.

This has led to the development of IBM's Deep Thunder which provides hyper-local weather forecasts to customers within a resolution of 0.2–1.2 miles [20].

11.2.4 Smart Earth

While AI offers significant potential to solve the environmental challenges confronting the Earth left undirected, it also can accelerate deterioration of the climate. This research focuses on leveraging AI systems today and providing maximum significant effect in pressing environmental challenges. This proposes strategies by which AI can effectively reinvent mainstream industries and processes tackling climate change, provide security for food and water, preserve ecosystems, and foster human well-being. This problem seems closely related to the ongoing issue of how to guarantee AI isn't detrimental to humans. The main aim for creating "safe" AI is proving that it is value-aligned – the vision of a positive scope based on the ideals of humanity, offering secure implementation of the technology to humanity. This implies that checks and balances built to ensure that emerging AI processes stay "secure" will include the well-being of the natural environment as a fundamental aspect [21].

In the 1940s, the first concrete steps were taken toward artificial intelligence. AI is extensively used in our daily lives and caused a revolutionary vision consisting of six factors:

- *Big data:* Machines give exposure where huge quantities of data, either organized (in databases and spreadsheets) or unstructured (such as text, audio, video, and images), can be accessed. All this evidence records our experiences and increases the understanding of the world for humans. Big data will only get larger, as trillions of sensors are installed in devices, products, shoes, autonomous vehicles, and elsewhere. AI-assisted analysis of this knowledge helps us to use this evidence to uncover historical patterns, to forecast more accurately, to render decisions more efficient, and more.
- *Power consumption:* Emerging developments such as cloud computing and graphics processing units have become simpler and easier for complicated AI-enabled applications to handle large volumes of data by parallel processing.
- *Globally linked:* Social networking sites have radically transformed the way people communicate. Such expanded openness has facilitated information exchange and enabled knowledge sharing, leading to the development of "collective intelligence," also comprising open platforms that involve devising AI software and app sharing.
- *Open-source applications and information:* Open-source applications and information facilitate the modernization and implementation of AI, as seen in the success of open-source machine learning frameworks and platforms such as TensorFlow, Caffe2, PyTorch, and Parl.ai.
- *Enhanced algorithms:* Scientists also achieved progress in various areas of AI, especially in "deep learning," involving layers of neural networks, constructed in a way that is inspired by the approach of the human brain to processing information. Another recent research field is "extreme learning," in

which the AI programmer learns by trials and errors guided by a reward feature, with little or no initial input data.
- *Accelerating returns:* Market pressures drive the growth of AI since companies use enhanced analytics and open-source tools to increase their market benefits and raise their yields by, for instance, rising consumer product personalization or using smart automation to boost productivity.

The combination of these variables helped to push AI from *in vitro* (in research laboratories) to *in vivo* (in daily lives). Known businesses, as well as start-ups, will now be exploring developments and innovations in AI. Many people have begun exploiting AI technologies to traverse towns, shop online, find tips for movies, screen out unwelcome emails, or share a journey to work, whether they know it or not.

Therefore, AI is being considered for its extensive potential. Fifty-four percent acknowledged making substantial investments in AI in a 2017 PwC study of global executives, while a shortage of technical expertise remains a major concern. The realized prices are expected to take off as companies continue to invest in software, technology management, assets, and AI-enabled innovations: rising from $1.4 billion in annual sales from AI-enabled applications in 2016 to $59.8 billion by 2025, according to one research study [22].

11.3   AI IN THE MONITORING ENVIRONMENT

Artificial intelligence is making its way through many domains like healthcare, automotive, finance and economics, cybersecurity, military, hospitality, and now video games, too. But the most important and positive impact it has is on its application in environmental monitoring, which is also the need of the hour. Real-time monitoring of soil, air quality, oceans, and lakes' health are some of the basic applications of AI. For example, IBM Research has been extensively exploring the application of AI in monitoring the environment and has come up with various prototypes for the same [6]. The best example for cleaning up of oceans using AI and Data Science is the Ocean Cleanup Project undertaken for clearing the debris in the Great Pacific Garbage Patch [23].

11.3.1   MONITORING SOIL

The agriculture industry consumes more than 70% of the world's water usage, annually. Small farms produce 80% of the food source for the world population, so ensuring the quality and safety of water supplied is essentially critical. IBM's AgroPad is a real-time enabled prototype that is being used for chemical analysis of soil or water samples using AI (Figure 11.3) [6]. It is a business card-sized paper device with a microfluidic chip that performs the analysis of the sample of water or soil and results are obtained in seconds [6]. Machine learning and vision algorithms are used to translate the measured colour of the composition and intensity into concentrations of chemical constituents present in the sample [6]. This is comparatively more reliable than human vision-based tests alone. The parameters that can be measured are pH,

FIGURE 11.3   IBM's AgroPad to analyze soil and water samples [24].

chlorine, magnesium, nitrogen dioxide, and aluminium [6]. This technology could revolutionize digital agriculture.

AI-enabled agriculture bots are helping farmers in protecting their crops from weeds and also combat the labour challenge. Monitoring of weed and timely spraying can be possible by incorporating computer vision.

Two research students have devised a smart farming system that works on the principles of AI and IoT (Internet of Things). The device called E-parirakshak has been designed to monitor agricultural fields based on factors like fertility, soil temperature, and moisture, controlling water levels, blowers, and sprinklers from a remote location. The data is pushed to the cloud from where it can be retrieved and analyzed later. It also helps in the detection of diseases in crops and updates the farmers in real time.

11.3.2 MONITORING WATER

Plankton is considered to be biological and natural sensors of aquatic health. These are a collection of organisms that serve as the basis of the oceanic food chain and the primary source of protein for billions of people. These organisms include bacteria, protozoa, archaea, algae, and other floating organisms. These organisms can alter their behaviour even after slight changes in the quality of the water are detected. Therefore, IBM researchers have developed small, autonomous microscopes in water for plankton monitoring, determining different species, and tracking their movement. [6]. These findings can be further used for their response to changes from temperature to oil spills to run-offs. The microscope consists of an imager chip to capture the shadow of plankton as it swims over, creating a digital image of its health [6].

FIGURE 11.4   The Interceptor [25].

The Ocean Cleanup Project based in the Netherlands is an initiative to extract plastic pollution from oceans, mainly the Great Pacific Garbage Patch. The Interceptor is a technology developed to avoid new plastic reaching the water, not only to clean up the plastic (Figure 11.4). It is solar-powered, autonomous, and remote-sensing enabled with scalability and cost-effectivity.

The very first Interceptor which is a prototype is currently deployed in the Cengkareng Drain in Jakarta, Indonesia (Figure 11.4). The local government of Malaysia has incorporated the Interceptor as an addition to their clean-up plans to stop the flow of plastic into rivers using barriers. It has been deployed in the Klang River which runs through Kuala Lumpur (Figure 11.5, left and right view).

FIGURE 11.5   The Interceptor deployed in Jakarta (left) and Klang River, Malaysia (right) [25].

11.3.3 Monitoring Air

Air pollution is of great concern in many cities around the world. It fluctuates greatly based on certain parameters like weather, humidity, wind, and temperature [6]. Certain pollutants interact with water molecules to create smog and while winds can disperse pollutants, they can also carry forward other pollutants with them. IBM has launched an initiative called the Green Horizon that utilizes the concept of the Internet of Things (IoT) and AI to predict air quality and bring down pollutant levels [6]. These systems can make sense out of the huge amount of data and make forecasts far more effectively. The Green Horizon initiative is making an impact in Beijing and will move to other cities like Johannesburg and New Delhi [6].

Blue Sky Analytics is a big data and AI start-up with a mission to provide actionable intelligence starting with air pollution [26]. In October 2019, it launched BreeZo, an app that provides real-time and contextual air quality data and helps people minimize exposure to air pollution. The company is also building an AI-enabled app called Zuri, for mapping farm and forest fires, and Zorro, for monitoring industrial emissions.

11.4 RISKS OF ARTIFICIAL INTELLIGENCE

Artificial intelligence, while being a field of vast successes and further scope for improvement, seems to slowly prove itself to be highly disruptive. It should be considered that since artificial intelligence uses algorithms, it cannot mimic the behaviour of humans from an emotional and realistic sense, thus creating discrepancies in the overall life of individuals. Algorithms cannot consider the various kinds of reactions a person could portray in response to a single stimulus, and this is where the issue lies, and branches out to several other problems [26].

11.4.1 Bias

Biases are a huge problem in an everyday human's life, and for a person with a bias to feed biased data, and train a system to follow these biased data sets, we are essentially curating a system that in turn logically produces results to match our sense of bias. This could mean bias towards, for instance, ethnic minorities, religious minorities, selective employment conditions, etc. [26].

11.4.2 Liability

Legal action has been a subject where artificial intelligence continually fails in, because as humans, for an error in the actions of a machine with artificial intelligence, who would we pass the blame to? Would it be the machine in its entirety or the person who made it? The fact to be considered is, whether we are ready to accept that in fields of high precision, medicine for instance, would humans be able to trust a machine to replace himself and consider in retrospect the small margin of error it could result in?

11.4.3 ASI (ARTIFICIAL SUPERINTELLIGENCE)

This would be a potential condition in which the artificial intelligence expressed by a machine eventually surpasses the general intelligence of humans as well. To prevent the possibility of this, ethical reforms and morals need to be transmitted. But technology is a little young for this, and in turn, this is not yet possible [26].

While AI allows everyone to effectively operate the effects of climate change and preserve the environment and corporation, banking, healthcare, medicine, law, education, etc., it's not without risks. Several notable people like the late physicist Stephen Hawking and Tesla CEO Elon Musk have cautioned of the potential risks of AI unchecked [26].

The six categories of risk associated with AI as per the World Economic Forum report are as follows [27]:

- *Performance:* The assumptions of the AI black box may not be available to people, which makes it difficult to determine whether they are right or beneficial. Deep learning can be dangerous when applications like early warning systems are required for natural disasters.
- *Security:* Potentially AI could be "taught," enabling hackers to introduce their data sets or manipulate existing systems for malicious reasons.
- *Control risks:* When AI systems communicate with each other autonomously, they can deliver unpredictable results. Two species, for example, came up with their language which humans could not comprehend.
- *Economic risks:* Companies that adopt AI at a slower level are economically impacted as their AI rivalry progresses. We can already see the closing of brick and mortar stores as the market is shifting to digitization.
- *Social risks:* As AI leads to automation, it is destroying the jobs in almost all the industries. Autonomous weapons systems could also speed up global conflicts and worsen them.
- *Ethical risks:* AI makes generalized conclusions in decision-making, this often leads to increased bias decisions. Data processing also questions the safety aspects.

To combat such risks, the WEF states "the stability, the accountability and the integrity of the AI system must be ensured for government and business." More collaboration between public and private entities, technologists, policymakers, and also philosophers and further innovation in science is required to avoid and consider the possible danger of artificial intelligence.

11.5 CONCLUSION AND FUTURE

Artificial intelligence is advancing at lightning speed, with discoveries and applications each day. While we cannot ignore the risk factor it has in our lives, integrating it with water and electricity systems to make a sustainable city is a future goal that many countries wish to approach quickly. In place of climate change, artificial intelligence technology-based products and discoveries could pave a path to at least lessen the effects of climate change, if not eradicate it as a whole. The current applications of artificial intelligence in environmental science have been discussed, whether it is in real-time testing of air pollution levels or to test the salinity of the ocean and seawater for the thriving of marine life, and while we can use these techniques and discoveries to make our lives easier, we should not focus on a complete reliance on artificial intelligence. Some artificial intelligence algorithms are rather opaque, and this, in turn, can affect us in several ways. For instance, we may not be able to identify how much of the information being provided is in turn being used. Making any form of assumption or prediction beyond the training that has been given can result in unpredictable results, even though the character of generalization that most machine learning algorithms possess entails it to give us a result.

In short, we can claim that artificial intelligence is helping us save the environment and thus improve our sustainability. Admittedly, it has adverse effects as well, as demonstrated above. The key is to find a balance between using and exploiting artificial intelligence. The techniques are not cheap either, so a lot of financial and computational power is required for this purpose. Having said that, it can be argued that the recently exploited idea of applying AI in environmental science has been a success, and the future seems quite prospective in this particular field.

ACKNOWLEDGEMENT

The authors listed in this chapter wish to express their appreciation to the RSST trust Bangalore for their continuous support and encouragement. As a corresponding author, I also express my sincere thanks to all other authors whose valuable contribution and important comments made this chapter to this form.

Conflict of Interest The authors listed in this chapter have no conflict of interest as known best from our side. There was also no problem related to funding. All authors have contributed equally with their valuable comments which made the chapter to this form.

REFERENCES

1. Russell, S.J. and Norvig, P. 2010. *Artificial Intelligence-A Modern Approach*, Third International Edition.
2. Poole, D.I., Goebel, R.G. and Mackworth, A.K. 1998. *Computational intelligence* (p. 142). New York: Oxford University Press.
3. McCarthy, J., Minsky, M.L., Rochester, N. and Shannon, C.E. 2006. A proposal for the Dartmouth summer research project on artificial intelligence, august 31, 1955. *AI Magazine*, 27(4), pp. 12–12.
4. Lieto, A., Lebiere, C. and Oltramari, A. 2018. The knowledge level in cognitive architectures: Current limitations and possible developments. *Cognitive Systems Research*, 48, pp. 39–55.
5. Tutorialspoint.com. 2020. *Artificial intelligence - research areas - tutorialspoint.* https://www.tutorialspoint.com/artificial_intelligence/artificial_intelligence_research_areas.htm [Accessed 16 Jan. 2020].

6. IBM Research - Energy and Environment. 2020. *IBM research - energy and environment*. https://www.research.ibm.com/energy-and-environment/ [Accessed 15 Jan. 2020].

7. Autonomous & Connected Vehicles. https://automotive.ricardo.com/hybrid-electronic-systems/autonomous-connected-vehicles [Accessed 21 Jan. 2020].

8. Darrah, S.E., Bland, L.M., Bachman, S.P., Clubbe, C.P. and Trias-Blasi, A. 2017. Using coarse-scale species distribution data to predict extinction risk in plants. *Diversity and Distributions*, 23(4), pp. 435–447.

9. Kroodsma, D.A., Mayorga, J., Hochberg, T., Miller, N.A., Boerder, K., Ferretti, F., Wilson, A., Bergman, B., White, T.D., Block, B.A. and Woods, P. 2018. Tracking the global footprint of fisheries. *Science*, 359(6378), pp. 904–908.

10. Mac Aodha, O., Gibb, R., Barlow, K.E., Browning, E., Firman, M., Freeman, R., Harder, B., Kinsey, L., Mead, G.R., Newson, S.E. and Pandourski, I. 2018. Bat detective—Deep learning tools for bat acoustic signal detection. *PLoS Computational Biology*, 14(3). https://doi.org/10.1371/journal.pcbi.1005995

11. Gorelick, N., Hancher, M., Dixon, M., Ilyushchenko, S., Thau, D. and Moore, R. 2017. Google earth engine: Planetary-scale geospatial analysis for everyone. *Remote Sensing of Environment*, 202, pp. 18–27.

12. Kranstauber, B., Cameron, A., Weinzerl, R., Fountain, T., Tilak, S., Wikelski, M. and Kays, R. 2011. The movebank data model for animal tracking. *Environmental Modelling & Software*, 26(6), pp. 834–835.

13. Google. 2020. *Using AI to find where the wild things are*. https://www.blog.google/products/earth/ai-finds-where-the-wild-things-are/ [Accessed 21 Jan. 2020].

14. partner.microsoft.com. 2020. *Gramener case study*. https://partner.microsoft.com/en-us/case-studies/gramener/ [Accessed 21 Jan. 2020].

15. Nature.com. 2020. *AI empowers conservation biology*. https://www.nature.com/articles/d41586-019-00746-1 [Accessed 21 Jan. 2020].

16. National Geographic Society Newsroom. 2020. *The California academy of sciences and national geographic society join forces to enhance global wildlife observation network iNaturalist: A growing community of iNaturalist users—and the artificial intelligence they help power—help observe and monitor more than 165,000 species around the world*. https://blog.nationalgeographic.org/2018/06/26/the-california-academy-of-sciences-and-national-geographic-society-join-forces-to-enhance-global-wildlife-observation-network-inaturalist/ [Accessed 21 Jan. 2020].

17. Medium. 2020. *How AI can save earth's biodiversity*. https://medium.com/@kesari/how-ai-can-save-earths-biodiversity-94555d57dd28 [Accessed 21 Jan. 2020].

18. McGovern, A., Elmore, K.L., Gagne, D.J., Haupt, S.E., Karstens, C.D., Lagerquist, R., Smith, T. and Williams, J.K. 2017. Using artificial intelligence to improve real-time decision-making for high-impact weather. *Bulletin of the American Meteorological Society*, 98(10), pp. 2073–2090.

19. Cho, R., 2018. *Artificial intelligence–a game-changer for climate change and the environment*. https://phys.org/news/2018–06-artificial-intelligence game-changer-climate.html [Accessed 09 March 2020].

20. McFadden, C., Young, C., Lang, F., English, T. and Lang, F. 2020. AI might be the future for weather forecasting. *Interestingengineering.com*. https://interestingengineering.com/ai-might-be-the-future-for-weather-forecasting [Accessed 21 Jan. 2020].

21. Bakker, K. and Ritts, M. 2018. Smart earth: A meta-review and implications for environmental governance. *Global Environmental Change*, 52, pp. 201–211. [Accessed 09 March 2020].

22. Pwc.com. 2020. https://www.pwc.com/gx/en/news-room/docs/ai-for-the-earth.pdf [Accessed 21 Jan. 2020].

23. https://theoceancleanup.com/. [Accessed 09 March 2020].
24. Chris Albrecht. September 6, 2018. *IBM AgroPad combines paper, AI, and the cloud to analyse soil and water.* https://thespoon.tech/ibm-agropad-combines-paper-ai-and -cloud-to-analyze-soil-and-water/. [Accessed 09 March 2020].
25. https://theoceancleanup.com/rivers/. [Accessed 09 March 2020].
26. Ambika Choudhury. March 1, 2020. *How blue sky analytics is utilizing ML & Geo-spatial data to minimize air pollution.* https://analyticsindiamag.com/this-gurgaon-ba sed-startup-is-utilising-ml-geospatial-dataset-to-minimise-exposure-to-air-pollution/. [Accessed 09 March 2020].
27. How to manage AI's Risks and Rewards. January 11, 2018. World economic forum. https://www.weforum.org/agenda/2018/01/how-to-manage-ais-risks-and-benefits/. [Accessed 21 Jan. 2020].

12 A Genetic Algorithm-based Artificial Intelligence Solution for Optimizing E-Commerce Logistics Vehicle Routing

Suresh Nanda Kumar and
Ramasamy Panneerselvam

CONTENTS

12.1   INTRODUCTION

Every manufacturing or service organization as well as e-commerce organization makes use of warehouse storage and distribution to other stores or to customers using transportation services of vehicles and carriers such as delivery trucks. Physical distribution is the movement of goods by means of the outbound logistics process, outward from the end of the assembly lines/shops of factories to the customers. Supply-chain management comprises logistics as one of its fields and links logistics with the firm's other functions like engineering and production and links closely with the communications network. It includes the procurement of raw materials and components, upstream, transforming them into finished products, storing them, and then transporting them to the customers efficiently and effectively. Logistics is the process of transporting, storage, handling, and packaging of goods and materials, from the beginning to the end of the manufacturing, sales, and finally the disposal in order to satisfy customer requirements and also to add competitiveness to the business.

12.2 TRANSPORT COSTS AND GOODS CHARACTERISTICS IN LOGISTICS

Transportation can be considered to be one of the key activities of economic importance to the organization, among the different activities of the constitute logistics for business applications. It is common knowledge that around 30–65% of the companies' logistics expenses are transportation costs (Chang, 1998). The costs of transportation consist of components such as the mode of transportation, transport infrastructure, packaging materials such as containers and pallets, transport terminals like railway stations, ports, airports, etc., labour, and time. It also depends on the property of the goods transported, i.e. the value, volume, weight, and stowability.

Hence, logistics and supply chain managers must understand the various aspects of transportation deeply. The transportation achieves the efficient movement of goods and provides delivery of products to the customers on time. Transport impacts logistics activities and also has a great impact on the production and sale of products. Hence, transportation cost is a necessary cost and ways must be found to minimize the transportation costs. Value of transportation costs varies with the type of industry. Transportation costs constitute a very big part of the selling price. Hence, it affects profits more, and therefore it needs to be minimized.

12.3 LOGISTICS CHALLENGES ON FESTIVE DAYS

With a sudden surge in demand during festive days, the logistics for the e-commerce players feel the pressure like never before. With home deliveries being the norm now, supply chain managers need to ensure that they adhere to their SLAs and avoid miss-routes to provide on-time deliveries to their customers.

With rising customer expectations, simple efficiency in operation no longer cuts the bill for e-commerce players. Companies now need to ensure that the customers are serviced in the time-window of their choice. Time-definite delivery is the new benchmark that companies need to follow.

Nearly 50% of the costs incurred by logistics companies come from executing the first- and last-mile delivery services. Logistics firms are trying to bring down costs by increasing efficiency, reliability, and speed of the transportation activity. With the increasing adaptation of digital technologies by the logistics industry, many companies these days are applying artificial intelligence (AI) to their logistics and supply chain activities with the hope of maximizing their resources by minimizing the time and cost of planning and executing the routes, the number of vehicles to be used, and the time of delivery of the goods and services to the customers.

Artificial intelligence is the simulation of human intelligence processes by machines, particularly computer systems. Artificial intelligence is extensively applied for improving logistics experience by increasing reliability, reducing the cost of transportation, faster processing, and deciding optimal routes for last-mile operations. Given the intense competition in the industry, customer satisfaction has become the main factor of competitive advantage and for business dominance; e-commerce firms like Amazon and Flipkart are investing hundreds of millions of dollars with a view to improve their delivery processes in terms of speed and efficiency and to achieve next-day or even same-day delivery. Artificial intelligence is the main technology used to realize these goals by finding the optimal route, calculating exact delivery times, and deciding the correct product mix in the warehouse in order to minimize logistics expenses.

12.4   AI IN LOGISTICS

The efficiencies gained using AI have achieved the greatest visibility in the fields of vehicle routing, network/route planning and optimization, and forecasting demand. Innovations that have been achieved using machine learning in these fields have led to the emergence of more agile businesses and also becoming more dynamic. As a result, food delivery aggregators such as Foodpanda, Zomato, Swiggy, and Uber are able to provide an excellent customer experience. AI can be applied in order to achieve improved planning of capacity, route optimization, dynamic pricing, optimizing resource allocation, and vehicles with agility and nimbleness to the areas where demand is high in order to decrease the waiting time of the customers.

The package carrier United Parcel Service (UPS) makes use of an application called ORION (On-Road Integrated Optimisation and Navigation), which is a GPS tool that used expansive fleet telematics and advanced algorithms to analyze vast quantities of data that enables UPS' drivers make deliveries on-time and at the same time being cost-effective. UPS has also implemented projects to improve efficiencies for its employees and to help reduce their environmental impact.

Dynamic route planning and optimization can be achieved based on traffic conditions and other factors. A reduction of even 1 mile for a driver each day in one year will lead to a savings of nearly $50 m for UPS. The deployment ORION will save for UPS nearly 100 million miles per year, which amounts to consumption of nearly 10 million gallons of fuel less, annually. It leads eventually to a decrease in carbon dioxide emissions by about 100,000 metric tonnes.

AI will eventually set a new standard of efficiency across supply-chain and logistics processes in the months to come. The scenario is changing quickly, creating a "new normal" in how global logistics companies manage data, run operations, and serve customers, in a manner that's automated, intelligent, and more efficient.

Irrespective of how these developments are viewed, AI and related technologies are going to radically transform the international supply-chain management and logistics industries.

12.5 GENETIC ALGORITHMS

Meta-heuristics combine different mechanisms to find solutions of good quality. In contrast to classical heuristics, they accept non-improving solutions during the search to overcome local optima. Normally, meta-heuristics produce better results than classical heuristics even though there is increased computation times (Gendreau et al. 2002).

Genetic algorithm is a technique in AI and was first developed by J. Holland at the University of Michigan in 1975. Solutions to a combinatorial problem are encoded as chromosomes. The chromosomes are evaluated for their fitness by an evaluation function and good properties of a generation of solutions are propagated to the following generations.

A genetic algorithm starts with a collection of chromosomes known as the initial population. Each of the chromosomes represents a solution to the given problem. There could be a random generation of the initial population, in which case it will take more time for the algorithm to converge to the solution or some heuristic method can be used for the generation of the initial population generated and since the population is now closer to the solution, it would take less time to converge. The GA's next step is known as selection. This mechanism involves selecting potential chromosomes as the parent chromosomes, which in turn depends on the fitness values of the parent chromosomes and computed by using the fitness evaluation function. The parent chromosomes thus selected will experience recombination by means of the crossover operations to generate the offspring chromosomes. Mutation occurs in a small number of newly produced offspring chromosomes. Mutation is carried out in order that a degree of randomness is introduced so that the GA is preserved and prevented from getting converged to a local optimum. The gene sequence undergoes a random swap in a mutation operation or randomly undergoes a bit negation in the event of the offspring being encoded in bits. The formation of new population is now made possible by substituting the parent chromosomes with the offspring chromosomes.

The genetic algorithm will continue through this process until a stopping criterion is met, which can be one of the following:

- Predefined number of generations has been produced.
- There was no improvement in the population, which would mean that the GA has found an optimal solution.
- A predefined level of fitness has been reached.

Genetic algorithms have been used extensively to obtain solution for the vehicle routing problem with time windows (VRPTW). The genetic algorithm is based on natural reproduction, selection, and evolution based on Darwin's theory.

12.6   VEHICLE ROUTING PROBLEM

Vehicle routing problem (VRP) involves the optimal transportation of goods by means of movement and delivery of goods. The VRP determines the least cost routes from a central depot to a set of customers. The VRP has been widely studied ever since its introduction in 1959. Many variants to the VRP have been researched and discussed by many researchers.

The principle of the VRP is described below.

The route of a vehicle to deliver goods must satisfy the following constraints:

- Every customer on the route must be visited exactly one time.
- All the routes begin at the depot and conclude at the depot. The vehicles must return to the depot, once all the customers on the route have been serviced.
- The sum of all customer demands on a route must not exceed the vehicle capacity servicing the route.

The vehicle routing problem is pictorially depicted in Figure 12.1. From this figure, one can note that each vehicle originates at the central depot, visits a set of customer or supplier nodes once, fulfil their demands or perform the pickups at the nodes, and then returns to the depot once again. The nodes can be either supplier sites in the case of e-commerce company supplier site pickups or customer sites where the e-commerce company delivers to the customer sites. It can also be different manufacturers to whom the suppliers deliver raw materials, components, parts, and other supplies of items. In this process, the number of customers served by all the vehicles is equal to the total number of customers under consideration/in the network. The various colours depicted in the figure represent the various routes of the vehicles to fulfil customer demand. This output was generated in the Heuristic Lab software environment, using a plugin developed in this research. If the number of vehicles under consideration is only one, then that vehicle has to start its journey from the depot and visit each customer node exactly once and return to the depot. In this

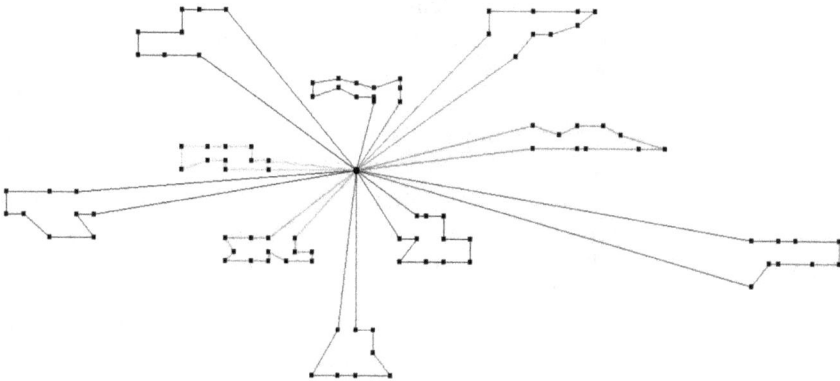

FIGURE 12.1   An example solution to a vehicle routing problem.

process, the total cost/total time of transportation is minimized. This is called as simple travelling salesman problem (TSP).

The vehicle routing problem has been widely studied in the operations research and optimization literature. The earliest works were done in seminal papers of the researchers Dantzig and Ramser (1959) and then later the VRP was discussed by another famous pair of researchers, Clarke and Wright (1964). Now, VRP can be solved by using different heuristic and meta-heuristics methods. The VRP is so widely studied because of its extensive application in a wide range of problems and its ability in identifying efficient methods for reducing distribution networks operational costs.

The total distance between any two locations in the network is known and the demand for the product is specified for each service location (node) in the route. In the time-dependent vehicle routing problem (TDVRP), the time to serve the customers vary along by considering traffic conditions like congestion, accidents, etc. in the route. In order to collect the items from various suppliers, the third party logistics service providers (3PL) or logistics transportation service provider must visit all scheduled suppliers during different hours required by the suppliers. Some suppliers may further request visits within certain time windows. But it is not easy to meet the time window constraints since the delivery processes are frequently affected by real-time traffic flow conditions. The traffic congestion occurring during the peak hours might cause severe delays of pickups and deliveries. Therefore, in order to pick up items from the suppliers more efficiently, it is necessary to address several challenging problems among customers and suppliers, in a 3PL setting, whereby the e-commerce fulfilment centre's vehicle visits the suppliers' sites to pick up the items ordered: (1) to satisfy suppliers' specific time windows; (2) to reduce the travel time and also the distance travelled, which in turn is affected by peak-hour urban traffic congestion, and also the travel speeds are not uniform.

An improved ACO algorithm was developed by Yu et al. (2009) for the VRP. The algorithm's effectiveness was tested by the data of 14 data points. Pisinger and Ropke (2007) proposed a general heuristic for the VRPTW, CVRP, MDVRP, site-dependent VRP, and open VRP. Schyns (2015) employed ACO algorithm to solve the VRP, including mixed batch arrival and dynamic capacity with time windows. Bell and McMullen (2004) compared the ant colony algorithm with the TS algorithm and genetic algorithm and found that the ant colony algorithm required less computation time. Meng et al. (2019) have developed a GA-based solution for what they call the customer-oriented vehicle routing problem considering energy conservation and emission reduction (C-VRPESER). The objective is to minimize the total cost based on the customers' group which can lead to an en-group distribution vehicle routing optimization. Planning the correct scheduling vehicles, which are of different kinds for each customer group, will lead to an improved utilization rate of the vehicles. The constraint of vehicle capacity can also be met. The optimization objective will include minimizing the vehicle fixed costs (minimizing the number of vehicles used), time penalty cost, fuel cost, and carbon emissions.

Most vehicle routing models assume constant travel times throughout the day. In reality, however, travel times fluctuate because of predictable events such as congestion during rush hours or from unpredictable events such as road accidents/incidents, unexpected weather conditions etc. (Ichoua et al. 2003) and also uncertainties like volatile demand for service. Many VRPs are based on the assumption that there is a constant and deterministic travel times between the various customer nodes and the central depot (Kok et al., 2012) or it is the same as the distance travelled between customer nodes. But in an actual scenario, the vehicles do not travel at a constant speed (due to factors like congestion, weather conditions, etc.), which impacts the transportation cost because of increased fuel consumption (Kuo 2010).

The TDTSP and the TDVRP were studied by Ehmke et al. (2012, 2016). The time-dependent travel times were used on the basis of Floating Car Data (FCD) from Stuttgart, Germany. The travel times so obtained were then converted into planning data sets using data-mining techniques. The authors applied two different kinds of travel-time planning sets. The time-dependent vehicle routing problem with time windows (TDVRPTW) makes the basic and important assumption that the travel times are a function of the present time. Traffic congestion affects travel speeds and hence it affects the total time to complete the route and the number of vehicles utilized and the total transportation cost can also be found. Travel time between customers (nodes) and depot has been identified as one of the major factors that increases the adverse impacts of congestion. Traffic congestion also has a direct effect on carriers' cost structure and the relative weight of wages and overtime expenses (Figliozzi, 2009).

TDVRPs satisfy a property which is known as the "non-passing property," and in simpler terms, the First-In First-Out (FIFO) property (Ichoua et al., 2003). According to Doerfler (2017), supply-chain management professionals face a lot of delivery challenges, especially in handling routing of the vehicles. She says that according to Hani Mahmassani, the William A. Patterson Distinguished Chair in Transportation and Director of the Northwestern University Transportation Center in Evanston, Illinois, routing is a form of roulette because traffic congestion, road construction, and weather conditions can adversely affect the time it takes to deliver packages to their destination. A civil and environmental engineering professor, He says "if you plan that a certain route should take two hours with travel time and it takes twice as long, you lose your schedule synchronization quickly". He also adds that "if you plan for the worst case or even the 95th percentile of travel time, the extra time may be unnecessary, and you've wasted the use of the vehicle. Thus, there is a loss of productivity."

The TDVRP takes the first aspect into account by assuming that travel times depend on the time of the day. All of these uncertainties if not addressed properly will result in an overall increase in logistics costs. Hence, vehicle routing is a complex logistics management problem and represents a key class of problems to solve in order to lower costs and optimize logistics resources. The main purpose of VRP is to find optimal routes for multiple vehicles visiting a set of locations by minimizing both time and cost.

12.7 PERFORMANCE MEASURES

The quality of solution obtained for the vehicle routing problem is judged in terms of the number of vehicles utilized and also the total distance travelled. If a single vehicle is used to satisfy the demands of all the customers, the cost of the number of vehicles as well as the cost of transportation will be the minimum. If the total demand exceeds the capacity of the vehicle, then the use of single vehicle becomes infeasible, which necessitates the use of multiple vehicles. This then gives rise to the multiple Traveling Salesman Problem (mTSP), which in other words is called the vehicle routing problem. In this case, the objective of the problem is to minimize the number of vehicles used and also the total distance travelled by the vehicle and thereby the total costs of transportation.

When multiple vehicles are used to satisfy the demands of the customers, the objectives are as follows:

- Minimize the total distance of travel of all the vehicles put together and hence the total cost of transportation of satisfying the demands of all the customers.
- Minimize the number of vehicles required to meet the demands of the customers.

12.8 THE TIME-DEPENDENT VEHICLE ROUTING PROBLEM WITH TIME WINDOWS

The time-dependent vehicle routing problem is an extension of the basic VRPTW. The TDVRPTW is being researched increasingly because of its practical importance to transportation problems. This is due to the fact that the traffic conditions are different and vary during different times of the day and as a result manufacturers, suppliers, and e-commerce retailers need to schedule their pickups and deliveries at the appropriate times of the day, considering the time windows of the customers and suppliers to make their order fulfilment needs efficient and faster. The methodology adopted by Kumar and Panneerselvam (2015) study was based on genetic algorithm.

In a discussion with a vehicle driver about delivery of prawn eggs for breeding in Andhra Pradesh from Tamil Nadu, he described how he had to change routes near Chennai East Coast Road to avoid traffic congestion during peak hours from morning 6 a.m. to 10 a.m. and then from evening 5 p.m. to 11 p.m. He had to deliver the prawn eggs to the customer in 10 hours' time with a maximum relaxation of 2 hours. An important real-life property found in transportation problems in the retail industry is time-dependent travel times, which is also known as dynamic travel times, where travel time depends on the time of departure and traffic conditions such as rush hours. Esmat and Mirmohammadi (2015) develop a tabu-search-based metaheuristic to solve the TDVRPTW. Here they present an Integer Linear Programming (ILP) and the tabu search algorithm to handle the problem in large-scale instances. Duygu et al. (2014) propose a tabu search and adaptive large neighbourhood search-based solutions to solve the time-dependent vehicle routing problem. The conducted experiments confirm that the proposed procedure is effective to obtain very good solutions to be performed in real-life environment.

12.9 APPLICATION OF ARTIFICIAL INTELLIGENCE IN ROUTE PLANNING AND OPTIMIZATION

According to a report by RedSeer (2019), sales during the festive season in September to October, 2019, in India by the major e-commerce retailers such as Amazon and Walmart-owned Flipkart generated a sales by the e-tailers, a whopping $3 billion of Gross Merchandise Value (GMV) with a large chunk of customers from Tier-2 cities or Bharat. RedSeer had forecasted a total sale of $3.7 billion in the first phase of the festive season. The actual sales of $3 billion are nearly 80% of RedSeer's forecasted sales. E-commerce companies such as Flipkart and Amazon could generate up to US$6 billion or Rs 39,000 crore in sales this festive season, according to consulting firm RedSeer (as compared to GMV of $30 billion for Singles Day in China, about $8 billion for Cyber Monday, and about $6 billion for Black Friday in the United States in 2019). However, these events don't come without its challenges.

12.10 DEVELOPMENT OF GA FOR TIME-DEPENDENT VEHICLE ROUTING PROBLEM WITH TIME WINDOWS

The genetic algorithm – Random Sequence Insertion-based Crossover (RSIX) (a detailed description on the development of the genetic algorithm-based solution for TDVRPTW) – is given. Traffic congestion is a common occurrence in many urban cities worldwide. This phenomenon of traffic congestion results in a significant variation in vehicle travel speeds during peak hours of traffic, especially during the morning and evening times. This situation is unlike other vehicle routing models, where a constant travel speed is assumed irrespective of congestion. Urban vehicle route designs, which do not take into consideration such significant travel speed variations, result in solutions that are inefficient and suboptimal. Routes that are designed without taking traffic situations into consideration will lead freight and passenger vehicles into congested arterial roads and streets and thereby not only increase supply chain and logistics costs but also contribute to environmental ill-effects that are associated with freight and passenger traffic in urban locations such as the emission of greenhouse gases, noise, and air pollution.

Travel time between customers (nodes) and depot has been identified as one of the major factors that increases the adverse impacts of congestion. Traffic congestion also has a direct effect on carriers' cost structure and the relative weight of wages and overtime expenses (Figliozzi, 2009). Dynamic vehicle routing problem is based on the time-dependent variations in travel times. A genetic algorithm is developed for solving the TDVRPTW. The VRP with pickup and delivery is used for picking up or to deliver to the various customers or from the various supplier sites. These sites specify a time window and the multiple vehicles with having various capacities and with real-time variations in the times of travel from one node to the other are considered.

To load and start the time-dependent VRP variant in HeuristicLab, a valid file to load the time-dependent VRP data is required. So, the data from the Solomon (CVRPTW) instances is used and the travel time matrix is added. The dimension of the travel time matrix must match the number of cities. For the travel times, random values between 0.5 and 1.0 were generated using random function in Excel.

So, the objectives of the research (Nanda Kumar and Panneerselvam, 2017) were to minimize the total distance travelled and to minimize the total number of vehicles used to cover all the customer nodes. The minimization of the total distance travelled is treated as the primary measure and the corresponding number of vehicles utilized to serve all the customer nodes is treated as the secondary measure.

In the genetic algorithm, a chromosome is a chain of integers (genes) and each of the integers represents a customer node. This chromosome represents all the customers. The customers on it are separated to several routes, each of them representing a sequence of deliveries that must be covered by a vehicle. The chromosome consists of information on the number of vehicles used to service the customer nodes, separately so as to minimize the number of vehicles used to service the customer nodes. This information is called as the vehicle-part. In the vehicle information, the number of vehicles is equal to the number of routes in the customer chromosome. The number on each of the vehicle genes represents the number of customers to service. The sum of these numbers in vehicle information should be equal to the total number of customers. The application of the RSIX crossover method to a pair of chromosomes yields a pair of offspring. The offspring chromosome is chosen based on the fitness value. Lower fitness value indicates a desirable solution, since it will lead to less distance travelled. For example, in Figure 12.2, there is an illustration of a possible solution of 3 vehicles for 8 customer nodes. There are 3 genes in the vehicle-part of chromosome, which means that the 8 customers are divided into 3 routes. The 3 on the first gene in vehicle-part represents route 1 that services 3 customer nodes: 1, 5, and 4. The 2 on the second gene means that route 2 services customers 2 and 8. The 3 on the third gene indicates that route 3 services customers 7, 6, and 3.

Vehicle travel times in the cities and urban areas vary due to various reasons and factors, like congestions in traffic, accidents, road repairs, VIP movement, and delays due to bad weather. If these travel time variations are ignored, when forming plans for the routes needed for pickup and delivery of goods by vehicles from customers and suppliers, one may end up in creating route plans that may make the vehicles end up in urban heavy traffic congestions. As a result of this, vehicles may be spending

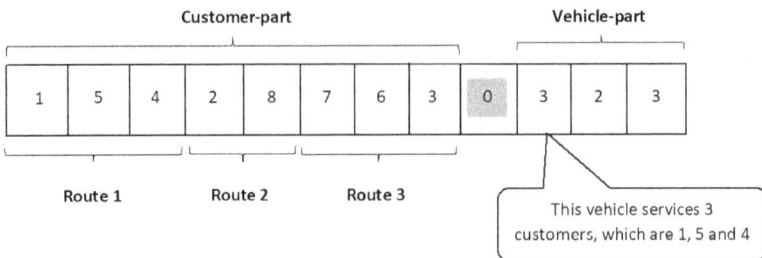

FIGURE 12.2   An example of the customer and vehicle chromosome.

unnecessary time in traffic jams and get delayed, resulting in long waiting times for the customers/suppliers and not having exact information regarding the times at which the vehicles arrive actually. Due to this, the time windows during which the demand or supply nodes have to be visited become difficult to be satisfied.

Consideration of time-dependent travel times and also the demand information arising during real-time to solve the VRP can help in reducing the costs of travel that occurs if the changes in traffic scenarios are ignored.

The total time taken for travel between two locations depends on the specific time of departure. Hence, to take these external influences into consideration, the VRPTW is extended and is studied as TDVRPTW. In this particular scenario, the driving time that changes with the time of the day is suitably represented by a time-dependent function.

12.10.1 The GA-based Crossover for the Time-Dependent VRPTW

The skeleton of a generalized genetic algorithm is presented below:

1. Generate an initial population, which consists of a set of chromosomes to represent the characteristic of all the individuals of the initially selected population.
2. Evaluate each of the chromosomes in the initial population using the fitness function.
3. Select a subpopulation for crossover operation.
4. Perform crossover operation between different pairs of chromosomes in the subpopulation to produce two offspring.
5. Perform mutation for each of offspring based on a mutation probability.
6. Replace the offspring in the current population along with their fitness function values.
7. Sort the chromosomes of the current population either in the descending order of their fitness function values in the case of maximization problem or in the ascending order of their fitness function value in the case of minimizing problem.
8. Repeat step 3 to step 7 for the specified number of generations.
9. Find the chromosome with best fitness function value for implementation.

The steps of the proposed genetic algorithm (SNRPGA) for the time-dependent vehicle routing problem with time windows are presented below. Kumar and Panneerselvam (2015) developed a crossover technique for the genetic algorithm, named Random Sequence Insertion-based Crossover (RSIX).

A schematic view of the crossover operation (RSIX) used in this algorithm is shown in Figure 12.3. A crossover operation is a major process which is applied on two chromosomes from the current subpopulation to produce two offspring. This operation modifies the arrangement of genes of the chromosomes considered and the resultant chromosomes are called offspring. There are many methods for crossover operation according to different problems. "Random Sequence Insertion-Based

P₁ 6, 1, | 4, 7, 2 | 3, 5

6, 1, | 3, 4, 5 | 4, 7, 2 | 3, 5

P₂ 1, 2, | 3, 4, 5 | 6, 7

1, 2, | 4, 7, 2 | 3, 4, 5 | 6, 7

(a)   (b)

C₁ 6, 1, | 3, 4, 5 | 7, 2

6, 1, | 3, 4, 5 | 4, 7, 2 | 3, 5

C₂ 1, | 4, 7, 2 | 3, 5, 6

1, 2, | 4, 7, 2 | 3, 4, 5 | 6, 7

(d)   (c)

C₁   6, 1, 3, 4, 5, 7, 2
C₂   1, 4, 7, 2, 3, 5, 6

(e)

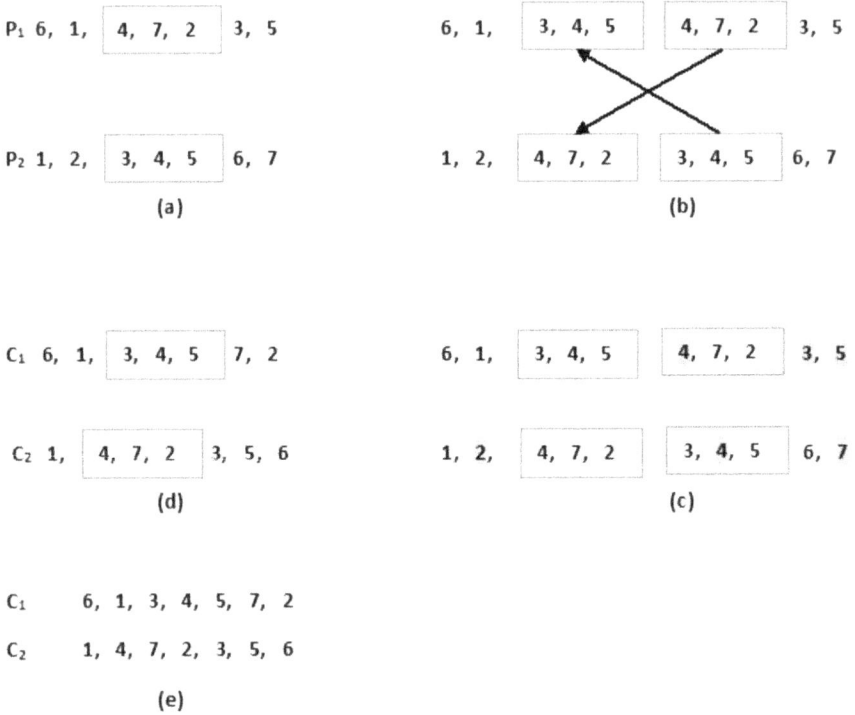

FIGURE 12.3   Crossover operation.

Crossover" method is proposed to perform the crossover operation between any two chromosomes, which is explained in detail below.

The steps of the Random Sequence Insertion-based Crossover (RSIX) method for the TDVRPTW are presented below.

Consider two parent chromosomes (P_1 and P_2) with seven genes in each of them as shown in Figure 12.3a. The gene elements in each of the parent chromosomes are from 1 to 7, but in some random order.

Step 1: The route chromosomes of the initial population are generated randomly by using a random seed. The random seed is used to initialize the new pseudo-random number generator for the time-dependent VRP problem instances. For the travel times, random values between 0.5 and 1.0 have been generated.

Step 2: Two chromosomes from the chromosome pool are randomly chosen as parents.

Step 3: Generate two crossover points, which will lead to three chromosome segments in each chromosome as shown in Figure 12.3a.

Step 4: Copy the middle segment of the chromosome P_1 in Figure 12.3a immediately after the middle segment of the chromosome P_2 in Figure 12.3a and

copy the middle segment of the chromosome P_2 in Figure 12.3a immediately after the middle segment of the chromosome P_1 in Figure 12.3a,
as shown in Figure 12.3b, i.e. swapping operation of the middle crossover
genes segment takes place, as in Figure 12.3b.

Step 5: In each chromosome in Figure 12.3b, check whether each of the
genes in the second segment is present in the first segment, third segment,
and fourth segment in it. If so, shade that gene as shown in Figure 12.3c.
Similarly, check whether each of the genes in the third segment is present
in the first segment, second segment, and fourth segment in it. If so, shade
that gene as shown in Figure 12.3c.

Step 6: Write the genes of each of the chromosomes in Figure 12.3c from left
to right after dropping the shaded genes to give the results as shown in
Figure 12.3d. This results in obtaining two new offspring having gene segments that underwent crossover and they are added into the next generation
as shown in Figure 12.3d.

Step 7: The final two offspring of the crossover operation are as shown in
Figure 12.3e.

- The new offspring are tested for fitness values.
- The smaller the "fitness-value," the stronger the road chromosome
  obtained.

12.10.2 EVALUATION OF FITNESS FUNCTION

Every solution has a fitness value assigned to it, which measures its quality. A route
for a vehicle with respect to a chromosome/offspring is obtained by assigning its
genes (customer nodes) serially from left to right and then copying its first gene
(customer node) at the end of the route for the vehicle routing problem. A vehicle is
assigned to each route to deliver the items to the customer nodes in the route and/or
pickup items from the customer nodes in the route. Every road or route chromosome/
offspring has its own fitness value, defined through a fitness function. The fitness
function value of a route chromosome is the sum of the distances of the arcs (pairs
of adjacent customer nodes) in the route of the route chromosome. In this case, the
smaller the fitness function value, the stronger the road chromosome.

The fitness function is formulated for this VRP, and the quality of a solution s
depends on the total cost or the distance travelled for all the vehicles:

$$f_s = \sum_r distance_{s,r}$$

where $distance_{s,r}$ denotes the cost or distance travelled in route r in solution s.

12.10.3 MUTATION

After performing the crossover operation, mutation operation is performed to further
perturb the arrangement of the genes in each of the offspring. This operation selects

two gene positions randomly and interchanges them for a given probability. This means that a random number is selected for each offspring. If this random number has a lesser value, or equal in value to specified probability of mutation, chromosome undergoes mutation. Else, mutation will not be carried out on that offspring.

Mutation helps to prevent the genetic algorithm from converging to a local optimum. Other genetic algorithms parameters can also influence the GA efficiency. Crossover probability which is specified in the GA determines the rate at which the crossover occurs.

In this section, a genetic algorithm with the modified random sequence insertion-based crossover (RSIX) method for the vehicle routing problem with time windows (TDVRPTW) is proposed.

This crossover technique (RSIX) is applied to develop a new genetic algorithm (SNRPGA) (Nanda Kumar and Panneerselvam, 2017)

The steps of the proposed genetic algorithm (SNRPGA) for the time-dependent vehicle routing problem with time windows are presented below.

Step 1: Input the following:
- Number of customer nodes (n)
- Number of vehicles (k)
- Capacity of the vehicles (a)
- Set Generation Count (GC) = 1.
- Maximum number of generations to be carried out (MNG) = 1000

Step 2: Generate a random initial population (L) of 100 (N) chromosomes (suitable solutions routes for the problem).

Step 3: Evaluate the fitness function $f(x)$ of each chromosome in the population L.

Step 4: Selection – sort the population L by the objective function (fitness function) value in the ascending order, since the objective of the study is minimization of the total distance travelled. Copy a top 30% of the population to form a subpopulation S rounded to the whole number. Smaller fitness value is preferred here.

Step 5: Randomly select any two unselected parent chromosomes from the subpopulation S. Let them be $c1$ and $c2$ using tournament selection.

Step 5.1: Perform two-point random crossover using the random sequence insertion-based crossover (RSIX) described in the earlier section among the chromosomes $c1$ and $c2$ to obtain their offspring $d1$ and $d2$ by assuming a crossover probability of 0.7.

Step 5.2: Perform mutation on each of the offspring using a mutation probability of 0.3.

Step 5.3: Evaluate the fitness function with respect to the total distance travelled and number of vehicles utilized value for each of the offspring $d1$ and $d2$.

Step 5.4: Replace the parent chromosomes $c1$ and $c2$ in the population with the offspring $d1$ and $d2$, respectively, if the fitness function of the offspring is less than that of the parent chromosomes.

Step 6: Increment the generation count (GC) by 1, i.e. $GC = GC + 1$.

Step 7: If $GC \leq MNG$, then go to step 4; else, go to step 8.

Step 8: The topmost chromosome in the last population serves as the solution for implementation. Print the tour along with the total distance travelled and the number of vehicles used.

Step 9: Stop.

The algorithm developed by Nanda Kumar and Panneerselvam (2017) was compared with some existing algorithms to find the superiority of their new algorithm. So, the SNRPGA (Nanda Kumar and Panneerselvam 2017) was applied to the TDVRPTW and was compared with another existing algorithm using a complete factorial experiment with two factors: Problem Size (Factor A) and Algorithm (Factor B) with four replications in each experimental combination.

In the first comparison of the new algorithm, SNRPGA with the existing algorithm developed by Demir (2012) for the TDVRPTW in terms of the total distance travelled, it was found by the authors that there was significant difference among them. Further, through Duncan's multiple rage test, they proved that the proposed SNRPGA is superior to the existing algorithm developed by Demir (2012) in terms of the total distance travelled.

The results of the factorial experiment in terms of the total distance travelled are shown in Table 12.1. The application of ANOVA to the data given in Table 12.1 gives the results as shown in Table 12.2.

From the ANOVA results shown in Table 12.2, one can infer that the factors "Algorithm" and "Problem Size" have significant effects on the total distance travelled. Since, there is a significant difference among the two algorithms compared in terms of the total distance travelled, Duncan's multiple range test is next conducted to identify the best algorithm by arranging the algorithms in the descending order of their mean total distance travelled from to right.

The standard error used in this test is computed as shown below using the mean sum of squares of the interaction terms (Problem Size × Algorithm) and the number of replications under each of the algorithms (24). One can notice the fact that the mean sum of squares of the interaction term AB is used in estimating the standard error (SE), because the F ratio for the factor "Algorithm" is obtained by dividing its mean sum of squares by the mean sum of squares of the interaction term AB_{ij} (Nanda Kumar and Panneerselvam, 2012):

$$SE = \left(MSS_{AB} \div n \right)^{0.5} = \left(21194.060 \div 24 \right)^{0.5} = 29.72$$

The least significant ranges (LSR) are calculated from the significant ranges of Duncan's multiple range tests table for $\alpha = 0.05$ and 5 degrees of freedom as shown in Table 12.3.

The treatment means for the Factor B (Algorithm) in terms of the total distance travelled are arranged in the descending order from left to right, as shown in Figure 12.4.

TABLE 12.1
Results of the Factorial Experiment in Terms of the Total Distance Travelled for TDVRPTW

Problem Class (Factor A)	Algorithm (Factor B)		
Class	Replication of Class	ALNS (Demir)	TDVRPTW (SNRPGA2)
Random 1	1 R1	971	831
	2 R1	932	675
	3 R1	948	717
	4 R1	1048	664
Clustered 1	1 C1	822	593
	2 C1	826	567
	3 C1	827	585
	4 C1	827	580
Random Clustered 1	1 RC1	1207	867
	2 RC1	1114	771
	3 RC1	1258	847
	4 RC1	1457	811
Random 2	1 R2	740	525
	2 R2	701	688
	3 R2	731	590
	4 R2	794	506
Clustered 2	1 C2	585	422
	2 C2	585	430
	3 C2	586	432
	4 C2	586	430
Random Clustered 2	1 RC2	777	762
	2 RC2	783	573
	3 RC2	923	732
	4 RC2	962	606

TABLE 12.2
Analysis of Variance for Total Distance Travelled

Source of Variation	Sum of Squares	Degrees of Freedom	Mean Sum of Squares	Calculated F Ratio	F Ratio (α = 0.05)	Remarks
Algorithm (B)	697352.302	1	697352.302	146.138	4.12	Significant
Problem Size (A)	1313751.984	5	262750.397	55.062	2.47	Significant
Problem Size × Algorithm (A × B)	105970.302	5	21194.060	4.441	2.47	Significant
Error	171787.094	36	4771.864			
Total	2288861.682	47				

TABLE 12.3
Duncan's Multiple Range Tests

No. of treatments – 1 (j)	Significant Range	Standard Error	LSR = Significant Range × Standard Error
2	2.872	29.72	85,356

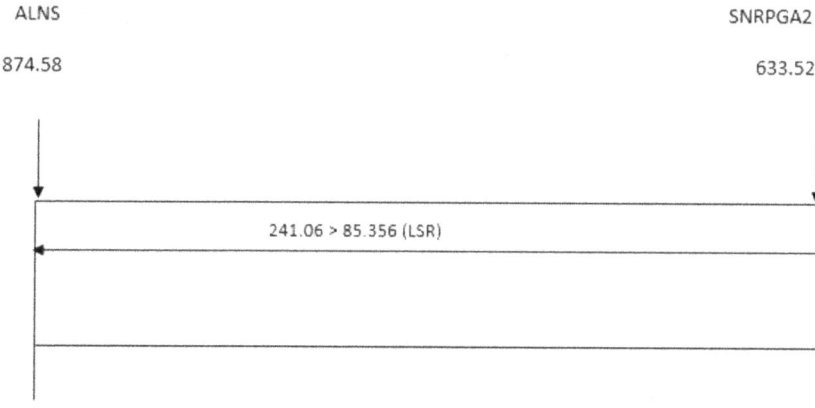

ALNS   SNRPGA2

874.58   633.52

241.06 > 85.356 (LSR)

FIGURE 12.4 Results of Duncan's multiple range test for distance travelled.

From the Duncan's multiple range test performed as shown in Figure 12.4, it is also clear that the TDVRPTW is superior in performance when compared to the existing algorithm used in this study for comparison, in terms of total distance travelled.

12.11 COMPARISON OF ALGORITHMS IN TERMS OF VEHICLES UTILIZED

In this section, a comparison is made among the proposed algorithm SNRPGA and an existing algorithm for the TDVRPTW in terms of number of vehicles utilized using a complete factorial experiment. The existing algorithm was an iterated local search algorithm proposed by Hashimoto et al. (2008). The number of factors in the experiment is 2, viz. Factor A (Problem Size) and Factor B (Algorithm). The number of levels for the Factor A is 6, Random 1, Clustered 1, Random Clustered 1, Random 2, Clustered 2, and Random Clustered 2. The number of levels for the Factor B is 2, viz. TDVRPTW and SNRPGA. The number of replications under each experimental combination is 4. The results in terms of the number of vehicles utilized as per this design are as shown in Table 12.4.

In this model, Factor A (Problem Size/Problem Class) is a random factor and the Factor B (algorithm) is a fixed factor. Since Factor A is a random factor, the interaction factor AB_{ij} is also a random factor. The replications are always random and the number of replications under each experimental combination is 4. The derivation of the expected mean square (EMS) is given in Nanda Kumar and Panneerselvam (2012). To test the effect of A_i as well as AB_{ij}, the respective F ratio is formed by dividing the mean sum of squares of the respective component (A_i or AB_{ij}) by the mean sum of squares of error. The F ratio of the component B_j is formed by dividing its mean sum of squares by the mean sum of squares of AB_{ij}.

The results of ANOVA of the data given in Table 12.4 are shown in Table 12.5.

From the ANOVA results shown in Table 12.5, one can infer that the factors "Problem Size," "Algorithm," and "Interaction of Problem Size and Algorithm" have significant effects on the response variable "Number of Vehicles Utilized." Since there are significant differences among the algorithms, the best algorithm is obtained

TABLE 12.4
Results of Number of Vehicles Utilized for TDVRPTW Problem

Problem Class (Factor A)	Algorithm ((Factor B)		
	Replication of Class	TDVRPTW (Hashimoto)	SNRPGA2
1. Random 1	1	9	9
	2	14	9
	3	13	10
	4	10	9
2. Clustered 1	1	10	8
	2	10	8
	3	10	8
	4	10	8
3. Random Clustered 1	1	10	10
	2	11	11
	3	11	11
	4	14	12
4. Random 2	1	4	3
	2	4	3
	3	3	3
	4	3	3
5. Clustered 2	1	3	3
	2	3	3
	3	3	3
	4	3	3
6. Random Clustered 2	1	4	3
	2	4	3
	3	3	4
	4	3	3

TABLE 12.5
Analysis of Variance for Number of Vehicles Utilized for TDVRPTW Problem

Source of Variation	Sum of Squares	Degrees of Freedom	Mean Sum of Squares	Calculated F Ratio	F Ratio (α = 0.05)	Remark
Problem Size (A)	551.604	5	110.321	260.430	2.47	Significant
Algorithm (B)	3.521	1	3.521	8.311	4.12	Significant
Problem Size × Algorithm (A × B)	5.604	5	1.121	2.646	2.47	Significant
Error	15.250	36	0.424			
Total	575.979	47				

using Duncan's multiple range test by arranging the algorithms in the descending order of their mean number of vehicles utilized from left to right.

The standard error used in this test is computed as shown below using the mean sum of squares of the interaction terms (Problem Size × Algorithm) and the number of replications under each of the algorithms:

$$SE = \left(MSS_{AB} \div n \right)^{0.5} = \left(1.121 \div 24 \right)^{0.5} = 0.216$$

The least significant ranges (LSR) are calculated from the significant ranges of Duncan's multiple range tests table for $\alpha = 0.05$ and 36 degrees of freedom as shown in Table 12.6. The results of Duncan's multiple range test are shown in Figure 12.5. In this figure, the algorithms are arranged as per the descending order of their mean number of vehicles utilized from left to right. From Figure 12.5, it is clear that there is a significant difference between the two algorithms in terms of the mean number of vehicles utilized and further the proposed algorithm SNRPGA utilizes the minimum mean number of vehicles compared to the other algorithm. Hence, the proposed algorithm SNRPGA is superior to the existing algorithm considered in this research for the TDVRPTW problem.

As already stated, any algorithm that is developed now is to be compared with some existing algorithms to find the superiority of the new algorithm. So, the SNRPGA is applied to the TDVRPTW and is compared with an existing

TABLE 12.6
Duncan's Multiple Range Tests

No. of Treatments – 1 (j)	Significant Range	Standard Error	LSR = Significant Range × Standard Error
2	2.872	0.216	0.6204

ALNS   SNRPGA2

7.0   6.0

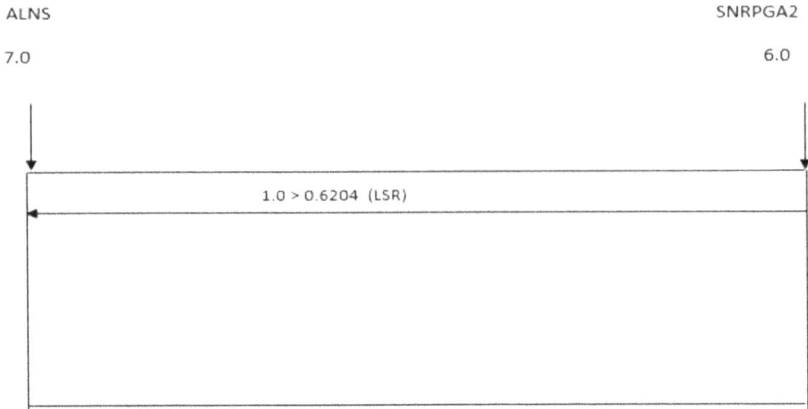

FIGURE 12.5   Results of Duncan's multiple range test for number of vehicles used.

algorithm through a complete factorial experiment with two factors: Problem Size (Factor A) and Algorithm (Factor B) with four replications in each experimental combination.

In the first comparison of the proposed SNRPGA with an existing algorithm developed by Demir (2012) for the TDVRPTW in terms of the total distance travelled, it is found that there is significant difference among them. Further, through Duncan's multiple rage test, it is proved that the proposed SNRPGA is superior to the existing algorithm developed by Demir (2012) in terms of the total distance travelled.

In the second comparison of the proposed SNRPGA with an existing algorithm developed by Hashimoto et al. (2008) for the TDVRPTW in terms of the number of vehicles utilized, it is found that there is significant difference among them. Further, through Duncan's multiple rage test, it is proved that the proposed SNRPGA is superior to the existing algorithm developed by Hashimoto et al. (2008) in terms of the number of vehicles utilized.

In the second comparison of the proposed algorithm, SNRPGA with an existing algorithm developed by Hashimoto et al. (2008) for the TDVRPTW in terms of the number of vehicles utilized, it was found that there was significant difference among them. Further, through Duncan's multiple rage test, it was proved that the proposed SNRPGA is superior to the existing algorithm developed by Hashimoto et al. (2008) in terms of the number of vehicles utilized.

12.12   CONCLUSION

The time-dependent vehicle routing problem (TDVRP) is a class of vehicle routing problems, where the time to serve the customers vary along with the consideration of the traffic conditions in the route. In order to collect the items from various suppliers or deliver items to customers, the 3PL logistics transportation service provider must visit all scheduled customers or suppliers during the different hours specified by the supplier or the customer. Some customers or suppliers may also need visits to happen during a certain time window. This variant of the vehicle routing problem is known as the vehicle routing problem with time windows. Time window constraints are difficult to meet given the real-time road conditions like traffic congestions which makes the vehicles to wait long hours in the traffic jams especially in urban environments. Peak hour traffic congestions lead to long delays. Hence, in order to pick up or deliver items from the suppliers and customers more efficiently, many challenges must be overcome and solved among customers and suppliers, in a 3PL kind of arrangement, whereby the 3PL vehicle visits the customers' or suppliers' sites to deliver or to pick up the items, in an e-commerce or manufacturing or other set-ups, to satisfy the customers and suppliers' specific time windows and to minimize the travel times severely impacted by heavy traffic conditions.

The vehicle routing problem with time windows is solved using the genetic algorithm with multi-chromosome representation. The genetic algorithm is an AI technique that finds near optimal solution to any combinatorial optimization problem which is of the NP-hard type. It is later used for finding a (near) optimal solution to a VRPTW variant, the TDVRPTW, which takes into consideration the various times of the day at which the vehicle travels, since the travel time varies during various times of the day such that the total distance travelled is minimized for the generated number of number of tours for the vehicles. So, the objectives are to minimize the total distance travelled and to minimize the total number of vehicles used to cover all the customer nodes. Hence, in the logistics sector, genetic algorithms are reducing delivery times and decreasing costs. In the logistics business, every mile and minute matter. Companies can use a route planner based on genetic algorithms to map out optimal routes for deliveries.

This study can be useful for planning the supplier site pickups by e-commerce companies, taking into consideration traffic conditions during different periods of the day with time window requirements of the suppliers.

REFERENCES

Bell, J.E. and McMullen, P.R. "Ant colony optimization techniques for the vehicle routing problem," *Advanced Engineering Informatics*, vol. 18, no. 1, pp. 41–48, 2004.

Chang, Y.H. *Logistical Management*. Hwa-Tai Bookstore Ltd., Taiwan, 1998.

Clarke, G. and Wright, J. R. "Scheduling of vehicle routing problem from a central depot to a number of delivery points," *Operations Research*, vol. 12, no. 4, pp. 568–581, 1964. doi:10.1287/opre.12.4.568.

Dantzig, G. and Ramser, J. "The truck dispatching problem," *Management Science*, vol. 6, pp. 80–91, 1959.

Demir, E. *Models and Algorithms for the Pollution-Routing Problem and Its Variations*. PhD Thesis, University of Southampton, Southampton, 2012.

Doerfler, S. "Delivering last-mile options," *Inside Supply Management*, vol. 28, no. 2, pp. 26–30, 2017.

Duygu, T., Jabali, O. and van Woensel, T. "A vehicle routing problem with flexible time windows," *Computers & Operations Research*, vol. 52, Part A, pp. 39–54, 2014.

Ehmke, J.F., Campbell, A.M. and Thomas, B.W. "Vehicle routing to minimize time-dependent emissions in urban areas," *European Journal of Operational Research*, vol. 251, no. 2, pp. 478–494, 2016.

Ehmke, J.F., Steinert, A. and Mattfeld, D.C. "Advanced routing for city logistics service providers based on time-dependent travel times," *Journal of Computational Science*, 2012, doi:10.1016/j.jocs.2012.01.006.

Esmat, Z.-R. and Mirmohammadi, S.H. "Site dependent vehicle routing problem with soft time window: modeling and solution approach," *Computers & Industrial Engineering*, vol. 90, pp. 177–185, December 2015.

Figliozzi, M.A. (2009) A Route Improvement Algorithm for the Vehicle Routing Problem with Time Dependent Travel Times. Proceedings of the 88th Transportation Research Board Annual Meeting, Washington DC, 11–15 January 2009, pp. 616–636.

Ganediwalla, S., Kumar, A., Vardhan Verma, A., Parasramka, T. and Jain, C. *Retail Technology – The Next Frontier*. RedSeer, 2019. https://redseer.com/reports/retail-technology-the-next-frontier/

Gendreau, M., Laporte, G. and Potvin, J.-Y. "Metaheuristics for the VRP," in Toth, P. and Vigo, D. eds, *The Vehicle Routing Problem, SIAM Monographs on Discrete Mathematics and Applications*, Philadelphia, pp. 129–154, 2002.

Hashimoto, H., Yagiura, M. and Ibaraki, T. "An iterated local search algorithm for the time-dependent vehicle routing problem with time windows," *Discrete Optimization*, vol. 5, pp. 434–456, 2008.

Ichoua, S., Gendreau, M. and Potvin, J.Y. "Vehicle dispatching with time-dependent travel times," *European Journal of Operational Research*, vol. 144, pp. 379–396, 2003.

Kok, A.L., Hans, E.W. and Schutten, J.M.J. "Vehicle routing under time-dependent travel times: the impact of congestion avoidance," *Computers & Operations Research*, vol. 39, no. 5, pp. 910–918, 2012.

Kumar, S. and Panneerselvam, R. "A time-dependent vehicle routing problem with time windows for e-commerce supplier site pickups using genetic algorithm," *Intelligent Information Management*, vol. 7, pp. 181–194, 2015. doi:10.4236/iim.2015.74015.

Kuo, Y. "Using simulated annealing to minimize fuel consumption for the time-dependent vehicle routing problem," *Computers & Industrial Engineering*, vol. 59, no. 1, pp. 157–165, 2010.

Meng, F., Ding, Y., Li, W. and Guo, R. "Customer-oriented vehicle routing problem with environment consideration: two-phase optimization approach and heuristic solution," *Mathematical Problems in Engineering*, vol. 2019, 2019, Hindawi. https://doi.org/10.1155/2019/1073609

Nanda Kumar, S. and Panneerselvam, R. "A survey on the vehicle routing problem and its variants," *Intelligent Information Management*, vol. 4, no. 3, pp. 66–74, 2012. doi:10.4236/iim.2012.43010.

Nanda Kumar, S. and Panneerselvam, R. "Development of an efficient genetic algorithm for the time dependent vehicle routing problem with time windows," *American Journal of Operations Research*, vol. 7, pp. 1–25, 2017. doi:10.4236/ajor.2017.71001.

Pisinger, D. and Ropke, S. "A general heuristic for vehicle routing problems," *Computers & Operations Research*, vol. 34, no. 8, pp. 2403–2435, 2007.

Schyns, M. "An ant colony system for responsive dynamic vehicle routing," *European Journal of Operational Research*, vol. 245, no. 3, pp. 704–718, 2015.

Yu, B., Yang, Z. and Yao, B. "An improved ant colony optimization for vehicle routing problem," *European Journal of Operational Research*, vol. 196, no. 1, pp. 171–176, 2009.

13 Application of Machine Learning for Fault Detection and Energy Efficiency Improvement in HVAC Application

Umashankar Subramaniam, Sai Charan Bharadwaj, Nabanita Dutta, and M. Venkateshkumar

CONTENTS

13.1 INTRODUCTION

Sustainability is a continuous process of maintaining climate change where the usage of natural resources for technological advancement has a neutral effect on the ecosystem [1]. Whereas sustainable development refers to "comprehensive and holistic approach of temporal process that will lead to an end point of sustainability."

Sustainable living is a socio-economic challenge. United Nations Development Programme (UNDP) has listed out 17 major sustainable development goals [2] and this forms the basis for leaders and policymakers across the world as a visionary framework for adopting and enforcing law of the land. Sustainable living describes the human lifestyle of using natural resources in such a way that it has a neutral effect on the ecology and ecosystem. The people who generally participate and practice a set style of living will often try to contemplate to reduce their carbon footprint by every means: transport, energy, water, etc. It is often described as a form of living that meets socio-economic, cultural, and ecological needs without jeopardizing Nature for future generations.

Today, in India, pollution has become a vital issue. Globally, CO_2 emission has reached 37.1 billion tons, increasing at a rate of 2.7% compared to the previous year. India is likely to add at least 6.3% on the estimate in 2017, i.e. 2.6 billion tons [3]. The buildings all over the world contribute significantly to carbon emissions and are the major stakeholders in the end user energy consumption. The construction sector accounts for 9.1% [4] of the country's gross domestic product (GDP) and it is contributing significantly to CO_2 emissions. Globally, the building sector consumed 125 EJ in 2016; building and construction industry together accounted for 36% of the total energy consumption. Now the onus lies on India to push the construction sector towards green and sustainable buildings.

13.2   SUSTAINABLE LIVING IN BUILDINGS

A sustainable building [5] definition says it is the building impact that will be reduced if the energy water used for building energy purpose increases with the increase in efficiency of the system. It will keep the bad impact on human health and environment should be better constructed for design operation and maintenance."

The most indispensable parameter for sustainable buildings is energy and water consumption and conservation. An ideal sustainable building comprises a power generation unit for generating power by renewable means such as solar photovoltaic (PV) and wind. This significant feature highlights that sustainable buildings virtually do not require any power from the grid and are isolated from the grid or from a local microgrid supplying power to other small buildings in its vicinity. In some parts of the world where harnessing renewable energy is not possible inside the buildings, they are connected to the grid purchasing or operating by renewable forms of energy. To reduce CO_2 emissions from the buildings, it is highly essential to optimize the energy usage inside the buildings.

Another chief objective of the green buildings is to conserve water. Seventy percent of India's population lives in rural areas, with each household consuming on an average 117–125 litres [6] of water per day – washing consuming the water most. Eighty-six percent of the residential buildings do not have rainwater harvesting pits in India. With the recent implementation of rainwater harvesting pits through National Rural Employment Guarantee Scheme (NREGS), there have been slight changes in the groundwater level in rural India. Most of the cities which are the driving engines of the economic growth in the country are facing acute water crisis. These include Delhi, Bengaluru, Hyderabad, Chennai, and emerging cities like Pune, Amravati, and Coimbatore. Today roughly about 0.5 billion people live in Bangalore, including more than 2 million information technology (IT) professionals; 40% of the

Per Capita Water Consumption per Day(Liters)

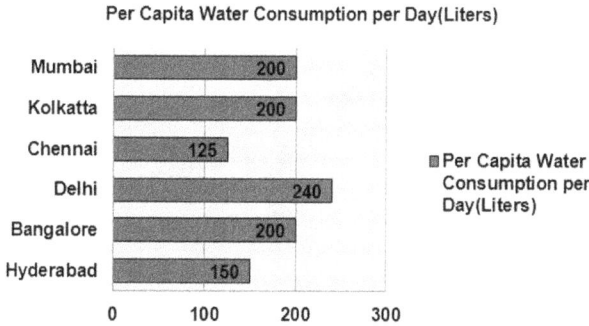

FIGURE 13.1 Per capita water consumption per day (in litres) in metros of India.

population is chiefly dependent upon groundwater[7]. Some experts say that at least 21 cities in India will reach zero groundwater level by 2020, which is quite alarming. In the following sections, we will look at various energy consumption scenarios in the buildings worldwide as well as in the Indian context and how technological advancements can achieve sustainable developmental goals and sustainable living inside the buildings. Figure 13.1 illustrates the per capita water consumption per day in major cities of India [8].

13.3 ENERGY SCENARIO IN BUILDINGS

The world is shifting its energy paradigm into a different ecosystem, but fissures are noticeable across its key structures. The key structures are as follows:

- Affordability
- Reliability
- Sustainability

The aforementioned are closely related to each other and require comprehensive understanding and a line of action for energy policy. In the modern era of human civilization with progressive industrialization, globalization, and digitalization and its policies, it is expected that cities will add a population of 1.7 billion by 2040. This is quite alarming for policymakers, engineers, think tanks, and people who are concerned about meeting the unprecedented energy demand. Figure 13.2 shows world energy consumption, and Figure 13.3 shows CO_2 emissions by sector across the world [9].

This astronomical growth of energy demand is dominated by emerging economies led by India and followed by other countries. Most of the people across the world spend 80% of their daily life inside buildings. The building and construction sector together account for 36% of global energy consumption (Figure 13.2). Forty percent of total CO_2 emissions which includes both direct and indirect emissions is from buildings (Figure 13.3). Energy usage in buildings continues to grow at a rapid pace. With the adoption of energy efficiency initiatives in the early millennium,

Worlds End Energy Consumption (World Bank 2015)

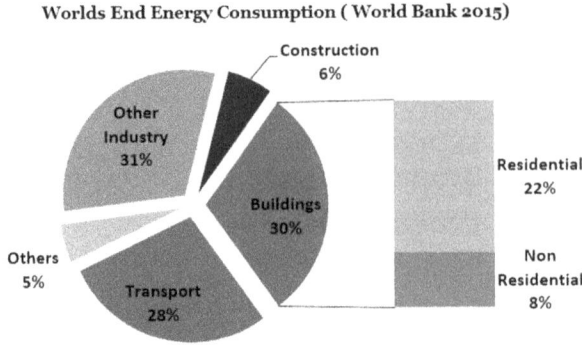

FIGURE 13.2   World's end energy consumption (World Bank 2015).

C02 EMISSIONS WORLD BANK DATA IN BUILDINGS WORLDWIDE

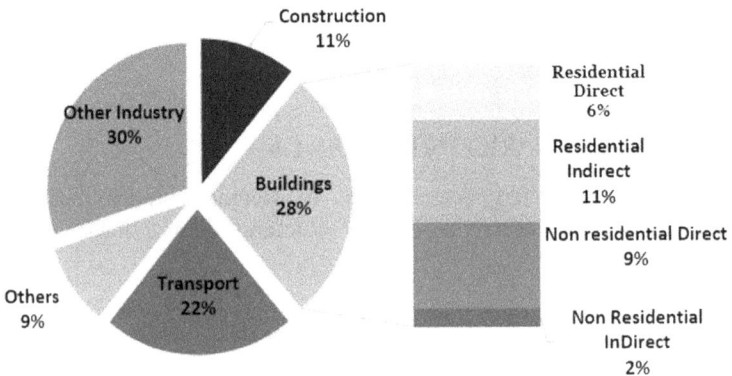

FIGURE 13.3   World CO_2 emissions by sector.

there has been a significant energy saving over the last decade and half; otherwise, the energy usage would have been 31% more by 2022 [10].

Coming to the Indian context, industry consumes the highest energy with around 40%. Buildings are the second largest consumers of end energy, followed by transport and others (Figure 13.4). Figure 13.5 represents the percentage share of end user energy consumption in commercial buildings of India, which illustrates that heating ventilation and air-conditioning (HVAC) systems consume about 55% of total energy consumption. Figure 13.6 provides a comparison of end user energy consumption among various units inside the buildings Energy usage in buildings is dominantly contributed by heating, cooling, lighting, pumping, and appliances, sometimes synonymously called HVAC systems (Figures 13.5 and 13.6) [11].

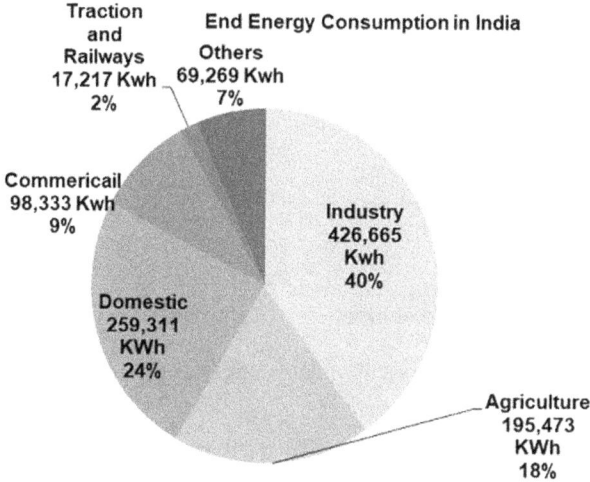

FIGURE 13.4 India's energy consumption in 2017.

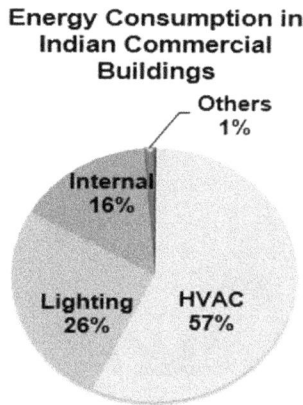

FIGURE 13.5 Energy consumption in commercial buildings in India.

Cooling: Energy usage for cooling the space inside the building grew three times faster than any other end usage inside buildings from 1996 to 2016. Space cooling is typically characterized by an electrical fan powered by an electrical source or an air-conditioner. Cooling is a major social challenge as out of 2.8 billion living in the hottest parts of the globe, only 8% have access to cooling facilities (Figure 13.7) [12, 13].

This increase in energy demand creates great thrust for power generation as well as distribution to meet the peak load demand of end energy usage as well as its driving agents, resulting in increased emissions due to power generation. Electrical energy is the driving force for the economic activity and development of infrastructure in the country. Energy demand is significantly influenced by the increase in population and economic growth in the hottest demographic regions of the world:

End Energy Usage Consumption- BEE 2017

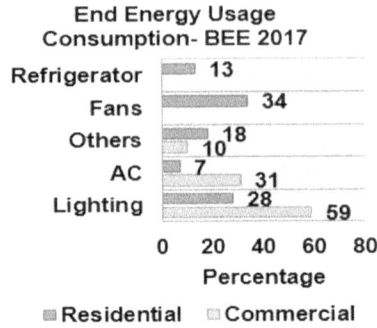

Refrigerator   13
Fans   34
Others   10 18
AC   7 31
Lighting   28 59

0   20   40   60   80

Percentage

■ Residential   □ Commercial

FIGURE 13.6   Energy consumption by various components in Indian buildings.

Room AC sales in India By star rating in Fiscal year 2017-18,

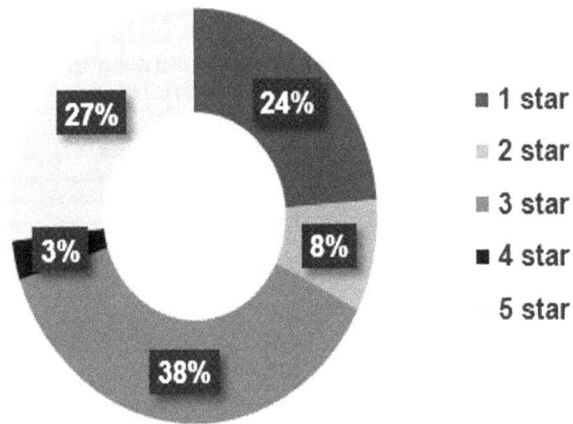

27%   24%

3%   8%

38%

■ 1 star
▪ 2 star
▪ 3 star
■ 4 star
5 star

FIGURE 13.7   AC sales in India 2017–2018.

India, China, Indonesia, and others. *Household fans:* Electrical fan is the simplest form of cooling. It is estimated that 2.3 billion residential fans have been under usage till 2016. Fifty-five percent of the global household has at least one fan in use. Fans consume about 10% less energy than the air-conditioners. Fans will continue to play a major role in the cooling demand in the developing countries as fans are much more affordable than air air-conditioners [13]. *Heating:* It is the largest component of energy uses in buildings, accounting for 36% of the total building energy consumption according to the 2018 IEA Report [14]. Pumps provide space heating, cooling, and hot water in buildings. Pumping is a predominant technology that has evolved over the years and is being used in AC's chillers and other systems. Heat pumps are highly efficient and are commercially proven. Over the decades, the typical pumps in the residential sector are

- Split or room AC
- Air water heater pump
- Water to water heater pump
- Ground source heat pumps

Lighting: Lighting applications in India accounts for about 59% of total energy consumption in residential buildings and 26% in commercial buildings. With the advent of compact fluorescent light (CFL), light-emitting diode (LED) lights, and their market-penetrative investments, savings of over 6000 PJ have been achieved in OCED countries. India is primarily leading the LED market. Sale of LEDs is expected to exceed by 15% across the world. India has already distributed 50 million LEDs until 2017.

In 2017, the lighting sector [15] had a demand of 2 EJ whereas the non-residential sector had a demand of 4 EJ. With the policy implementation and effective investment into the market, LED account for 25% of the installed capacity.

Appliances: Appliances alone have consumed 30% of global fuel energy and accounts for 17% of the world's total end energy consumption [16]. As previously mentioned, India is an emerging economy associated with meteoric urbanization and increasing purchasing capacity of the individuals. These are some of the striking factors for mounting up the floor area of appliances. Energy consumption by appliances has also swelled to a whopping level of 58% since 2000. It is estimated that appliances will contribute 9% of the total energy in buildings with the implementation of energy-efficient policies.

It is worth mentioning that India has taken a major step forward with its first ever building energy code for residential buildings named Energy Conservation Building Code (ECBC) in 2017 [17]. To sum up, the energy usage will now tend to remain flat until 2040 despite the increase in the floor capacity of 60%, making buildings 40% efficient. India has saved 6% additional energy since 2000. China and India are the major energy consumers, with a stake of 82%, among the six emerging economies in 2017; they contribute 5% of fossil fuel and 145 MT CO_2 emissions (Figure 13.8)] [18].

13.4 DIGITALIZATION AND ARTIFICIAL INTELLIGENCE FOR ENERGY EFFICIENCY IN BUILDINGS

Early computers used to occupy a lot of space and perform small calculations via vacuum tubes, but in later stages Central Processing Unit (CPU) and other modifications were made. Over the years, man has achieved the capacity and capability of developing computer systems with reduced size and yet increasing computational capacity and performance speed in all dimensions (Figure 13.9).

13.4.1 ARTIFICIAL INTELLIGENCE (AI)

It is the branch of computation science which aids computers and machines to think like humans. It is the science and engineering that make intelligent machines

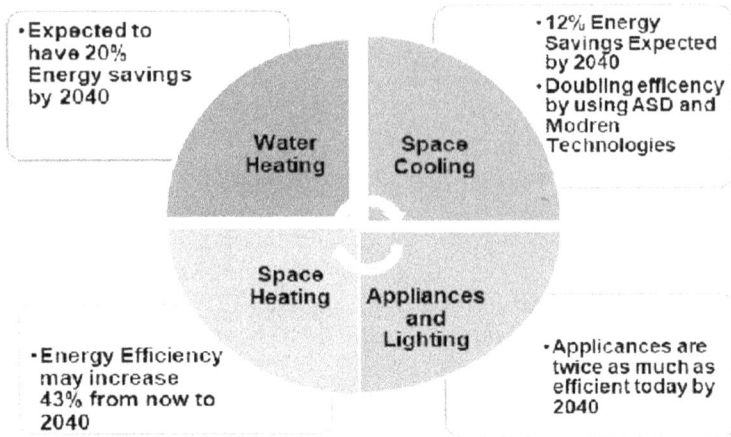

- Expected to have 20% Energy savings by 2040
- 12% Energy Savings Expected by 2040
- Doubling efficency by using ASD and Modren Technologies

Water Heating   Space Cooling

Space Heating   Appliances and Lighting

- Energy Efficiency may increase 43% from now to 2040
- Applicances are twice as much as efficient today by 2040

FIGURE 13.8   World's energy savings projected for 2040.

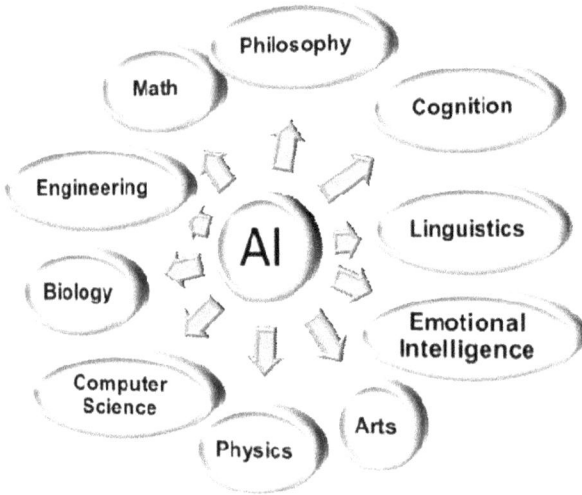

Philosophy
Math
Cognition
Engineering
AI
Linguistics
Biology
Emotional Intelligence
Computer Science
Physics
Arts

FIGURE 13.9   Illustration of AI subjects.

especially intelligent. Integral philosophy and motto of this is "can a machine think and behave like humans" [19].

This makes a loud statement that it is a technology of creating machines that exhibit characteristic features like thinking, behaviour, sense, learn, explain, and adhere. Thus, artificial intelligence is a multidimensional, cross-functional, and multidisciplinary subject involving computer science, physics, biology, psychology, linguistics, arts, sociology, engineering, mathematics, and others, as depicted in Figure 13.10. AI systems will adopt the changes without explicitly being programmed, rather than being specifically intended to perform a task (Figure 13.11).

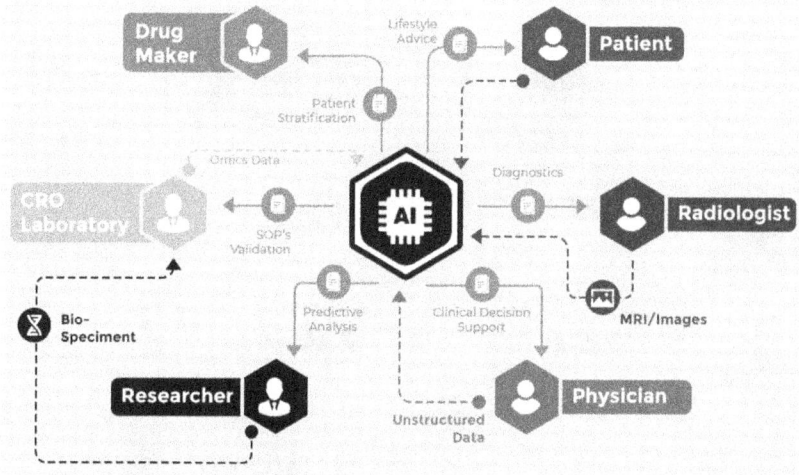

FIGURE 13.10   Some applications of AI.

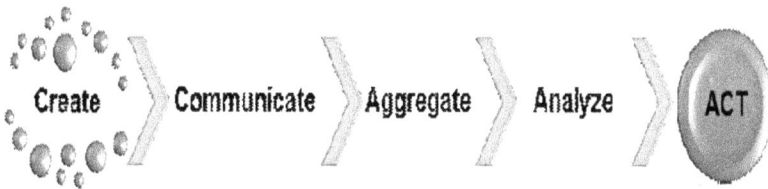

FIGURE 13.11   Various stages in typical IoT system.

13.4.2   INTERNET OF THINGS (IoT)

The term Internet of Things was first coined by Kevin Ashton [20] in early 2000, which laid the foundation for the IoT at Massachusetts Institute of Technology (MIT) labs subsequently. [21] Recently, the word IoT has been added to *Oxford Dictionary*, which says "that it is an interconnection via internet of computing devices embedded in everyday objects enabling them to send and record data."

IoT systems generally include artificial intelligence communication protocols, census data, acquisition systems, and actuators in smart device for usage and active engagement (Figure 13.12) [22].

The various stages and building blocks of an IoT system are as follows:

- *Create:* Use of sensors to record the physical event or a state.
- *Communicate:* Transfer of recorded event for a communication protocol.
- *Analyse:* Analysing the patterns behaviour redundancies of the aggregated data to provide predictive, presumptive, prescriptive forecasting insights: transfer of recorded event for a communication protocol (Table 13.1)
- *Act:* Changing and maintaining initiative state or event upon the analysis.

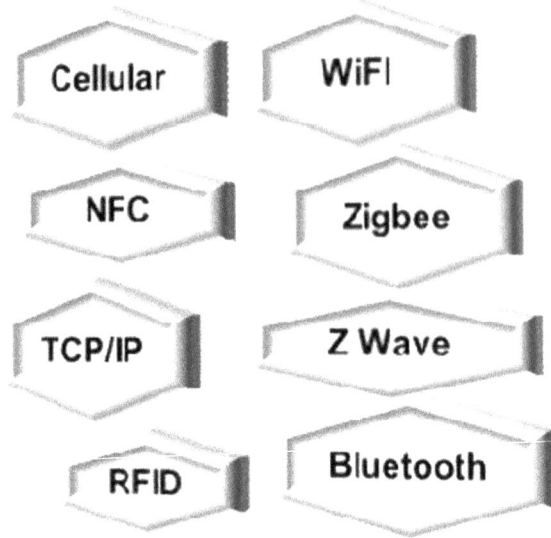

FIGURE 13.12 Various communication protocols used in IoT.

TABLE 13.1
Some Commonly Used Sensors Inside Buildings

S. No.	Sensor Type/Name	Comments
1	Temperature and RH	Used for measuring temperature and humidity of room say DHT11 or LM35
2	CO_2 and smoke sensor	Useful in detection of gas leakage for home as well industry, e.g. MQ2 smoke sensor
3	Water flow rate sensors	Measure water flow from the sensors, e.g. YF-S201
4	Motion detect sensors	Used to detect the human activity, e.g. HC-SR501 PIR sensor
5	Acoustic sensor	Used for sound detection, e.g. KY-038
6	Light sensors	Deployed for luminosity and light measurement TSL2561 or LDR
8	HVAC and smart energy meters	Used for measuring power and energy, e.g. smart energy meters
9	Imaging	Thermal IR- or CMOS-based sensors to detect and analyse occupancy patterns

13.4.3 MACHINE LEARNING (ML)

It is a subset and a sister branch of AI which is used to analyze the data which learns and runs through data without being explicitly told. It performs exceptionally well if the data is very large (big data). Data can be in numerous forms such as analogue sensors, images, acoustics, human activity, and others.

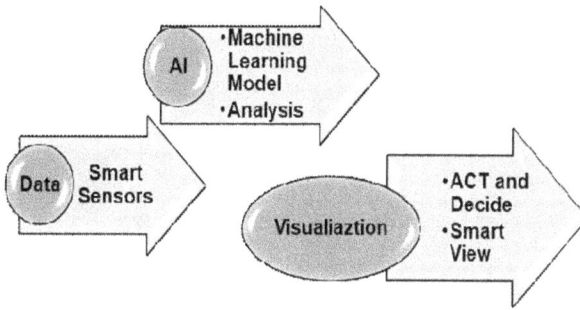

FIGURE 13.13 Simple machine learning block diagram.

Machine learning [23] has been basically divided into three types: supervised learning, unsupervised learning, and reinforcement learning. It is a form of cognitive theory and science. This machine learning models can be used inside buildings for predictive modelling of the energy and for fault detection of the various equipment and appliances inside buildings.

The basic and simple flow block diagram of machine learning is shown in Figure 13.13.

13.4.4 Influence of Digitalization on HVAC Systems

The data collected by sensors are stored in the server and the next step is to automate the further processes to start analyzing the data trends related to the cooling or heating pattern, energy usage, peak load demands [24], occupant behaviour [25], and overall thermal comfort inside the space of the building. These insights provide more degrees of freedom and operational flexibility to regulate chiller or heating equipment to achieve energy saving (Figure 13.14).

13.4.5 Energy Optimization and Scheduling

Let's say a building has 100 data points or nodes that are being connected to a central hub where data has been stored in the server. The thermal comfort inside the building is a function of human activity, flooring, walls, space area, equipment, solar radiation, and the presence of electronics. In order to cool a space in a building, chiller has to blow cold air into the space to decrease the temperature. Over the course of the day, there will be various levels of requirements of cooling and heating pattern in accordance with the different nature of humans and activities. The role of IoT and AI-based systems is to record, analyze, and interpret the data to provide meaningful outcomes of energy usage at peak times, flat time, and no load periods by blending occupant behaviour into the model to achieve localized control of thermostat [26].

13.4.6 Predictive Maintenance and Fault Diagnosis

As previously mentioned, machine learning models which are part of artificial intelligence can be used for predictive maintenance and fault diagnosis of systems.

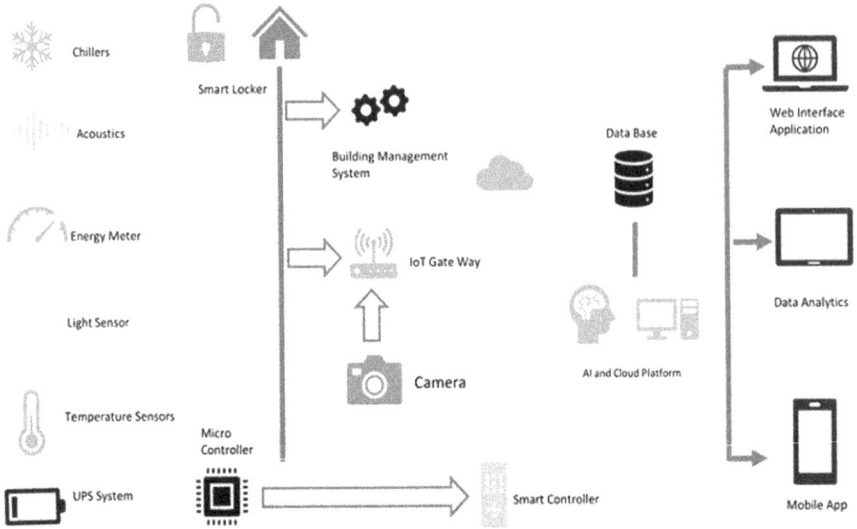

FIGURE 13.14   Typical IoT–AI system for smart building.

Pumps, which are an integral part of the buildings, also contribute significantly to energy consumption as most of the time it is driven by an induction motor, which is subjected to various faults (Figures 13.15 and 13.16) [27].

Data from vibration sensor smart flow rate meters and pressure sensors will allow us to predict the pumping faults well in advance to save unnecessary excessive energy usage and sudden collapse of the pump inside the buildings. Such a kind of predictive fault diagnosis [28] can be extended to heating, cooling, and ventilation

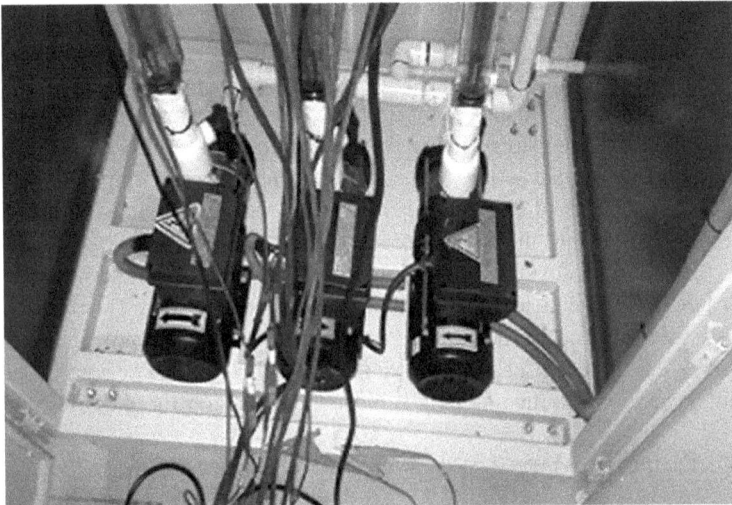

FIGURE 13.15   Water pump monitoring inside a building.

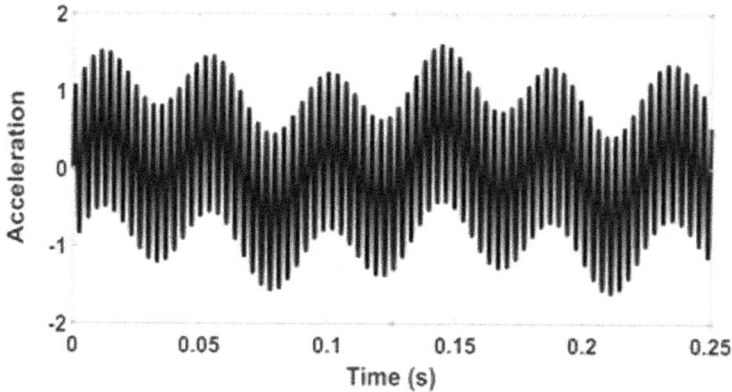

FIGURE 13.16 Vibration sensor data collected to analyse real-time faults in pumps.

as well. If any fault happens in the motor or water pumping system, the trained data of the motor or pumps will be fed to the simulation and predictive control software which will classify the faults using training and testing data with the help of machine learning software. Support vector machine algorithm is one of the machine learning algorithms which will identify the faults in the system in very less time. So for fault classification, SVM algorithm has been used which classified the faulty and no faulty points from unclassified data (Figure 13.17).

Other interesting aspects of artificial intelligence are listed below:

- Sending in app notifications of problematic equipment.
- Notifying service provider about the equipment repair at a particular location on a particular building.
- Moral consolidation and energy consumption demand generating reports and forecasting models to grid authority's for effective scheduling.

To achieve this in greater efficacy in the system, we need to monitor each and every facility management services and deployment of energy harvesting methods to provide power to sensors and AI-powered devices to conserve energy requirement to carry out this service.

13.5 CONCLUSION

Buildings account for 36% of the total energy consumption and 28% of the total carbon emissions. The construction sector in India accounts for 9% of the GDP. On an average, an Indian citizen consumes around 125–150 litres of water per day for his essential needs. To address this challenge, India needs to push towards the green and sustainable buildings incorporating digital technologies like IoT in fusion with data-driven machine learning models to conserve energy. There are numerous applications that are being relied upon, including artificial intelligence. AI can be implemented even in old buildings using retrofit methods. To achieve low carbon

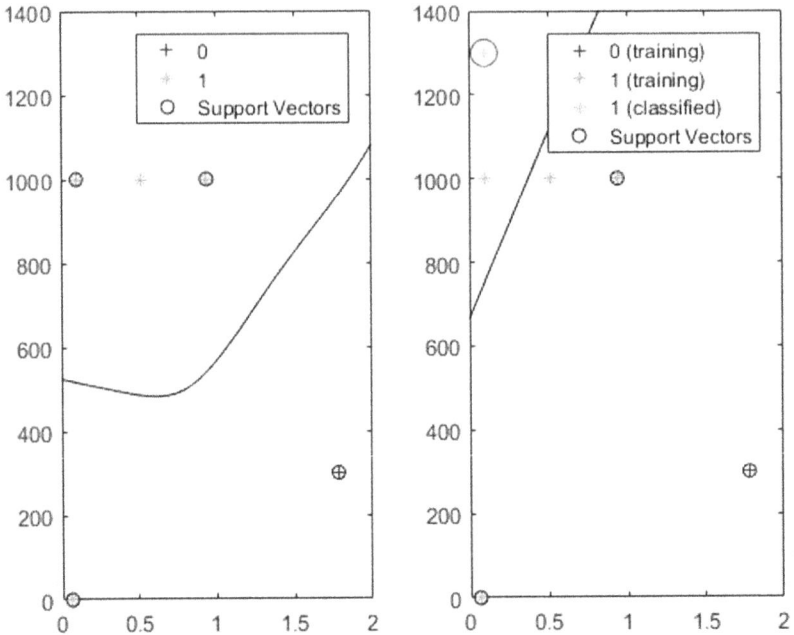

FIGURE 13.17   Classification of faults using SVM algorithm.

emissions and higher energy efficiency, we need to adapt and implement new technologies that will conserve energy in buildings. Buildings provide a huge potential for energy savings and reducing the carbon emission and thereby achieving a 2°C reduction in global temperature. A 30% improvement in the global average of the building energy intensity by 2030 is needed to meet the ambitious goals of the Paris Accord 2015. Machine learning and deep learning technology have brought remarkable solutions by enabling both continuous monitoring and ground-level fault detection. The prediction control app-based device based on the hybrid technology combined with machine learning, deep learning, and IoT, GPU-based human–machine interface can be helpful to predict the fault condition before the system is totally shut down. So whenever any abnormal condition is detected, the machine will indicate through the alarm that the system is in danger, so it is possible for the system operator to identify the faulty part easily and take necessary action. This predictive controller GPU-based device is applicable not only for pumping system but also for any heavy industry. Machine learning and deep learning technology have been extensively applied in the biomedical sector, but in the industrial application and mainly in the pumping system it is very rare. The predictive control hybrid model is the new point of study where the researcher is planning to reduce the energy loss and the time of the process and thus making the system flawless. This study is trying to find out the possibilities of anomalies in the HVAC system and its solution by using an AI technique.

CONFLICT OF INTEREST

The authors declare no conflict of interest.

REFERENCES

1. U. S. E. P. Agency, Learn *About* Sustainability, United States Environmental Protection Agency, [Online]. Available: https://www.epa.gov/sustainability/learn-about-sustainabi lity#what. [Accessed 16 February 2019].
2. U. N. D. Programme, *Sustainable Development Goals*, UNDP, [Online]. Available: http://www.undp.org/content/undp/en/home/sustainable-development-goals.html. [Accessed 15 February 2019].
3. U. Goswami, "India's CO2 Emissions Forecast to Increase by 6.3% This Year," *Economic Times*, 06 December 2018. [Online]. Available: https://economictimes.ind iatimes.com/news/politics-and-nation/indias-co2-emissions-forecast-to-increase-by-6 -3-this-year/articleshow/66963109.cms. [Accessed 13 February 2019].
4. C. India, *India in Figures*, MOSPI, 2018. [Online]. Available: http://www.mospi.gov .in/sites/default/files/publication_reports/India_in_figures-2018_rev.pdf. [Accessed 15 February 2019].
5. J. C. Howe, & M. Gerrard (2010). *The Law of Green Buildings: Regulatory and Legal Issues in Design, Construction, Operations, and Financing.* American Bar Association.
6. O. &. T. Singh, "A Survey of Household Domestic Water Consumption Patterns in Rural Semi-Arid Village, India," *GeoJournal*, vol. 78, no. 5, pp. 777–790, 2013.
7. A. Shukla, "Alarming: 21 Indian Cities Will Run Out of Water by 2030," *Business World*, 19 June 2017. [Online]. Available: http://www.businessworld.in/article/ Alarming-21-Indian-Cities-Will-Run-Out-Of-Water-By-2030/19-06-2017-120383/. [Accessed 9 February 2019].
8. S. Mungara, "Five Major Metros Staring at Severe Water Crisis: Expert," *Times of India*, 5 October 2018. [Online]. Available: https://timesofindia.indiatimes.com/city/ hyderabad/five-major-metros-staring-at-severe-water-crisis-expert/articleshow/660801 51.cms. [Accessed 8 February 2019].
9. B. D. a. J. D. Thibaut Abergel, *UN Environment and International Energy Agency (2017): Towards a Zero-Emission, Efficient, and Resilient Buildings*, United Nations Environment Programme, Paris, 2017.
10. C. India, *Energy Statistics 2018*, Ministry of Statistics and Programme Implementation, New Delhi, 2018.
11. D. C. P. Commission, *The Final Report of the Expert Group on Low Carbon Strategies for Inclusive Growth*, Planning Commission GOI, New Delhi, 2014.
12. 2. OECD/IEA, *The Future of Cooling*, OECD/IEA, Paris, 2018.
13. T. E. a. R. Institute, *Improving Air Conditioners in India*, TERI, New Delhi, 2018.
14. OECD/IEA, *Energy Efficiency: Heating*, IEA, [Online]. Available: https://www.iea .org/topics/energyefficiency/buildings/heating/. [Accessed 1 February 2019].
15. OCED/IEA, *Lighting: Tracking Clean Energy Progress*, IEA, 2018. [Online]. Available: https://www.iea.org/topics/energyefficiency/buildings/heating/. [Accessed 16 February 2019].
16. OCED/IEA, *Energy Efficiency: Appliances*, IEA, 2018. [Online]. Available: https://ww w.iea.org/topics/energyefficiency/buildings/appliances/. [Accessed 16 February 2019].
17. B. G. Ministry of Power, *Energy Conservation Building Code*, 2017. [Online]. Available: https://beeindia.gov.in/sites/default/files/BEE_ECBC%202017.pdf. [Accessed 16 February 2019].

18. OCED/IEA, *Energy Efficiency: Buildings*, IEA, 2018. [Online]. Available: https://www.iea.org/topics/energyefficiency/buildings/. [Accessed 16 February 2019].

19. T. Point, *Artificial Intelligence - Overview*, Tutorials Point, [Online]. Available: https://www.tutorialspoint.com/artificial_intelligence/artificial_intelligence_overview.htm. [Accessed 16 February 2019].

20. Wikipedia, *Kevin Asthon*, Wikipedia, [Online]. Available: https://en.wikipedia.org/wiki/Kevin_Ashton. [Accessed 16 February 2019].

21. K. Asthon, "That 'Internet of Things' Thing," *RFID Journal*, pp. 97–114, 2009.

22. M. M. E. R. C. Jonathan Holdowsky, *Inside the Internet of Things*, Deloitte University Press, UK, 2015.

23. S. a. S. B.-D. Shalev-Shwartz, *Understanding Machine Learning: From Theory to Algorithms*, Cambridge University Press, Cambridge, 2014.

24. H. M. R. a. K. T. C. Zhou, "Artificial Intelligence Approach to Energy Management and Control in the HVAC Process: An Evaluation, Development and Discussion," *Developments in Chemical Engineering and Mineral Processing*, pp. 42–51, 1993.

25. S. Salimi & A. Hammad (2019). Critical review and research roadmap of office building energy management based on occupancy monitoring. *Energy and Buildings*, 182, 214–241.

26. H. B. Gunay, W. Shen & G. Newsham (2019). Data analytics to improve building performance: A critical review. *Automation in Construction*, 97, 96–109.

27. D. S. U. a. P. S. Nabanita, "Mathematical Models of Classification Algorithm of Machine Learning," in *International Meeting on Advanced Technologies in Energy and Electrical Engineering (IMAT3E'18)*, Fez, Morocco, 2018.

28. N. Dutta, "Centrifugal Pump Cavitation Detection Using Machine Learning Algorithm Technique," in *2018 IEEE International Conference on Environment and Electrical Engineering and 2018 IEEE Industrial and Commercial Power Systems Europe (EEEIC/I&CPS Europe)*, Palermo, 2018.

14 Smart City Using Artificial Intelligence Enabled by IoT

P. Srividya and Sindhu Rajendran

CONTENTS

14.1 INTRODUCTION

A city is said to be smart if the residents are able to access various services and networks efficiently. The services must be sustainable, flexible, and easily accessible. The goal is accomplished by the use of information, telecommunication, and digital technologies. Traditional cities are transformed into smart cities by the use of information and communication technology (ICT). Emerging technologies like IoT,

BD, blockchain, and robotics help to make a city smart. Developing a smart city is a collaborative work that involves many government and private organizations, public sectors, and citizens themselves.

The intention of building a smart city is to overcome the problems arising from rapid urbanization and increasing population. Even though the initial cost involved in developing a smart city is high, in the long run it helps in reducing energy consumption, environmental pollution, city wastes, water consumption, traffic congestion, and other problems. Different transportation forms, pollution management, Wi-Fi systems, green energy, universal identification, and local commerce promotion are some of the initiatives of building a smart city.

Some of the facilities that can be included in a smart city are public Wi-Fi systems, mobile apps for citizens to report congestion of roads, systems to manage and monitor electricity and water usage, systems to intimate contamination of water resources, systems to identify potholes in the roads, systems to manage parking systems, and many others.

A city is said to be smart if it possesses the following features [1]:

- Usage of digital technologies and electronics in building the city infrastructure.
- Usage of ICT in transforming living conditions and working environment.
- ICT-embedded government systems.
- Usage of ICT to bring about innovation and in enhancing the knowledge offered.

The infrastructure of smart city includes the physical infrastructure, installed sensors, software, and firmware.

This chapter covers the various aspects of the smart city which include the structure of a smart city, requirements to build a smart city, various components involved, challenges faced, and technologies involved in building a smart city. This chapter also highlights the challenges faced in making a city smart.

14.2 STRUCTURE OF A SMART CITY

Structure of a smart city is shown in Figure 14.1. It involves the following three-layered structure:

1. *Technology base:* Includes smartphones and different sensors to collect the data in real time. High-speed communication networks and open-data portals are used to connect the sensors.
2. *Specific applications:* The raw data acquired from the sensors are to be translated into alerts and actions. This can be done by application developers and technology providers.
3. *Public usage:* The applications so developed must be adopted by the public to change the lifestyle.

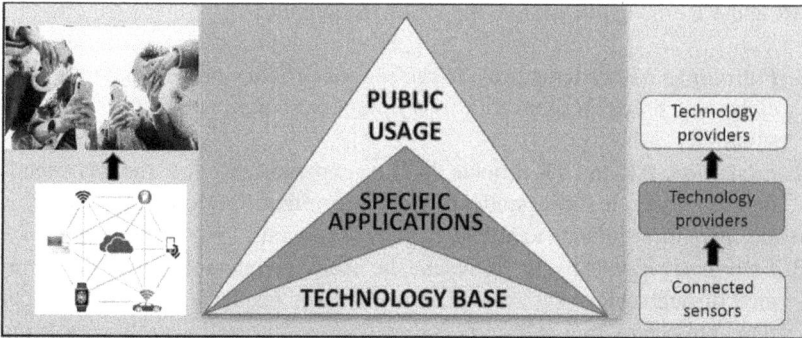

FIGURE 14.1   Structure of a smart city.

14.3   REQUIREMENTS TO BUILD A SMART CITY

The process of building a smart city is an iterative process, with processing and analysis at each iteration step. It forms the city's backbone.

IoT provides seamless connectivity to all devices like wearable, entertainment, home appliances, public mobility, medical devices, connected vehicles, buildings, agriculture, and services that go beyond machine-to-machine communication. Any IoT platform requires the following things to build a smart city:

- Smart things like sensors, actuators, cameras, etc. to gather the data.
- Tools to analyse the data collected by sensors.
- Cloud gateways to collect the data from IoT devices, store them, and forward them to cloud in a secured manner.
- Data processors to collect the data, store it, and distribute it to control applications.
- A data warehouse to clean and organize the collected data.
- Algorithms to automate city services and to improve the performance of various control applications based on data analysis.
- Applications to send the data to the actuators.
- Applications to connect smart things and users.

14.4   AUGMENTED REALITY AND VIRTUAL
REALITY IN BUILDING A SMART CITY

Augmented reality (AR) uses digital technology to enhance the images captured using smartphones. Additional digital information in the form of text or image can also be added to the captured image using apps.

Virtual reality (VR) uses computer technology to create altogether a different reality than the one that is in front of us.

AR and VR find application in the following streams:

1. *Patrolling:* AR systems provide relevant environmental details to the officials, which enable them to be well prepared before entering treacherous situations.
2. *Education:* AR and VR provide thrilling experiences to the students attending the classes. It allows students to explore unfamiliar things in depth. This attracts more students to attend classroom sessions.
3. *Emergency management:* To rescue the needy, an interactive map is distributed through AR to the responders to help the residents to plot their location. This helps in locating the safest rescue route and also in identifying unsafe areas.
4. *Mental health services:* AR and VR help in treating certain mental health problems like anxiety.

14.5 MAJOR COMPONENTS REQUIRED TO BUILD A SMART CITY

1. *Technology:* AR and VR play a vital role in the development of smart cities. Both VR and AR need an enormous amount of information to be delivered quickly in order to provide their experience. This demands a powerful network that can be established using 5G technology. With the ability to provide data in real time, AR and VR help in transforming government services.
2. *IoT:* It is the nerve of the city that keeps all the devices connected. It offers connectivity to all smart devices, connected vehicles, smart home appliances, smart buildings, smart public mobility, and other services [2, 3].
3. *Network of smart things:* This includes sensors, actuators, and cameras for collecting data. Every physical device which makes up IoT ecosystem has a sensor embedded in it. Sensors collect and transmit the data to the cloud. IoT interconnects all the devices and makes them work together.
4. *Cloud gateways:* It is helpful in collecting data from low-power IoT devices, store the data, and forward the data securely to the cloud.
5. *Data processor:* It is helpful to collect the data from various data streams and distribute it to data lake.
6. *Data lake:* It is required to store all the raw data.
7. *Data warehouse:* It is essential to clean the data and structure it.
8. *Geospatial technology:* It plays a vital role in providing the location and the necessary framework in collecting the data and analyzing it.
9. *Artificial intelligence:* The vast amount of data collected in the smart city becomes useless if it is not properly processed to generate information. AI algorithms and techniques help in processing and analyzing the data obtained from machine-to-machine interaction.
10. *Control applications:* It is required to send commands to IoT actuators.
11. *User applications:* It is helpful in connecting smart things and citizens.

14.6 CHALLENGES IN BUILDING A SMART CITY

Some of the challenges faced when a smart city is being built are data collection, security issues, technology involved, and requirement for expertise.

1. *Data collection:* The problems involved are as follows:
   - Overloading of data – noise and heterogeneous data
   - Issues due to interoperability
   - Managing open data
2. *Security issues:* Some of the security issues involved are as follows:
   - Cyberattacks on interconnected critical regions which are complex and enormous
   - Significant implications due to the attack
   - Sharing responsibility to shield the city from attacks
3. *Technology involved:* The main challenges involved in technology usage are as follows:
   - Providing network coverage throughout the city
   - Managing network capacity
   - Reusing existing infrastructure
4. *Requirement for expertise:* The main setback in expertise involve the following:
   - Lack of training
   - Insufficient funds for training

14.7 TECHNOLOGIES INVOLVED IN BUILDING A SMART CITY

Various technologies like AI, IoT, blockchain, big data, and robotics can be used to build a smart city as explained in the following sections.

14.7.1 INTEGRATING AI AND IoT TO BUILD A SMART CITY

The goal behind developing a smart city is to optimize costs, reduce wastages, and provide better living standards for the residents by incorporating various aspects of technologies in the system. The numerous fields of AI and IoT together have a greater potential in transforming the traditional city into a smart one.

Various branches of AI can be used to analyzse how the city is being used by the people. The AI's pattern recognition technology can be used to analyze a huge amount of raw data, patterns, and hidden correlations collected from different sensors installed all over the city for various purposes like road congestion monitoring, weather monitoring, etc. Analysis of sensor data using AI tools helps to create predictive models that can be used to control applications to send commands to IoT devices. Deep learning can be used to track the speed of the vehicles, read the license plate on the vehicles, count the number of vehicles passing by and the number of pedestrians walking on the road, and track the vehicles parked in the parking lot and establish patterns.

A smart city built with AI powered by IoT can be used in many application areas like smart environment, smart buildings, smart traffic management, smart grid, smart farming, smart waste management, smart health sectors, smart security systems, smart transport, smart parking, drone delivery, smart postal services, smart disaster and calamity managements, and many more.

14.7.2 BLOCKCHAIN IN BUILDING A SMART CITY

Even though the usage of blockchain to build a smart city is a pretty new concept, it assists in connecting the services of the smart city at the same time and increases security and transparency. Blockchain system provides security and privacy to data by encoding the information captured and storing it on cloud. This also reduces paper work, pollution, and wastage. Blockchain network can also be used in healthcare service to store health information required in emergency situations. This proves to be helpful in accessing relevant information. It can also assist in billing and transaction processes as well as in sharing smart grid energy.

14.7.3 BIG DATA IN BUILDING SMART CITY

Big data, IoT, and smart city are strongly correlated as one entity is interdependent on other two entities. The data generated in a smart city can be in the form of text, videos, and images from various sensors, websites, social media, RFID data, atmospheric data, and various data bases. This entire fertility of data generated in a smart city constitutes big data. The data collected from different sources must be processed and analyzed using various algorithms to retrieve information hidden in them. The challenges involved in this are manifold. This includes data searching, data capture, visualization, data analysis, data storage, and data sharing. Big data analysis requires new approaches like enhanced decision-making capacity, in-depth analysis, and process optimization.

14.7.4 ROBOTICS IN BUILDING A SMART CITY

Collaboration of humans and robots can work out wonders in all the sectors of smart cities. Robot integration is transforming technologically advanced cities into smart cities. This transformation only depends on how humans are efficiently using the robots in bringing about a change. Robots can be used in various applications like surveillance and policing, transportation, traffic management, hotel service maintenance, and others.

14.8 COMPONENTS OF A SMART CITY

The main components involved in building a smart city is illustrated in Figure 14.2.

14.8.1 SMART ENERGY

As energy is an important requirement for our day-to-day activities, hence conserving energy is a vital factor. Energies are classified as sustainable, renewable, and

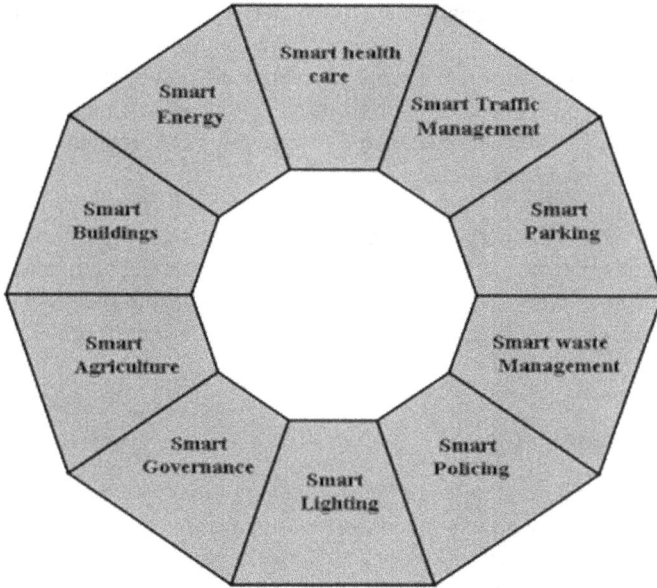

FIGURE 14.2   Components involved in building smart city.

non-renewable energies. Non-renewable energy is of greater concern as this type is not replenishable.

The term smart energy refers to the utilization of ICT in energy utilization and generation [4]. The smart energy model is based on developing smart power grids, smart power generation, smart storage, and smart consumption. Smart energy aims at reducing the carbon emission, distributing the energy efficiently to the end users, and optimizing the energy utilization, as shown in Figure 14.3.

Carbon emission can be reduced by using solar, wind, hydroelectric, or photovoltaic energies. Smart distribution can be achieved by using smart grids and smart infrastructures enabled by ICT. Smart grid facilitates to integrate different sources of energies like fossil fuel, photovoltaic, wind, and other energy forms onto a single

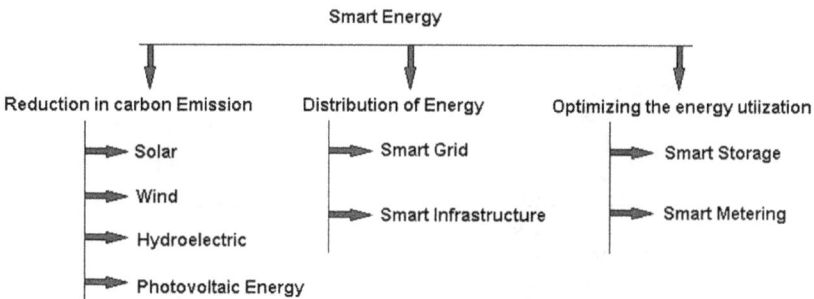

FIGURE 14.3   Smart energy system.

grid. The energy from various sources is synchronized by the grid to deliver electricity of specified frequency and voltage without any fluctuations. The grid ensures efficient, secured, and sustainable energy system by integrating the source of energy to end users. Smart infrastructure is the crucial component in a smart energy system. It is responsible for gathering the energy consumption information and for sharing the unit rate information to the users.

Optimization of energy utilization can be achieved through smart storage and smart metering. Smart energy storage can be obtained by using lithium-ion batteries or fuel cells that have a longer life and prove to be effective in energy storage and its efficient delivery. The smart energy meter records the energy consumed for the specified time duration and sends the reading to the central office for monitoring and billing.

14.8.2 SMART HEALTHCARE

Healthcare sectors have many voids like the ratio of growing population to the number of medical practitioners is unmatchable, mistakes by the hospitals in diagnosing diseases, patients receiving wrong treatments, patients not getting immediate treatment in some emergency situations, and many more. To meet the demands of the growing population with limited resources, the healthcare sectors must become smart, intelligent, and efficient. Smart healthcare sector integrates the use of biosensors, wearable devices, ICT, cloud computing, mobile apps, emergency services like ambulance with the traditional healthcare [5].

Some of the smart healthcare applications include periodical intimation to the nurses and doctors regarding the patient's health condition, transferring the real-time data of the patients to an expert doctor who resides in a far-off place for his expert opinion, automatic intimation to the pharmacist in case of medicine requirement to a patient, and telemedicine. The idea of the smart healthcare sector is shown in Figure 14.4.

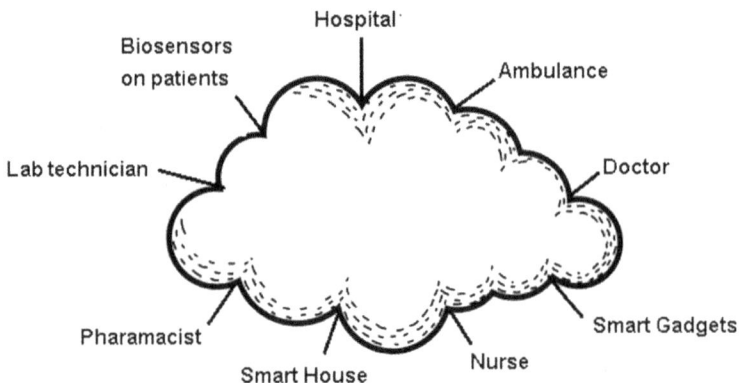

FIGURE 14.4 Smart healthcare.

FIGURE 14.5 Smart traffic management.

14.8.3 SMART TRAFFIC MANAGEMENT

With the growing population and digitalization, AI technique is playing a vital role. The ever-booming IoT can be used to implement smart traffic solutions to ensure that residents of a smart city commute in the city safely and quickly [6]. There exists a central traffic management system that receives real-time data about the flow of traffic via road surface sensors and cameras, the real-time data captured from camera is analysed and it notifies the user regarding the signal malfunctions and congestion, as shown in Figure 14.5.

14.8.4 SMART PARKING

Due to narrow roads and increase in traffic, parking is always a problem, especially during holidays which is a real struggle. IoT smart parking solutions can be implemented using road surface sensors embedded in the parking area, which can determine whether the slots are occupied or free and provides a map for parking on a real-time basis. A mobile app can also be developed to locate the available parking slot and also pay for the slots, as shown in Figure 14.6. Smart parking helps with congestion reduction and reduces carbon emissions.

14.8.5 SMART WASTE MANAGEMENT

Disposal of waste is one of the major issues to be dealt with since it emphasizes on the hygiene and eruption of diseases across the place. Proper disposal and collection of waste is an important service as there is an increase in population. A sustainable waste management system can be provided by implementing AI for smart recycling and waste management. Figure 14.7 shows the waste management system comprising sensors and devices attached to bins, which sends notification to municipal authorities to clear the waste collection as soon as it is about to be filled. Separate bins for paper, plastic, glass, and waste food items can be maintained in every locality.

FIGURE 14.6   Smart parking system.

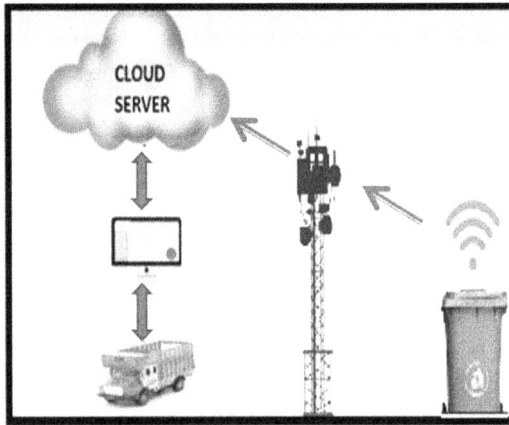

FIGURE 14.7   Smart waste management.

14.8.6 SMART LIGHTING

Smart lighting emphasizes on less consumption of energy. Street lights are necessary, but they consume a lot of energy. This can be reduced by the use of smart lighting. Since street lights are necessary, sensors are attached or there are Wi-Fi hotspots which automatically adjust the brightness depending upon the presence of pedestrians and automobiles. It employs a mesh network in real time to trigger adjoining lights and to create a safe circle of light around the localities. Figure 14.8 illustrates smart lighting system.

FIGURE 14.8 Smart lighting system.

14.8.7 SMART GOVERNANCE

A smart city aims at providing sustainable high-quality amenities to the citizens. This necessitates all the decision-makers to line up their goals and to work together in planning all the activities. Smart use of the available resources demands the interactions between different government agencies. Best results can be achieved when citizens are also allowed to participate in urban planning activities. These changes can be brought about by adopting smart governance. Smart governance involves the use of ICT sensibly in order to improve decision-making. It involves the process of refining the democratic process and changing the methods of delivering public services. This is done by better collaboration among different stakeholders, including government and citizens. Compliance towards the needs of people and improved decision-making using data, evidence, and other resources are emphasized through smart governance.

The following are the main aims of smart governance:

1. Simple laws, rules, and regulations and comprehensible government.
2. New system of governance with moral values.
3. Transparency in governance. This brings about a great transformation in the system of governance by reducing corruption.
4. A responsive administration to attend the difficulties faced by the citizens as quickly as possible.

14.8.8 SMART AGRICULTURE

The worldwide increase in population has increased the demand for food. In traditional farming methods, farming is usually done based on predications. Many times it leads to failure of crops. Due to this, farmers incur huge losses. For efficient cultivation, soil

FIGURE 14.9   Smart agriculture.

moisture, soil quality, air quality, sufficient irrigation, and crop weeds, proper fertil-
ization plays an important role. If these parameters are monitored automatically, and
suitable information is passed on to the farmers, quality and quantity of cultivation
increase. This necessitates for smart agriculture that uses IoT, big data, GPS, and connected devices. Smart agriculture assists in automated farming and also in collecting
and analyzing the data from the field by installing sensors, cameras, and actuators [7].
The analyzed data is then sent back to the farmer. This helps in growing high-quality
crops. Figure 14.9 illustrates smart agricultural system.

Usage of IoT in agriculture can help farmers in the following ways:

1. Monitoring weather conditions, crop development, and soil quality can help
   in improving the cultivation.
2. Internal processes are better controlled and this reduces the risk of inferior
   production and also allows for better distribution of the products.
3. If the crop yield is known in advance, planning to sell the product can be
   made.

14.9   DRAWBACKS IN IMPLEMENTING SMART CITIES

Inappropriate design of the smart city structures will result in few drawbacks:

1. Building a smart city requires huge data acquisition. In this process, there
   are chances that the information about the individual might get revealed.
2. Huge investment is required to build a smart city. This investment must be
   borne by the tax payers. This will also raise the cost of living.
3. The whole process involves the usage of sensors and other IoT devices
   placed at various places. These devices operate with battery. This causes a
   problem when battery replacement is required.

14.10   CONCLUSION

With rapid population explosion in urban areas, the challenges faced by the cities have drastically increased in an uncontrolled manner. Problems like resource scarcity, traffic management, pollution level, and many more shoot up with a rising population. To provide sustainable prosperity to inhabitants, cities now face economic, political, and technological threats. This demands the establishment of smarter systems in cities to optimize the utility of the exhaustible resources and to provide sustainable development without disturbing the environment. Thus, improvement of cities powered by digital technologies is set at the highest priority worldwide by the 21st century. Even though the day-to-day living standards of the people in urban areas is increased drastically due to digitization, the transformation of the traditional city to smart city has barely begun.

In the development of smart cities, the roles and responsibilities of the participants involved become significant. The key participants involved in building the smart city include the government bodies that aim at transforming the lives of public, their safety, and well-being; various private sectors that assist the government bodies by funding, constructing, and managing the urban infrastructures; and the stakeholders. The stakeholders include the citizens themselves and other non-profit organizations.

REFERENCES

1. https://hub.packtpub.com/how-ai-is-transforming-the-smart-cities-iot-tutorial/ `
   Hands-On Artificial Intelligence for IoT'.
2. Husam Rajab, Tibor Cinkelr, "IoT Based Smart Cities", *2018 International Symposium on Networks, Computers and Communications (ISNCC)*, June 2018.
3. Badis Hammi, Rida Khatoun, Sherali Zeadally, Achraf Fayad, Lyes Khoukhi, "Internet of Things (IoT) Technologies for Smart Cities", *ET Research Journals, the Institution of Engineering and Technology*, 2015.
4. Subba Rao, Sri VidyaGarige, "IOT Based Smart Energy Meter Billing Monitoring and Controlling the Loads", *International Journal of Innovative Technology and Exploring Engineering (IJITEE)*, March 2019.
5. Rustem Dautov, Salvatore Distefano, Rajkumaar Buyya, "Hierarchical Data Fusion for Smart Healthcare", *Journal of Bigdata*, Feb 2019.
6. Sabeen Javaid, Ali Sufian, Saima Pervaiz, Mehak Tanveer, "Smart Traffic Management System Using Internet of Things", *2018 20th International Conference on Advanced Communication Technology (ICACT)*, Feb 2018.
7. Sjaak Wolfert, Lan Ge, Cor Verdouw, Marc-Jeroen Bogaar, "Big Data in Smart Farming – A Review", *Agricultural Systems*, May 2017.

15 AI Emerging Communication and Computing

N. Girija and T. Bhuvaneswari

CONTENTS

15.1 INTRODUCTION

The current century is recognized as Industrial Revolution 4.0 (I.R. 4.0). I.R. 4.0, which includes concepts such as Internet of Things (IoT), blockchain technology, and artificial intelligence (AI), fulfils a substantial role and is recognized as a flourishing model. This chapter presents the role of AI in emerging communication technology. This emerging AI technology is affecting not only the communication and computing environment but also the emerging employment skill requirements in the industry. There is no denying the fact that industry requirements and academic sectors are interrelated. Therefore, in the coming years, it is inevitable that AI technology-related programming languages, applications and, projects will be part of the student curriculum. Deloitte is a UK origin freelancing auditing multinational company. Its report says that robotics and artificial intelligence influence nanotechnology, drones, sensor technology, and computer vision and hence are becoming vigorous in the employment market. That is a reason why BRICS is focusing on vocational education training and world skill development in their nations.

This chapter has two parts. The first part provides an introduction, history of the Industrial Revolution, and the stages of AI. It also provides an overview of how

deep these technologies spread in the mortal world. In the world, every country recognizes the impact of the industrial revolution which enriching its communication
technologies areas like mobile networks, internet-based social media. AI development is categorized into four stages. This chapter also discusses how each stage of
the AI-based algorithm has boosted and targeted the vulnerable brain and its sym-
pathetic expression.

The second part covers topics on AI-based technology diversification. In this part,
the AI technology is classified into artificial narrow intelligence (ANI), artificial
general intelligence (AGI), and artificial super intelligence (ASI) models. Each stage
of AI such as reactive AI, decision-making capability, emotional intelligence and
replication of the human brain, etc., does not allow anyone to guess the saturation
phase of AI. Each AI classification communication task with humane society is also
briefed in this chapter. The significance of machine learning and deep learning concepts and their algorithms in AI are also discussed. The machine learning algorithms
such as the non-maximum suppression algorithm and YOLO have been implemented
in various industry sectors like the health sector and agriculture-based AI-based
applications. AI researchers have also focused on computer vision and hybrid thinking in AI communications. These are also discussed in this chapter.

In conclusion, emotional intelligence, ethics, and the law of machine–human
attitudes, and Way of the Future, i.e. Tech Church, are discussed. These combinations based on technology design which will aid the human society reform are also
discussed.

15.2   INDUSTRIAL REVOLUTION 4.0

The 21st century is witnessing a revolution in industry, and this period is hence
referred to as Industrial Revolution 4.0 (I.R. 4.0). In every century, at least one industrial revolution changes the technology. That technology plays a vital role in human
culture and their skill set and economy, resulting in new industry initiation and more
business opportunities worldwide. Since the advent of industrial revolution, steam
power, coal, electricity, and many machinery like spinning industry have proved
helpful to humans as well as in employment generation. Initially dependent on animals like horses and buffalo for travelling, we have now gradually shifted to train,
airline, and automobile. Each industrial revolution reduces the gap between automation and humans.

The best and precious contribution of I.R. 2.0 is electrical energy. Telegraph, telephone, and railways are other contributions that have tremendously helped the society by enabling a rapid way of communication, fast raw material delivery from and
to ports, and comfortable travelling. It created direct and indirect job opportunities
and undoubtedly contributes to the current industry in society.

I.R. 3.0 is a precious digitization revolution in communication and technology.
Industries like computers and telecommunications played a vital role. I.R. 3.0, especially network and internet technology, has brought the globe under one distinctive
roof. Communication has become very rapid. Various types of communication like
wireless phones, mobile, email, messengers, social media, and messaging apps help

to connect people in any distinct region in the world. Nevertheless, I.R. 4.0 is converging into the replication of superhuman. These technologies are moving towards concepts like smart cities, smart houses, and secured transactions like blockchain and deep learning. AI technology is omnipresent not only in the modern workplace but also in a remarkable part of the lifestyle of the community. Just as I.R. 3.0 introduced modern concepts like computer science, information technology, and telecommunication, I.R. 4.0 seamlessly combines AI, machine learning, and deep learning in all the fields ranging from medical to automobile.

15.3 STAGES OF AI

In the earlier stages, AI could perform tasks with human proficiency and functionality. However, AI has limited capacity due to a lack of learning skills. Learning skills develop with experience. For gaining experience, memory-based learning considered representing an essential aspect. In this stage, AI response is based on the combination of various inputs groups. This stage is called reactive AI. In 1997, IBM developed *Deep Blue* as a reactive AI machine. Deep Blue beat the chess Grandmaster Garry Kasparov. This incident was an eye-opener for the human community to turn and explore AI machine proficiency.

In the second stage, AI continues with memory and decision-making. Storing a massive volume of data needs memory. AI takes a decision based on an archive of data stored in the memory. For the decision-making feature, the AI needs analytical skills. In this stage, AI is smarter in recognizing the thousands of images and scanned objects in image processing, similar to Google Assistants like Chatbot.

Until the third stage, AI was still not considered equivalent to a human as human beings learn various emotional aspects and can recognize other people's emotions. This feature in AI is called emotional intelligence (EI). Understanding the user's emotions and respond accordingly is very much needed for virtual assistants. An example of an EI-based AI application is Siri. Siri is Apple's voice-controlled personal assistant for Apple product users. Siri is a combination of AI and Natural Language Processing (NLP). The three critical tasks of Siri are service tasks, awareness of the situation, and communication interface.

The fourth stage is the replication of the human brain and self-awareness. AI understands emotional intelligence but also like human primitive emotions. It barely exhibits emotions like Sophia robot, which was developed by Hong Kong-based Hanson Robotics research laboratory. Dr. David Hanson described Sophia as capable of generating expressions of joy, grief, curiosity, confusion, mediation, sorrow, frustration, among other feelings. It means that AI can now concentrate even more on human muscle expression which is typically connected with the human brain.

The above-mentioned stages clearly state that in the future, most of the human communications will be with AI-based support instead of actual human beings. Each stage of AI releases physical stress to human society. The target of AI haunts the human brain. Human brains include sentiments- and emotions-based components. So, human brain characteristics might become more attached and dependent on emotional intelligence technology. As human beings are most emotionally

FIGURE 15.1   Stages of AI. *Source:* worldmarket.com-only ladder image.

attached to pets, likewise advanced features of Sophia model aims to enhance one of the family members. This bondage constructs a fashionable relationship between human–machine or human–robot (Figure 15.1).

15.4   CLASSIFICATION OF AI

AI technology is classified into three categories: Artificial Narrow Intelligence (ANI), Artificial General Intelligence (AGI), and Artificial Super Intelligence (ASI) (Figure 15.2).

15.4.1   ARTIFICIAL NARROW INTELLIGENCE

ANI is mostly used for some specific tasks, for example developing a self-driving car. This type of technology is disciplined to excel in some particular task only. The ANI programs are designed to get information from specific data set or predefined data ranges. Based on the data set, it will accomplish the given task like checking the weather forecast, chess playing like Deep Blue, and writing reports. ANI does not possess emotional intelligence nor cognizant sentiments like

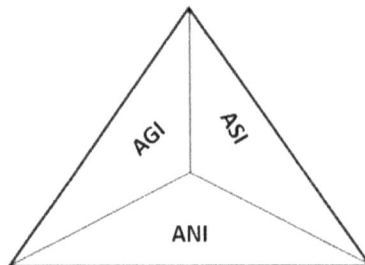

FIGURE 15.2   Classification of AI.

humans. Currently, ANI-based technologies are used in Google Assistant, Google Translate, and Siri. The advantage of ANI is that it improves the mechanical and tedious human lifestyle to a productive and efficient lifestyle. For instance, a self-driving car provides extra leisure to a person and relieves from being stuck in traffic. Keeping a personal secretary is impossible for every person, but Siri supports all secretarial jobs. Siri is a personal digital assistant. It sends a text while you are in drive, orders pizza, sets your alarm, creates reminders, and plays your favourite song when asked. Answering and sorting emails are tedious assignments. ANI technology takes care of forwarding recent emails, checking emails, and replying to emails. The development of ANI technology has not merely replaced the low-skilled human job opportunity but also given space to people to develop more intellectual skills than mere clerical skills.

Lee Se-Dol is one of the leading players of the "GO" Game. Alpha GO program defeated the player Lee-Se-Dol. The "DeepMind" algorithm using the *Alpha GO* program was developed by UK-based company DeepMind Technologies which was a part of Google (up to 2015). ANI at present is also doing the tasks of Inbox cleaning, auto sorting for Spam filters based on customer browsing history, adding things to your shopping cart during online shopping, and recommending videos for viewers to watch; these need a huge amount of quality data sets and rule-based machine learning algorithms.

15.4.2 Artificial General Intelligence

The AGI model is precisely defined as it will act appropriately like human-level AI. The other terms used to scientifically describe AGI model are "computational intelligence," "natural intelligence," "cognitive architecture," and "biologically inspired cognitive architecture" (BICA). AGI can accomplish a variety of objectives and carry out a diversification of tasks in various circumstances and surroundings like a human being. The person who claims to know diverse fields. For example, Isaac Newton was an English mathematician, prominent physicist, astronomer, theologian, and scientist. In the AGI community, the prominent researcher Goertzel (2014) expressed about the core AGI hypothesis as given below:

> *Core AGI hypothesis: the creation and study of synthetic intelligence with sufficiently broad (e.g. human-level) scope and strong generalization capability are at the bottom qualitatively different from the creation and study of synthetic intelligence with a significantly narrower scope and weaker generalization capability.*

The upgrade from one specific field to another, i.e. from ANI to AGI, requires general intelligence. That is the reason AGI aims to concentrate on machine learning. The combination of AGI and machine learning can recognize and acquire any type of intellectual task. This combination is called Strong AI.

AGI models use human cognitive skills to solve efficiently an unfamiliar task. According to Stephen Hawking, mortal beings are slow biological evolutionary models. Suppose innovative Strong AI redesigns itself and constructs unlimited cognitive skills, it is challenging for humans to ride AI. Currently, DeepMind Technologies owned by Alphabet Inc. (from 2015) is extensively focusing on AGI models.

The Google AI research team continuously focuses on machine learning for combining AI apps like *Google Lookout App*, especially for visually impaired people. Approximately 253 million people are visually impaired across the world. Lookout App works on portable pixel devices. Wherever the visually challenged people move around, these apps will accurately identify people, objects, and texts. Google has powerfully built *Pixel Night Sight* with ML advancements, which will capture the photo shot even in the dark.

AGI models focus to faithfully implement learning similar to capable humans. The human brain naturally contains 90 billion excitatory neurons and synapses. Excitatory synapses merely connect the neurons in the brain. Hundred trillion synapses are intentionally used to form the neural net in the human brain to accumulate unlimited information. AGI models plan to train not only storing and connecting information like the active brain but also in common sense, intuition, and reasoning skills like the brain.

AGI models are still under empirical research. The AGI researchers are focusing on this efficient model which is continuously undergoing various levels of cognitive tests to attain human intelligence. The first ultimate test attends the *Loebner Prize* competition. This prestigious annual competition checks the standard of turning the test of the model. The intelligence level pertains to audio-video conversation and responses.

The next level of the test represents intellectual tasks such as in the home like layman the AGI model prepares the coffee. The co-founder of Apple, Steve Wozniak, believes that the AGI model passes the coffee test. For the coffee test, AGI enters any kitchen and finds the ingredients like sugar, water, coffee powder, and coffee cup and perfectly mix all. The third test is the *Robot College Student test*, which will enrol like a student and attend course work and move around the campus. The Chinese AI robot AI-MATHS is introduced for taking college entrance math's test. This robot has been developed by Chengdu Zhunxingyunxue technology. This robot has completed two versions of Math test and Chinese test.

The fourth test is an economically important employment test that will prove the performance of AGI in the job market. AI researcher Nils J. Nilsson has proposed that the turning test be replaced by the employment test.

Across the globe, every scientist's dream is to receive a Nobel Prize. This level of the test considered the *Artificial Scientist Test*. AGI model pursues creative scientific research, problem-solving, literature review, and thesis publishing to receive Nobel Prize. The Australian physicists team experimented with the help of AI and performed the Bose–Einstein condensate. The team cooled the group of atoms to absolute zero temperature to a few hundred nanokelvins i.e. a billionth of a second. The team, which won the Nobel Prize in 2001, firmly believes the general-purpose research units called "graduate students" in upcoming research will move towards applying more flexible automation.

The Global Catastrophic Risk Institute published "A Survey of Artificial General Intelligence Projects for Ethics, Risk, and Policy" in 2017. As per this survey, most of the AGI R&D projects are in industry or academic institution. Figure 15.3 shows several AGI projects in various nations.

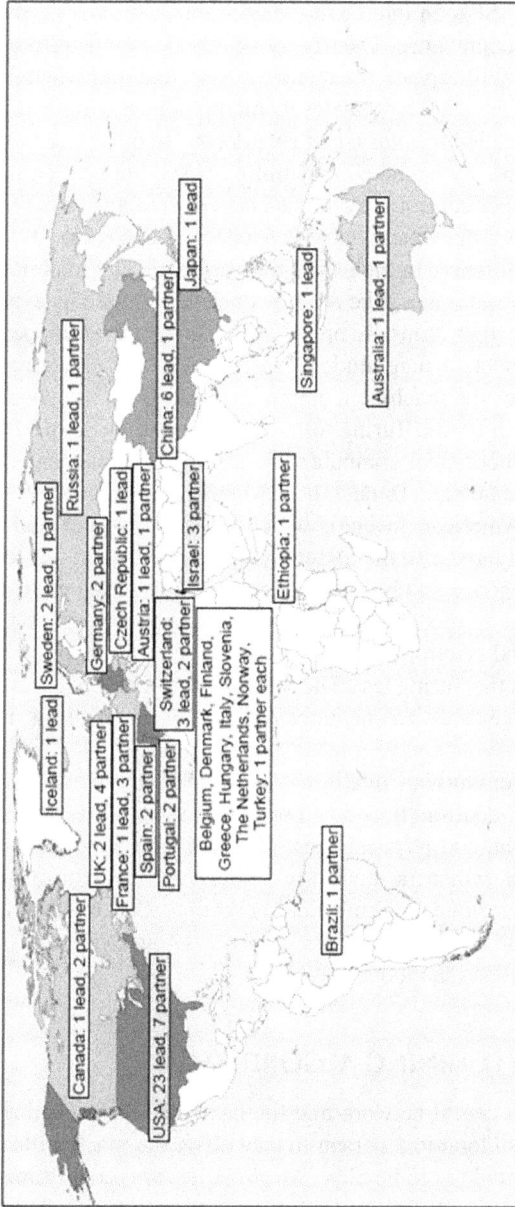

FIGURE 15.3 Nationwide AGI project. *Source: A Survey of Artificial General Intelligence Projects for Ethics, Risk, and Policy, 2017.*

15.4.3 Artificial Super Intelligence (ASI)

ASI models surpass humans. The multi-tasking AI is ASI. Artificial intelligence journey is moving towards the next level: Artificial Super Intelligence. The researchers have stated that ASI is an intelligence explosion. ASI aims to overtake humans with AI-empowered cognizance. The triggering question in the future regarding ASI would be whether it will replace the human intellectual and whether it might help sharpen human intelligence. The peak level or hypothetical point of this type of intellectual exposition is called *technological singularity*. In other words, singularity can be defined as machine autonomous. Any online-based application is monitored by humans and also proper control direction is given by humans. Whereas the machine autonomous develops a machine to make a decision and execute the operation without any human interference. In biological terms, the genetic material is developing human growth in several stages: like AI growth stages entitled by evolutionary algorithms called genetic algorithms are applied to an artificial neural network.

Turing test represents a magnitude of testing AI. The objective of this test is to investigate whether the machine (computer) intelligence behaviour is capable of thinking like a human or not. Turing test was developed by Alan Mathison Turing. He is the father of theoretical computer science and artificial intelligence, and he is an English mathematician. To date, no AI model has passed the Turing test. Dr. Ray Kurzweil, an American inventor who is the Co-founder and Chancellor of Singularity University and also the director of engineering, Google team, has developed machine intelligence and Natural Language. He is a part of AI field programs such as optical character recognition (OCR), text-to-speech synthesis, speech recognition technology, and electronic keyboard instruments. He hopes that in the future Singularity will pass the Turing test. The Staffordshire University research team has taken a step ahead to establish a contemporary way of turning tests, i.e. Multimodal Turning test.

Amazon Alexa remains an intelligent agent. Alexa represents a multiskill to handle smart home control, financial portfolio information, fitness control like sleep tracking, weight, and physical exercise, etc. Using Speech Synthesis Markup Language, the Alexa whispers mode use gentle voice without disturbing anyone around in a quiet environment. Figure 15.4 shows the subset of artificial intelligence, machine learning, and deep learning. It also displays how each subset of AI is associated with each classification model (Table 15.1).

15.5 MACHINE LEARNING ALGORITHMS

The brain of AI is a neural network and the heart of AI is machine learning. Just as the heart supplies blood to a person to stay alive, the machine learning supplies algorithm to AI technology to be alive. Machine learning algorithms and subset of machine learning, i.e. deep learning algorithms, accumulated with AI technology to advance to the next tremendous elevation.

Predominantly, recommended videos on YouTube and recommended products on Amazon have been done using supervised learning algorithms techniques. Face

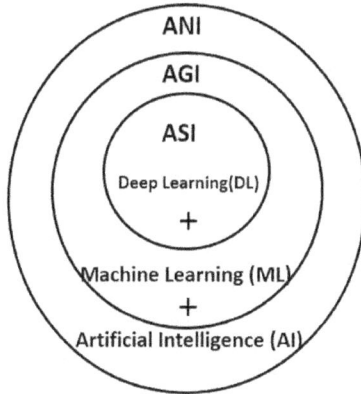

FIGURE 15.4 Subset of AI.

recognition and speech recognition are done using *deep neural network* and *k-means techniques* (Figure 15.5).

Classification and regression represent the two prominent features of supervised learning. The below-mentioned algorithms are the traditional algorithm techniques of machine learning. For Artificial Narrow Intelligence-based AI technology execution, the machine learning algorithms such as *YOLO* (You Only You Once) and *Sliding Window* are used to supervise learning technique and are also applied in the image processing. These types of algorithms are used in self-driving cars. *Non-maximum Suppression algorithm* is used in the chessboard for object detection. In Figure 15.6, the popular ML algorithms used in the real-time application are depicted.

15.6 AI EMERGING COMMUNICATION

The key aspect of AI is self-learning. The self-learning potential is based on machine learning for AI. Self-learning and neural network may aid the AI to meet the utmost knowledge. At the same time, supreme Knowledge, Emotional Intelligence, and Singularity are not enough for humans. Effective communication is one of the essential features of gaining knowledge. Effective communication needs to be typically categorized into either verbal or non-verbal communication. Non-verbal communications involve careful observation, monitoring, and searching. AI-based non-verbal communication technology such as *computer vision* and *hybrid thinking* in common are future AI technology contributions to modern society. How the vulnerable brain instantly communicates, for example, when a cautioning message is forwarded while driving, the driver automatically slows down the vehicle using speed breakers and pedestrians crossing. In the same way, *Waymo* is a computer vision technology properly used in a self-driving vehicle. It correctly identifies the pedestrians crossing beforehand for the safe driving.

Computer vision also plays a role in nursing *massive obstetric haemorrhage* (MOH) bleeding during Caesarean cases using *Gauss Surgical* computer vision

TABLE 15.1
Summary of Classification of AI-based Application and Test

	ANI	AGI	ASI
Application	• Deep Blue • Google Assistant • Google Translate • Siri • Alpha Go	• Google Lookout App • Pixel Night Sight	• Amazon Alexa
Test		• Loebner Prize • Coffee Test • Robot College Student Test • Employee Test • Artificial Scientist Test	• Multi-model Turning Test

healthcare technology. Aware mostly of the Oil and Gas industries in remote spots, *Osprey's intelligent visual monitoring* is prominently used in the oil and gas industry to detect leaks, safely monitor, and bolster security.

Worldwide, sustainable agriculture continues to be a crucial industry. Persisting farms to be more profitable in business needs *aerial phenotyping* ("AP") technologies,

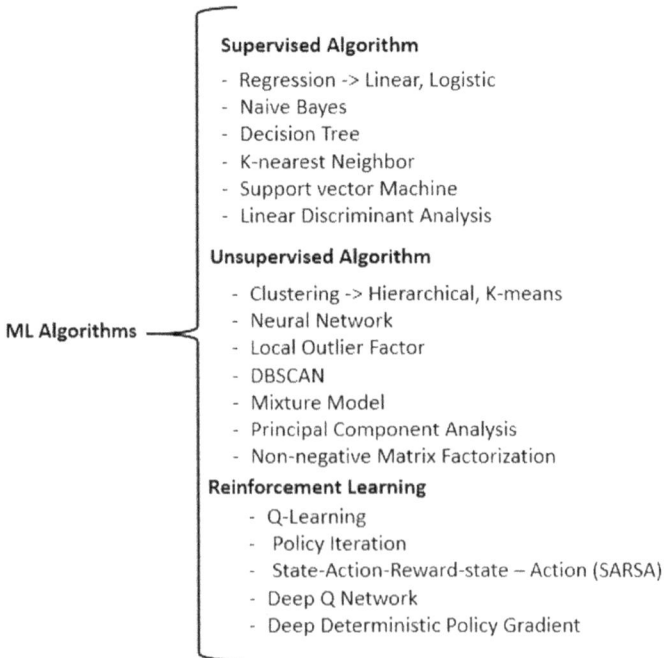

Supervised Algorithm

- Regression -> Linear, Logistic
- Naive Bayes
- Decision Tree
- K-nearest Neighbor
- Support vector Machine
- Linear Discriminant Analysis

Unsupervised Algorithm

- Clustering -> Hierarchical, K-means
- Neural Network
- Local Outlier Factor
- DBSCAN
- Mixture Model
- Principal Component Analysis
- Non-negative Matrix Factorization

Reinforcement Learning

- Q-Learning
- Policy Iteration
- State-Action-Reward-state – Action (SARSA)
- Deep Q Network
- Deep Deterministic Policy Gradient

ML Algorithms

FIGURE 15.5   Machine learning algorithm.

Machine Learning Algorithm

Convolutional Neural Networks ⟶ **Sentence Classification**

Artificial neural networks ⟶ **optimization problems**

Feed-Forward CNN ⟶ **Image colorization process**

Convolutional deep Neural
Networks (CNNs) ⟶ **Automatic speech recognition (ASR)**

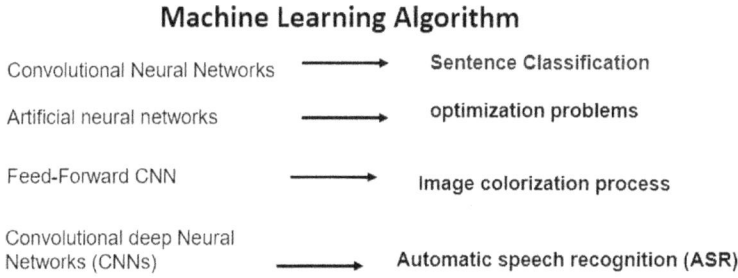

FIGURE 15.6 Real-time application using ML algorithms.

which is used in *Slant Range* for crop development for successful breeders. The above-discussed applications realistically are computer vision-based technology. Undoubtedly, in a short time, computer vision apps will play a vital role in assuring safety and risk awareness in routine life.

Based on searching continually and gaining information from the browser and search engine, the machine learning understands the surroundings and also acts accordingly. For browsing and seeking reliable information, any mechanical or electronic devices like the mobile are genuinely needed. The following promotion of AI technology racks the vulnerable brain as a capable device, i.e. hybrid thinking. Hybrid thinking has been intimately connected with the human brain. The neocortex is a part of the human brain. The neocortex has multilayered prominent parts involved in high-order brain functions like vision, audition and olfaction, cognition, and spatial-visual intelligence, especially for imagination. The neocortex is bigger in humans than in animals.

Nano-robotics adequately represent an imminent technology in the medical nanotechnology field. Nanorobot or Nanobot is intimately connected to the neocortex part of the human brain. Whenever a Nanobot allied, the ideal person instantly thinks for specific information without traditionally using any electronic devices. The Nanobot automatically connects to the cloud directly. This specific type of modern thinking is properly designated *hybrid thinking*. This intelligence technology provides not only detailed information but also a solution. This modern superior intelligence is addressed by *non-biological intelligence*. Ray Kurzweil, Director of Engineering at Google, has authored the book *How to Create a Mind*. He correctly stated in this book that for the next 10 years, hybrid thinking looks promising. He also believes that in the next two decades non-biological intelligence would be more potent than intelligent humans.

15.7 CONCLUSION

This chapter began with the description of I.R. 4.0 revolution. The key aim of I.R. 4.0 remains intellectually based automation, the implementation of which requires AI technology. In the earlier stage, computer-based technology was adopted for accumulating information, which was called a database management system for managing databases. Besides, a database system was used for storing historical and pertinent information needed for decision-making, which was called data warehousing and data mining. Data analytics is most demanding by industry for not only storing and decision but also for predicting. The demand of human society and industry is altering every industrial revolution based on technology development. This alteration is explained in various stages of AI and the classification of AI topics. The machine learning algorithm represents the core of AI application and AI emerging communication, and this chapter focuses on AI and ML with regard to various fields like agriculture, healthcare, and nano-robotics.

To conclude, when human automation is emerging, then ethics and the law of automation and religious-based moral discipline are also compulsory. A crucial aspect that enables human society to act is *ethics and law*. In 2017, Google DeepMind gathered the group as *"DeepMind Ethics & Society"* (DMES). The DMES' group members are philosophers, economists, risk analysis experts, political advisors, and AI technology experts. This group of experts designs the ethical procedure to be followed by the AI. Verity Harding and Sean Legassick are co-leads of Ethics & Society, who believe that AI technology must take the ethical and social impact responsibility. DeepMind has identified the key ethical challenges for AI: privacy, transparency, and fairness to economic impacts, governance, liability, and many more.

The next aspect is law. Connecting humans and AI involves appropriate legal regulations. Human faults are disciplined by legal decree; similarly, the Baltic nations of Estonia has developed a constitutional status called *"Kratt Law"*, i.e. algorithmic liability law for AI. Estonia is one country in which e-governance is done through advanced AI applications. The abbreviated two letters AI instantly flash at laymen as technology. But AI represents not only the emerging technology, it is also considered as the next-century religion. The technology patronized in the more rapid industrial revolutions. Way of the Future (WOTF) is a contemporary AI religion's tech church. WOTF religion focuses on the worship of a Godhead based on artificial intelligence. Anthony Levandowski is a man behind self-driving car technology in Uber and also part of Google light detecting and ranging (lidar) engineering team. Levandowski worked as CEO in the start-up company Pronto AI, which focused on developing the CoPilot project. The CoPilot project goal is to develop Advanced Driver-Assistance Systems (ADAS) for truck drivers, which will find lane detection like whether zebra lines exist, two lines whether white or yellow are parallel, collision warning, and switching lanes for determining length and width of the lane. This automated system is most useful for the entire globe to prevent vehicles-caused accidents. He started the WOTF, and he believes super intelligence is inevitable. Artificial Super Intelligence is much smarter than superhuman. Therefore, WOTF believes, it is clear to integrate human computational power like a singular model that will be the most supreme on this planet. Anthony Levandowski has confidence that God is immeasurable or can't ensure or control, but tech god (AI god) also has immeasurable smartness even though it is possible to interact with tech god and recognize it is listening. Religion is a significant regulatory for the moral values of the humane society. In the same way, the forthcoming era of AI religion has also influenced human society. Just like how god or spontaneous creations exist, the AI's journey will also virtually exist from ANI to AGI to ASI with more and more progress!

BIBLIOGRAPHY

1. https://www.bgp4.com/2019/04/01/the-3-types-of-ai-narrow-ani-general-agi-and-super-asi/ (15/10/2019)
2. https://www.britannica.com/event/Industrial-Revolution
3. https://www.forbes.com/sites/cognitiveworld/2019/06/19/7-types-of-artificial-intelligence/#7f0cc0cc233e
4. https://interestingengineering.com/the-three-types-of-artificial-intelligence-understanding-ai
5. https://www.hansonrobotics.com/news-meet-sophia-the-robot-who-laughs-smiles-and-frowns-just-like-us/
6. https://www.pocket-lint.com/apps/news/apple/112346-what-is-siri-apple-s-personal-voice-assistant-explained
7. https://www.nuffieldfoundation.org/sites/default/files/files/Ethical-and-Societal-Implications-of-Data-and-AI-report-Nuffield-Foundat.pdf
8. https://futurism.com/artificial-intelligence-officially-granted-residency
9. https://www.searchenginejournal.com/yandex-artificial-intelligence-machine-learning-algorithms/332945/#close
10. https://towardsdatascience.com/ai-machine-learning-deep-learning-explained-simply-7b553da5b960
11. https://9to5google.com/2019/03/18/deepmind-agi-control/
12. https://www.scientificamerican.com/article/will-china-overtake-the-u-s-in-artificial-intelligence-research/
13. https://www.mckinsey.com/industries/technology-media-and-telecommunications/our-insights/an-executives-guide-to-machine-learning
14. https://gmisummit.com/wp-content/uploads/2018/10/What-is-Hybrid-Thinking.pdf
15. Burden, D, Baden, MS, (2019), *Virtual Humans: Today and Tomorrow* (1st Ed.), Chapman & Hall/CRC. https://doi.org/10.1201/9781315151199
16. Chalmers, D, Appleyard, A, (2016), *The Singularity: Could Artificial Intelligence Really Out-Think Us* (1st Ed.), Imprint Academic.
17. Goertzel, B, (2014), Artificial General Intelligence: Concept, State of the Art, and Future Prospects, 5(1), 1–46, https://doi.org/10.2478/jagi-2014-0001
18. Shabbir, J, Anwer, T, (2015), Artificial Intelligence and Its Role in Near Future, *Journal of Latex Class Files*, 14(8), 1–11.
19. Baum, S, (2017), *A Survey of Artificial General Intelligence Projects for Ethics, Risk, and Policy*. Global Catastrophic Risk Institute Working Paper 17-1. https://ssrn.com/abstract=3070741.
20. Adriano, M, Althaus, D, Erhardt, J, Gloor, L, Hutter, A, Metzinger, T, (2015), Artificial Intelligence. Opportunities and Risks. In: *Policy Papers of the Effective*, Altruism Foundation (2), S.1–16. https://ea-foundation.org/files/ai-opportunities-and-risks.pdf
21. Watson, EN, (2019), The Super Moral Singularity—AI As a Fountain of Values, *Big Data and Cognitive Computing*, 3(2), 23. https://www.mdpi.com/2504-2289/3/2/23
22. Carlson, KW, (2019), Safe Artificial General Intelligence via Distributed Ledger Technology, *Big Data and Cognitive Computing*, 3(3), 40. https://www.mdpi.com/2504-2289/3/2/23

23. Bhasin, H, Bhati, S, et al, (2011), Application of Genetic Algorithms in Machine learning, *International Journal of Computer Science and Information Technologies*, 2(5), 2412–2415.
24. Wang, Yu, (2018), Reconfigurable Processor for deep learning in autonomous vehicles, Special Issue 1- The Impact of Artificial Intelligence on Communication Networks and Services, *ITU Journal*, 1(1), 9–21.

Index

For Product Safety Concerns and Information please contact our EU
representative GPSR@taylorandfrancis.com
Taylor & Francis Verlag GmbH, Kaufingerstraße 24, 80331 München, Germany

www.ingramcontent.com/pod-product-compliance
Lightning Source LLC
Chambersburg PA
CBHW060815220326
41598CB00022B/2622